Food Emulsions and Foams

Interfaces, Interactions and Stability

Food Emulsions and Foams

Interfaces, Interactions and Stability

Edited by

E. Dickinson
Procter Department of Food Science, University of Leeds, UK

J. M. Rodríguez Patino
Universidad de Sevilla, Spain

The proceedings of the conference Food Emulsions and Foams: Interfaces, Interactions and Stability organized by the Food Chemistry Group of the RSC together with the Group Tecnologia de Alimentos of the University of Seville, held on 16–18 March 1998 in Seville, Spain

Special Publication No. 227

ISBN 0-85404-753-0

A catalogue record of this book is available from the British Library.

Published by The Royal Society of Chemistry,
Thomas Graham House, Science Park, Milton Road, Cambridge CB4 0WF, UK

For further information see our web site at www.rsc.org
Typeset by Paston PrePress Ltd, Beccles, Suffolk
Printed by Athenaeum Press Ltd, Gateshead, Tyne and Wear, UK

Preface

The underlying objective of this edited book is to record current progress in the development of fundamental understanding of the stability and rheological properties of food dispersions containing particles, droplets and bubbles. Examples of complex multiphase foods of this type are yoghurt, ice-cream, mayonnaise, *etc*. The properties of manufactured food emulsions and foams depend on the processing techniques used in their formulation and on the nature of the interactions involving the various constituent molecular ingredients—proteins, lipids and hydrocolloids. The structural and compositional complexity of food colloids generally necessitates a consideration of simpler model systems in order to elucidate the key mechanisms and principles contributing to texture, taste and shelf-life. Of particular importance are surface chemical properties of adsorbed protein layers and the nature and strength of the interactions between proteins and other molecular components, especially lipids and polysaccharides.

Every two years since 1986 an international conference in the area of food colloids has been held in Europe under the auspices of the Food Chemistry Group of the Royal Society of Chemistry (UK). The latest such conference, entitled 'Food Emulsions and Foams—Interfaces, Interactions and Stability', was held in Seville, Spain, on 16–18th March 1998. The three main themes of the Seville meeting were (i) dispersions, (ii) fluid–fluid interfaces, and (iii) rheology of food colloids. The programme included invited overview lectures, contributed oral presentations, and an exhibition of over 90 posters. The conference was attended by 190 participants from 19 different countries. Most of the invited lectures and contributed oral presentations are recorded in this volume. Research papers based on some of the poster presentations are being published separately in a special issue of the journal *Colloids and Surfaces B: Biointerfaces*.

The lecture programme of the conference, and hence the selection of contributions for this volume, was the responsibility of the International Organizing Committee comprising Dr Rod Bee (Unilever Research, Colworth Laboratory), Dr Björn Bergenståhl (Stockholm), Prof. Denis Lorient (ENSBANA, Dijon) and Prof. Pieter Walstra (Wageningen Agricultural University), as well as the editors of this volume. Detailed arrangements for the conference were made by the members of the Local Organizing Committee: Prof. J. M. Rodríguez Patino (Chairman), Prof. J. de la Fuente Feria (Secretary), Prof. M. Ruíz Domínguez (Treasurer), Prof. M^{a}. R. Rodríguez Niño and C. Carrera Sánchez.

E. Dickinson (Leeds)
J. M. Rodríguez Patino (Seville)
May 1998

Contents

Rheology of Food Colloids

Interfacial Structures and Colloidal Interactions in Protein-Stabilized Emulsions

By D. G. Dalgleish

CENTRE INTERNATIONAL DE RECHERCHE DANIEL CARASSO, 15 AVENUE GALILEE, 92350 LE PLESSIS-ROBINSON, FRANCE

1 Introduction

It is approximately 20 years since the publication of the important studies of Graham and Phillips,[1,2] which described for the first time many of the details of the behaviour of proteins adsorbed to oil–water and air–water planar interfaces. A few years later came the description by Oortwijn and Walstra[3] of the formation of emulsions from milk fat and the proteins of milk. Since that time there has been a considerable increase in our understanding of protein adsorption, and this paper will describe some of the developments during succeeding years, to demonstrate the progress which has been made, and perhaps also to suggest where additional research may be required.

Functionally, proteins are used in food emulsions because they confer stability. The fact that they are food emulsions means, of course, that the protein also provides nutritionally valuable material, as well as exercising its functional role in the formation and stabilization of the emulsion. At the same time, the fact that these proteins exercise their function (in real foods) in the presence of a number of other food components must also be taken into account, and it may be that the major challenge to the scientist of food emulsions today is to understand the behaviour, not just of simple systems which contain a protein and some lipid and an aqueous phase, but also the more complex mixtures which are found in real food preparations. Moreover, the manner in which these 'real' emulsions are processed affects the behaviour of the proteins in solution and on the interface, with consequent effects for the structure and stability of the product. For example, the properties of the emulsion in ice-cream has been extensively investigated and described,[4] but there are also products such as cream cheese, where milk-based emulsions are formed, heated, acidified and further processed, and yoghurts, in which similar processes occur. The additional complexity of the presence in these products of casein micelles, rather than simply monomeric proteins, adds factors to the texture, and, it must be admitted, a certain spice to the study of such complex systems.

We may conveniently divide the study of simple and complex food emulsions into a number of topics. First, we need to know what is on the interface, how it gets there, and what in its surroundings affects this interaction. Secondly, there is the need to know as exactly as possible the state of this adsorbed material; for proteins, this requires descriptions of their molecular conformations and how the adsorbed proteins interact. Accordingly, the discussion which follows will be divided into these two sections.

2 The Adsorption and Competition of Proteins at Oil–Water Interfaces

Exchange Reactions

It is evident that many proteins adsorb spontaneously to oil–water interfaces. However, the simple fact of adsorption conceals a number of important questions. Although it is easy to make an emulsion using oil and a single purified protein, it is evident even at the most simple level that different proteins have different emulsifying capacities (however one chooses to define this rather elusive quantity). That is, some proteins make finer or more stable emulsions than others. However, attempts to predict emulsifying capacities from some properties of the proteins—for example, from the surface hydrophobicity,[5] or by defining a scale based on the abilities of different proteins to displace one another from the interface[6]—seem to be doomed to failure. This is because, in the first case, the adsorption of a protein proceeds not simply via surface sites, and, in the second, free exchange between adsorbed and non-adsorbed proteins is rather difficult to achieve. Simply making an emulsion with one protein and adding a second protein afterwards does not ensure that the most surface-active protein actually ends up dominating the interface. With the two purified casein fractions, α_{s1} and β-caseins, there is indeed replacement of the former by the latter, suggesting that the β-casein is the more surface active of the two.[6] However, even for the native mixture of caseins (α_{s1}, α_{s2}, β, κ) found in sodium caseinate, the adsorption of the individual proteins is not as simple as the model systems suggest.[7]

It has never been unequivocally established whether an equilibrium exists between adsorbed and non-adsorbed protein in food emulsions, especially because the emulsion is formed under conditions which are far from equilibrium, using a homogenizer. In addition, as has been pointed out often, proteins adhere to the interface through many points of contact, and so the spontaneous desorption of the whole protein requires a number of events to happen simultaneously, which is unlikely. For caseins, and presumably for other proteins as well, the protein is used as efficiently as possible during emulsion formation; when emulsions are made using only small amounts of casein, the latter spreads across the surface of the emulsion droplets to cover as much of the surface as possible. Thus emulsions made with caseinate are stable[8] with a surface coverage as low as 1 mg m^{-2} although the surface coverage associated with a saturated monolayer is generally found to be *ca.* 3 mg m^{-2}. This spreading occurs, but to a lesser extent, with the more rigid molecules of the whey proteins. Evidently, emulsions may be stable with less than maximal amounts of protein covering the surface, because it is possible for stable emulsions to exist with gaps between the adsorbed protein molecules (this would not be possible with any other type of surfactant, but the sheer size of the proteins and their capacity for steric stabilization may over-ride the undesirability of leaving gaps in the protein layer). This being the case, if an emulsion is prepared using less than saturating amounts of one type of protein, and then a second type of protein is subsequently added, it is possible to include a considerable quantity of this material in the interfacial layer without necessarily displacing any of the originally adsorbed protein.[8] In most of the experiments which have been described, there is little evidence that there is equilibrium between adsorbed and non-adsorbed protein; there may be changes in surface composition, but true equilibrium is a rare occurrence, if it occurs at all.

The exchange of protein between interface and bulk is often facilitated by the presence of a surfactant. The mechanism is the assisting of desorption by removing one by one the different points of contact of the protein with the interface, and replacing them with surfactant. This process is much more likely than the spontaneous simultaneous desorption of all of the points of contact of the protein. However, the surfactant must be chosen with some care, since not all surfactants behave in the same manner. For example, although water-soluble surfactants generally can remove proteins totally from the interface,[9] oil-soluble surfactants have often been found to remove only some of the protein,[10] presumably because the oil-soluble material is sufficiently hydrophobic that the protein may adsorb to it, even at the surfaces of the emulsion droplets. Even water-soluble surfactants may differ in their effects when different proteins are used; although we need to make allowances for the results from different laboratories, it appears more difficult for polyoxyethylene surfactants to displace whole casein than isolated β-casein.[9,11] This is similar to the competitive displacement of purified α_{s1}- and β-caseins, and suggests that the other caseins (α_{s2}- and κ-casein) may have an effect on the adsorption and structure of the adsorbed caseinate. The surfactants do not necessarily simply displace the protein; at low concentrations of surfactant, it is possible rather to loosen the protein on the interface, so that lateral diffusion may become easier.[12,13] Clear indications of this secondary effect of Tween-60 on caseins in adsorbed caseinate have been observed, and the suggestion may be either that the surfactant binds to the proteins, or that it allows the protein to take up a different conformation at the interface.[12]

In some cases, surfactants and protein do not compete at all. We have seen[14] that in oil-in-water emulsions made using caseinate and phospholipid as surfactants, caseinate is only slightly displaced by the phospholipid; indeed it may be considered uncertain whether the phospholipid adsorbs at all in some of these emulsions,[15] although it certainly enhances the emulsifying capacity of the caseins. It is possible that this latter effect arises from the formation of specific complexes between phospholipid and caseinate, either on or off the surfaces of the emulsion droplets; that such complexes are at least possible has been demonstrated by measuring direct interactions between specific phospholipids and individual caseins.[16] This type of action is discussed in one of the other papers in this volume (Singh *et al.*), where it is shown* that the stabilizing action of peptides is enhanced by the presence of certain phospholipids, although there is no direct adsorption of the phospholipid to the interface, except when the emulsion is heated.

Effects of Heat on Surfactant/Protein Behaviour

The surface composition of emulsion droplets is defined not simply by the proteins which are present, but by other surfactant materials as well. However, it is also necessary to consider the effects of processing on these systems, since nearly all food preparations are processed, especially by subjecting them to heating of greater or lesser severity. With the exception of studies on homogenized milks,[17] few studies have been published on the effects of heat on the composition of the adsorbed layers of emulsion droplets. Some recent work in my own laboratory demonstrates that this may have some rather unexpected effects. It is well known[18] that heating milk causes the whey proteins α-lactalbumin and β-lactoglobulin to denature, and to interact, by the formation of disulfide bonds, with the κ-casein and possibly also the α_{s2}-casein of the micelle. By extension we would expect an emulsion droplet stabilized by whey proteins to interact with added caseins when the mixture was heated. This, however, has proved not to be the case;[19] the addition of sodium caseinate to the emulsion was found to cause no interaction at all, apart from the adsorption of some of the casein to saturate the interface, as described above. Surprisingly, there was no interaction between the sulfhydryl-containing caseins and the adsorbed whey proteins, either before or after heating. This seems to suggest that the adsorbed whey proteins were incapable of forming disulfide bonds with molecules from solution, although it is known that slow interactions between adsorbed whey proteins do occur via disulfide formation. The implication of these findings seems to be that the sulfhydryl groups of adsorbed β-lactoglobulin molecules are inaccessible to other proteins approaching the interface from solution.

On the other hand, the addition of whey protein to an emulsion initially made with caseinate has a considerable effect, in addition to the simple adsorption to ensure saturation of the interface.[19] During heating (85 °C, 2 minutes or more), β-lactoglobulin adsorbs to the oil droplets, and simultaneously some of the α_{s1}- and β-caseins are observed to desorb. Evidently the heating has the effect of facilitating the exchange of proteins. This is shown in detail in Figure 1, where the displacement of the α_{s1}- and β-caseins is balanced by the adsorption of β-lactoglobulin and α-lactalbumin. The greater the amount of added whey protein, the greater was found to be the observed effect. Originally, we believed that this effect resulted from the denaturation of the serum proteins, but subsequently we were able to demonstrate that the phenomenon occurred at

* See p. 117.

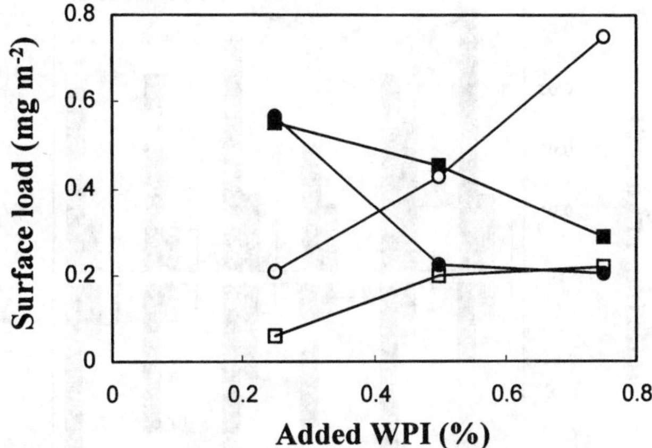

Figure 1 *Changes in the composition of adsorbed layers of protein in an emulsion (20% w/w soybean oil, 1% w/w sodium caseinate) during heating for 2 minutes at 85 °C in the presence of different amounts of added whey protein isolate (WPI), expressed as changes in the surface loads of the individual proteins:* ●, α_{s1}*-casein;* ■, *β-casein,* ○, *β-lactoglobulin;* □, *α-lactalbumin*

temperatures in the region of 40–50 °C, well below the normal temperature of denaturation of the whey proteins. This is illustrated in Figure 2, where the rates of adsorption of β-lactoglobulin and of desorption of α_{s1}-casein are shown for these moderate temperatures. Not only do the reactions occur at relatively low temperatures, they also do not depend on disulfide bond formation. A possible facilitating reaction may be the dissociation of the dimer of β-lactoglobulin, which takes place in this temperature range.[20] Therefore, it is not necessary for denaturation of the β-lactoglobulin to occur to cause the molecules be more surface-active, but just relatively small changes such as are defined by dissociation of multimers. This may also explain why earlier experiments at room temperature[21,22] failed to detect the phenomenon.

If we try to translate this observation into an industrial context, the implications may be that the surfaces of emulsions may not be at all stable during processing. Although an emulsion may be formed with one set of proteins on its interface, the influence of heating may be such as to change the proteins which adsorb, with some effects on either the stability or the more general properties of the emulsion. It may be partly for this reason that milks which are homogenized before heating have different behaviour from those homogenized after the milk has been forewarmed. The experiments described above were carried out on relatively simple emulsions made from caseinate; we do not know what is the effect if the casein is in micellar form—that is, if homogenized milk is heated in the presence of added whey proteins. Such evidence as exists[17] suggests that whey proteins in homogenized milk are less capable of displacing the casein micelles from the emulsion interface than are whey proteins in simple model emulsions.

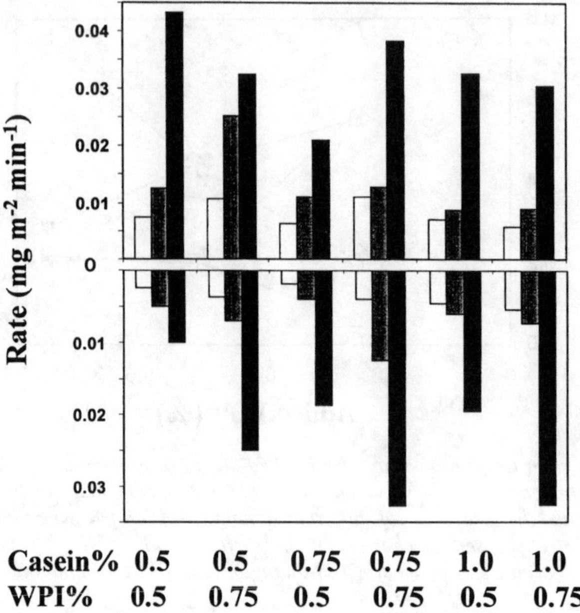

Figure 2 *Rates of the adsorption of β-lactoglobulin (upper plot) and of desorption of α_{s1}-casein (lower plot) during heating of emulsions (20% w/w soybean oil) originally made with sodium caseinate at different concentrations (0.5, 0.75, 1%), to which were added defined amounts of WPI (0.5, 0.75%). The rates are expressed in terms of the changes in surface loads of the different proteins during the heating. Open bars, heating at 40 °C; grey tone, heating at 45 °C; filled bars, heating at 50 °C*

Thus, heating may have the effect of creating or enhancing the exchange of proteins; another is simply to alter the emulsifying capacity of the proteins by denaturing them. Specifically, we may cite the case of the membrane fraction of buttermilk. By comparing the properties of membrane fractions isolated from buttermilk prepared from creams which had either been heat-treated or not, we have shown[23] that the heat-treated creams yield buttermilk membrane fractions in which the proteins are denatured, and which have low emulsifying capacity. Emulsions prepared using this material showed bridging flocculation, because the membrane material was present in the form of large particles, which evidently would not spread over the interface (Figure 3). On the other hand, the membrane fraction isolated from unheated buttermilk was much more functional in terms of its emulsifying capacity, and gave fat globules surrounded by thin membranes (Figure 4), so that evidently the native membrane structure is capable of being disrupted in the presence of lipid and to spread around the surfaces of the oil droplets.[24] Thus, although buttermilk is often considered to be an emulsifying agent, it is in general the casein micelles in it which provide the emulsifying capacity, rather than the membrane fraction, which denatures at a relatively low temperature (60–65 °C).

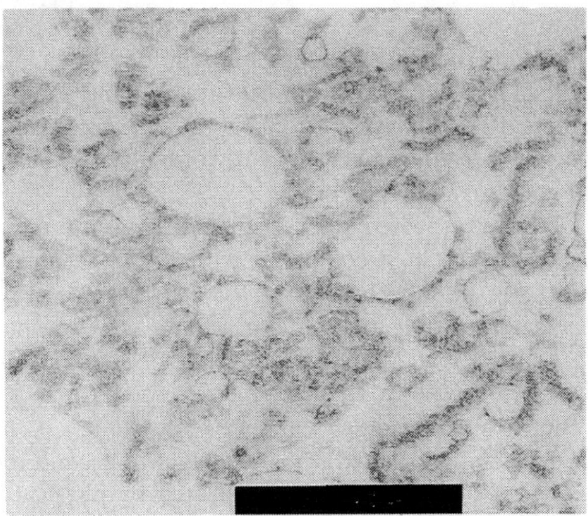

Figure 3 *Transmission electron micrograph of an emulsion (10% w/w soybean oil, 4% w/w milk fat globule membrane isolate). The membrane material was isolated from a commercial buttermilk derived from cream which had been pre-treated at a temperature of 85 °C. Scale bar = 1.1 μm*

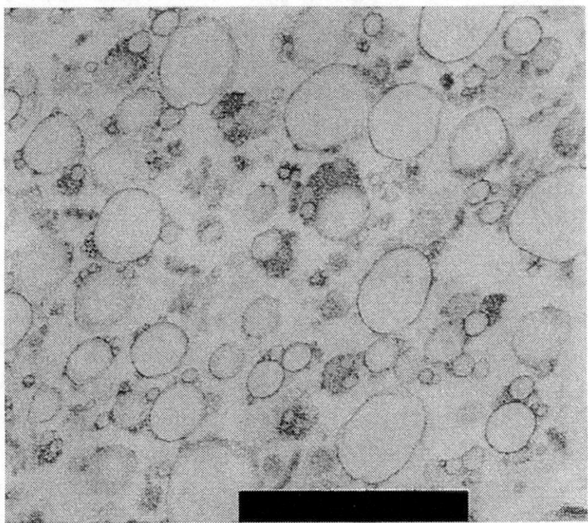

Figure 4 *Transmission electron micrograph of an emulsion (5% w/w soybean oil, 3% w/w milk fat globule membrane isolate). The membrane material was isolated from buttermilk derived from cream which had received no prior heat treatment. Scale bar = 0.7 μm*

Exchange and Changes During Processing

If, therefore, we consider the behaviour of emulsions during processing, we will find that the final composition, and therefore the structure and reactivities, of the oil droplets may not be those which we expect. This is, of course, well understood in products such as ice-cream and whipped toppings, but it is less well defined in other products which are heat-treated and stored for extended periods of time thereafter. Quite apart from changes in composition, these types of emulsions may also change with time because the proteins change their conformation or react together.

It should always be remembered that the effective concentration of proteins in the adsorbed layer is very high compared with the concentrations typically regarded as saturating in solution. Thus, for example, because of viscosity considerations, it is not possible to dissolve caseinate to a concentration of, say, 25%; however, a surface coverage of $2.5\,mg\,m^{-2}$ is more than sufficient to create such a concentration in the interfacial layer. At such extremes of concentration, there will be a much enhanced probability of intermolecular chemical reaction, which may cause the protein layer to become more rigid with time. While this behaviour has been demonstrated by physical and chemical means to occur in β-lactoglobulin,[25,26] it is also apparently possible even in such unstructured proteins as the caseins. Different experiments have shown that the interfacial viscosity of adsorbed α_{s1}-casein increases slowly with the age of the interface;[25] that β-casein becomes less susceptible to displacement by surfactant from an emulsion interface;[27] and that the calcium-sensitivity of caseinate-stabilized emulsions decreases with the age of the emulsion.[28] However, not enough is yet known about the causes of these different changes. Since neither α_{s1}-casein nor β-casein possess sulfhydryl groups, there is no possibility of the formation of disulfide bridges between the molecules, as happens with adsorbed β-lactoglobulin. However, it is not impossible that simply the very high local concentration of the protein at the interface is a factor, allowing some slow structural rearrangement and perhaps the formation of salt bridges to occur, especially if multilayers of protein are adsorbed to the interface. Certainly in emulsions it is quite possible that the speed of the non-equilibrium adsorption process may enforce some of the proteins to adsorb to non-optimal conformations for maximum interaction with the interface. A slow conformational re-equilibration is quite likely as a result. More recently (paper by Leaver *et al.*, this volume*) it has been demonstrated that a time dependent effect may also arise from chemical modification of the adsorbed protein by enals derived by peroxidation of polyunsaturated fatty acids in the oil phase. Clearly this covalent modification may have both functional and nutritional significance.

In all of these adsorption processes there is also an effect of the non-aqueous phase. The proteins adsorb slightly differently to interfaces of triglyceride oil or hydrocarbons.[29] Presumably, the hydrocarbons, being more hydrophobic than triglycerides, force stronger interactions with the hydrophobic portions of the proteins. However, there is another factor which makes the two non-aqueous phases difficult to compare, namely that they have different viscosities, and therefore produce emulsions with different particle sizes for the same homogenization pressure and the same concentrations of protein and non-aqueous

* See p. 258.

phase. Since casein, at least, is known to spread over the interface, we may expect that the caseins do indeed have different conformations on the two oil–water interfaces, but this may be partly due to the fact that the surface coverage, for defined overall contents of protein and non-aqueous phase, is different.

3 Conformations of Adsorbed Proteins

Spectroscopic and Related Measurements

There can be little doubt that the act of adsorption changes the conformation of proteins. The evidence comes from many types of experiments which in general reinforce one another. Some recent studies have demonstrated that enzymes (chymosin or lysozyme) adsorbed to oil–water interfaces are not only inactive when adsorbed, but their activity is not recovered on desorption.[30] Even though the conformational changes may be small, they are still sufficient to render the proteins denatured, if we take denaturation to mean some change, however small, from the native state. Analogously, adsorbed β-lactoglobulin has been found to be very much more susceptible to attack by trypsin than is the native protein in solution.[31] It is probable that this results from conformational change making the labile bonds of the protein more accessible to the enzyme, but it is perhaps simply an accelerating effect of having locally high concentrations of the protein on the emulsion interface.

One of the interesting aspects of recent research on adsorbed proteins is that a number of studies have been made using different spectroscopic techniques, namely NMR, circular dichroism (CD) and Fourier-transform infra-red (FTIR) spectroscopies. These allow some direct measure of at least the broad outlines of the conformation of adsorbed proteins. Thus, NMR studies have confirmed, as was already believed from other evidence, that the phosphoserine groups of adsorbed β-casein are not immobilized, but are in a flexible part of the protein remote from the site of adsorption.[32]

Perhaps surprisingly, the conformational changes resulting from adsorption do not necessarily result in a complete disordering of the macromolecular structure, as is suggested for the action of heat. The measurement of CD spectra of a protein (subtilisin) adsorbed to teflon microspheres has shown[33] that the amount of α-helix structure increases on adsorption; in other words adsorption causes an ordering of the polypeptide chain. Depending on the primary structure of the protein, it may be possible to form amphipathic helices, in which one face of the structure is rich in hydrophobic residues which aid adsorption. Possibly the increase in the α-helix structure of the adsorbed protein arises from such a cause. Of course, this cannot happen for every protein; there is no evidence, for example, that caseins, despite their flexible structures, form ordered structures when they adsorb.

For other proteins, FTIR shows evidence for the retention of some ordered structure, even though there is no evidence for increased structuration as a result of adsorption. The simplest observation is that when β-lactoglobulin is adsorbed to an oil–water interface, the structure is altered, as evidenced by changes in the amide I region of the FTIR spectrum.[34] However, although the change in the spectrum suggests that the intramolecular β-sheet of the protein undergoes some loss of structure, there is no change to the type of spectrum typical of a totally unordered polypeptide (Figure 5). A further observation partially confirms the hypothesis that there is no equilibration between adsorbed and soluble protein. Because the adsorbed protein has an altered spectrum from the protein in its native state, and because it retains this structure after desorption, then we would expect, if there is equilibration between the adsorbed and soluble protein, that the protein in the unadsorbed state should gradually assume the conformation typical of the adsorbed state. This does not seem to occur, as can be seen from the FTIR spectrum of the non-adsorbed protein in Figure 5, which resembles that of the native protein. Thus, spontaneous adsorption–desorption equilibria may not occur. When the adsorbed protein is removed from the interface by surfactants, its conformation does not revert to that of the native protein; indeed, there seem to be further changes as a result perhaps of complex formation with the surfactant (Figure 5).

The infra-red spectroscopy of β-lactoglobulin is in agreement with observations made using DSC. This technique has shown[35] that adsorbed β-lacto-

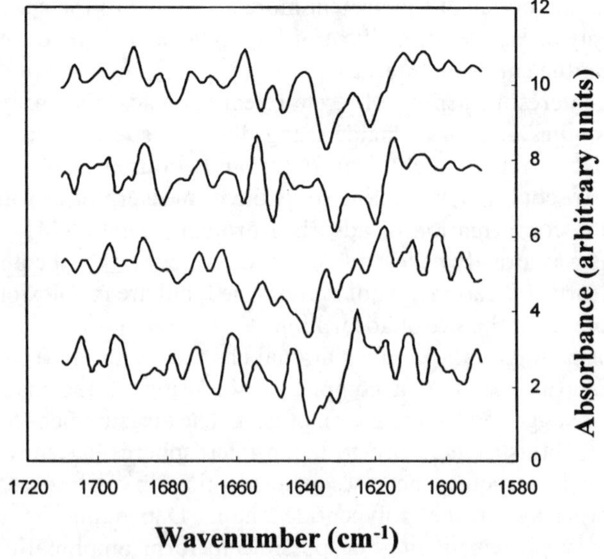

Figure 5 *Second derivative FTIR spectra of adsorbed and unadsorbed β-lactoglobulin. Top spectrum, native β-lactoglobulin in D_2O solution; second spectrum, β-lactoglobulin remaining in the aqueous phase after an oil-in-water emulsion was made and stored overnight. The similarity of the spectra in the region 1610–1650 cm^{-1} is indicative that no changes of conformation have occurred in the non-adsorbed protein. Third spectrum, β-lactoglobulin adsorbed to oil droplets isolated from the emulsion; profound changes in the spectral region of interest demonstrate conformational change. Fourth spectrum, β-lactoglobulin after desorption from the emulsion interface by Tween 20; this shows that the desorbed protein does not recover its original conformation after desorption*

globulin apparently has no thermal transition; that is, the protein appears to be denatured while on the surface of the oil. Moreover, displacing this protein from the interface with a surfactant (Tween 60) did not cause the thermal transition to reappear, and a separate study showed that unadsorbed protein was not itself denatured by the surfactant. Thus it may be argued that, whatever the conformation of the adsorbed (or desorbed) protein, it represents some minimal energy state which cannot be denatured further by heat. Therefore, we are correct in applying the term 'surface denaturation' to this protein.

In contrast, the companion whey protein, α-lactalbumin, shows considerable difference in its behaviour. Adsorption leads again to a change in the structure of the protein, but the change appears to be more reversible to adsorption by the surfactant, and the adsorbed protein retains a good deal of structure (according to FTIR spectra). However, FTIR studies suggest that α-lactalbumin itself does interact with the surfactant, since mixtures of Tween 60 and α-lactalbumin show different spectra. These observations are only partly confirmed by the DSC measurements. As with β-lactoglobulin, adsorbed α-lactalbumin gives hardly any thermal transition, but the protein does apparently recover after desorption by surfactant, to give a thermal transition very close in denaturation temperature and magnitude to that of the original protein. This is the case even though the FTIR spectra of protein which has never been adsorbed and of protein which has been adsorbed and then desorbed are not the same, suggesting that a permanent conformational change has occurred. It is rather difficult to be certain of the interpretation of the DSC results. It is possible that the proteins when they become adsorbed are held firmly enough so that they do not undergo a sharp thermal transition in the temperature range 30–100 °C; this would explain the observed behaviour, and is the implication of the work presented by Green *et al.* (this volume*) who have suggested that a continuous conformational change occurs in adsorbed lysozyme as the temperature is increased, rather than the sharp conformational change typical of the protein in solution. These results may suggest that there is a range of conformational states available to the adsorbed protein, each possessing a different temperature of denaturation, so that the DSC shows no thermal transition, the effect having been smeared out over a large temperature range. Thus the act of adsorption causes the loss of some tertiary structure of α-lactalbumin, which is perhaps imperfectly recovered when the protein is desorbed. However, if the adsorption does not cause any change in the secondary structure of the protein, it may still be desorbed in such a way as to cause a transition in the DSC. This altered structure is not, however, the molten globule form of α-lactalbumin, since that is characterized by a much lower transition temperature in the DSC. It is likely to be a further semi-denatured form of the protein, not corresponding to any of the defined forms which have been so far identified.

* See p. 285.

Scattering Measurements

Less detailed information on the conformations of adsorbed proteins can be obtained from scattering of electromagnetic radiation *i.e.*, light, X-rays, or neutrons. Essentially, these techniques give information on the distribution of mass close to the interface to which the proteins are adsorbed. Ideally, X-ray and neutron scattering experiments require the particles being studied to be monodisperse; light scattering (in the shape of dynamic light scattering) does not require this condition. Neutron reflectance, as an alternative to neutron scattering, occurs at planar interfaces, and cannot be used to study emulsion droplets. Only light scattering offers the possibility of rapid measurement, the other methods being rather slow. Although neutron reflectance and X-ray scattering give more detailed assessments of the distribution of mass close to the interface, it is the measurements of the hydrodynamics of the adsorbed material by light scattering which are perhaps more close to reality. This is because the hydrodynamic layer thicknesses of the adsorbed material may be closer to the observed behaviour of the particles as colloids, since the stability or otherwise of the colloid depends on the shape and interactions of the outer adsorbed layer.

The scattering experiments have provided valuable information on the conformations of adsorbed proteins, in particular the caseins.[36,37] It appears from the research that the adsorbed caseins form a layer of material close to the interface, but that they also have protruding tails of material away from the interface and into solution. Thus, although as much as 80% of the mass of the protein is to be found within 1–2 nm of the interface, the remainder may protrude into solution to a considerably greater distance.[37] These experiments confirm the hydrodynamics experiments, which clearly demonstrate that this extended layer exists.[38] Although there are differences in the conformations of the individual caseins, they all generally behave similarly, presumably because they are relatively flexible molecules. On the other hand, β-lactoglobulin is very different; adsorption of this protein produces only a thin layer of material which projects much less into solution;[39,40] but β-lactoglobulin is a more rigid protein than the caseins. Regrettably, none of these methods seems capable of resolving the question as to whether adsorbed proteins are either on or partially within the surface of the oil in an emulsion droplet. Although hydrophobic proteins exist which are capable, for example, of penetrating membranes to form pores, these appear to be specialized in their conformations. It is not established whether food proteins, such as the milk proteins which are often used to form edible emulsions, actually penetrate the surface of the oil or simply lie on the surface.

The adsorbed layers of caseins are susceptible to changes in their conformation depending on circumstances. The hydrodynamic thickness of the adsorbed layer, and hence the interactions of the emulsion droplets, depend upon the surface coverage of the protein; the more protein is adsorbed, the more the adsorbed layer protrudes,[8] possibly simply because the crowding of the molecules on the surface pushes the tails outwards (these, it should be remembered, are generally quite highly charged). Also, because of the charges on the protruding portions of the molecules, the structure of the adsorbed layer is altered by changes in ionic strength, since this decreases the repulsion between the charged portions of the chains. Ions which bind more specifically, such as Ca^{2+}, also tend to collapse the adsorbed layers,[41] as does ethanol,[42] although this latter component also appears to dissolve the oil from emulsion droplets and its precise effect on the adsorbed layers is not totally clear. However, since ethanol is a poor solvent for proteins, it will be expected to cause the adsorbed layers to collapse.[43]

It was mentioned above that protein on an interface that is depleted of protein (*i.e.*, having less than its maximal monolayer coverage) protrudes less into solution than when the interface is saturated. However, if a saturated interface is treated with surfactant, so that some of the protein is removed from the interface, it is found that the thickness of the adsorbed layer decreases in a different way; it seems likely that the adsorbed surfactant allows the protein molecules to retain their extended conformation, and there is a simple linear relationship between the thickness of the adsorbed layer and the amount of protein adsorbed.[12] This was not the case when the surface coverage of protein was changed simply by incorporating less protein into the emulsion.[8] In addition, there are probably interactions between the protein and the surfactant which may also help to maintain it in an extended position.

At least some of the behaviour of the caseins can be calculated from models, using the Scheutjens–Fleer model for adsorption of flexible polypeptides, and taking account of the distribution of charged groups in the protein molecules.[44] These reproduce well the observed conformational properties of α_s- and β-caseins, and also the highly phosphorylated protein from egg yolk, phosvitin.[38,45,46] These all form rather extended layers, whose general conformation seems to be described by the calculations. This extends to a possible description of the different sensitivities to added salt of emulsions made using the different caseins.[47] However, this approach may only be used for flexible proteins like the caseins, where there is little or no secondary structure. For globular proteins, the calculations will be much more difficult because of the existence of the original tertiary structures of these proteins, so that they cannot be treated as flexible chains, and other energy factors must be taken into account in calculating their interactions.

4 Conclusion: Surfaces and Emulsion Stability

Caseins are considered to stabilize emulsion droplets both by steric and by charge interactions, because they form a highly charged and extended layer around the fat droplets in an emulsion. The breakdown of the adsorbed layer does not, however, necessarily cause the immediate destabilization of the emulsion. Even trypsin treatment of emulsion droplets stabilized by individual or mixed caseins, which completely breaks down the adsorbed layer of protein, does not destabilize the colloids. This is the basis of the estimation of the dimensions of the adsorbed layers of protein.[8] If aggregation of the emulsion droplets occurred, it would not be possible to make these measurements.

Interestingly, replacement of the adsorbed casein layer using Tween 60 does not allow the hydrodynamics of the adsorbed layer to be measured, simply because the replacement of the adsorbed protein by the small-molecule surfactant destabilizes the emulsion just sufficiently so that slight aggregation occurs; this results in a slight increase in the apparent sizes of the emulsion droplets due to coalescence since the new surfactant layer is rather weak.

The casein on emulsion droplets seems to be more than sufficient to confer stability. This is also made clear when emulsions, prepared using whole caseinate rather than individual fractions, are treated with trypsin. Not only do the emulsions not aggregate after moderate treatment, they are in fact stabilized towards the action of calcium ions, which typically destabilize the emulsion.[48] This action may be explained by the general observation that the first peptides released by the action of trypsin on adsorbed caseins are the phosphorylated peptides, which are also the primary sites by which calcium binds to the caseins. Removal of these peptides by trypsin creates droplets which are much more stable to calcium than are the original droplets. At moderate extents of trypsinolysis, it is the α_s- and β-caseins which are degraded, leaving the κ-casein intact on the emulsion surface. This is sufficient to confer stability; but, once the κ-casein begins to be degraded, the emulsion becomes much less stable to the addition of calcium.

It is unfortunate that most of the operations designed to alter the steric stabilization properties of the caseins (*e.g.*, trypsin treatment) also remove much of the charge. However, the changes in the geometry of the protein seem to have relatively small effects upon the ζ-potentials of the emulsion droplets. This is presumably because the adsorbed layer is flexible and can change its conformation to maintain a constant charge density, so that the effects of modification are minimized. Thus, it is questionable whether the measured ζ-potential can give much information on the stability. An example is that when the ζ-potential is reduced by changing pH, then the expected aggregation occurs. However, calcium addition, which neutralizes the charges of the proteins, causes aggregation to occur at a much larger (absolute) value of the ζ-potential. Thus, there is more to the destabilization than simply charge neutralization.

However, calculations have shown that emulsions coated with α_{s1}-casein are more likely to be destabilized by salt than those stabilized by β-casein.[49] This calculation has been confirmed by experiments, and so it is evident that at least in this case the behaviour of the emulsion approximates to the expectations of classical colloid science. However, prediction for emulsions stabilized by globular proteins such as β-lactoglobulin is difficult. For example, what is the reason for the rather similar susceptibility to calcium of emulsions stabilized by caseinate (which binds a great deal of calcium) and β-lactoglobulin (which does not)?

A direct understanding of the implications for stability of the specific structural features of adsorbed proteins is elusive, except perhaps for the caseins, which are hardly typical proteins. The challenge for the future is to estimate the details of the conformations of adsorbed multimeric or globular proteins which will permit the stability behaviour of emulsions to be predicted.

References

1. D. E. Graham and M. C. Phillips, *J. Colloid Interface Sci.*, 1979, **70**, 415.
2. D. E. Graham and M. C. Phillips, *J. Colloid Interface Sci.*, 1979, **70**, 427.
3. H. Oortwijn and P. Walstra, *Neth. Milk Dairy J.*, 1979, **33**, 134.
4. K. G. Berger, in 'Food Emulsions', 3rd edn, ed. S. E. Friberg and K. Larsson, Marcel Dekker, New York, 1997, p. 413.
5. S. Nakai, *J. Agric. Food Chem.*, 1983, **31**, 676.
6. E. Dickinson, S. E. Rolfe, and D. G. Dalgleish, *Food Hydrocolloids*, 1987, **2**, 397.
7. E. W. Robson and D. G. Dalgleish, *J. Food Sci.*, 1987, **52**, 1693.
8. Y. Fang and D. G. Dalgleish, *J. Colloid Interface Sci.*, 1993, **156**, 329.
9. J.-L. Courthaudon, E. Dickinson, Y. Matsumura, and A. Williams, *Food Struct.*, 1991, **10**, 109.
10. E. Dickinson and S. Tanai, *J. Agric. Food Chem.*, 1992, **40**, 179.
11. D. G. Dalgleish, M. Srinivasan, and H. Singh, *J. Agric. Food Chem.*, 1995, **43**, 2351.
12. J. Chen and E. Dickinson, *Food Hydrocolloids*, 1995, **9**, 35.
13. J.-L. Courthaudon, E. Dickinson, Y. Matsumura, and D. C. Clark, *Colloids Surf.*, 1991, **56**, 293.
14. Y. Fang and D. G. Dalgleish, *Colloids Surf. B*, 1993, **1**, 357.
15. J.-L. Courthaudon, E. Dickinson, and W. W. Christie, *J. Agric. Food Chem.*, 1991, **39**, 1365.
16. Y. Fang and D. G. Dalgleish, *J. Am. Oil Chem. Soc.*, 1996, **73**, 437.
17. S. K. Sharma and D. G. Dalgleish, *J. Agric. Food Chem.*, 1993, **41**, 1407.
18. H. D. Jang and H. E. Swaisgood, *J. Dairy Sci.*, 1990, **73**, 900.
19. J. Brun, M.Sc. Thesis, University of Guelph, 1998.
20. S. P. F. M. Roefs and C. G. de Kruif, *Eur. J. Biochem.*, 1994, **226**, 883.
21. D. G. Dalgleish, S. E. Euston, J. Hunt, and E. Dickinson, in 'Food Polymers, Gels and Colloids', ed. E. Dickinson, Royal Society of Chemistry, Cambridge, 1991, p. 485.
22. J. A. Hunt and D. G. Dalgleish, *Food Hydrocolloids*, 1994, **8**, 175.
23. M. Corredig and D. G. Dalgleish, *J. Agric. Food Chem.*, 1997, **45**, 4595.
24. M. Corredig and D. G. Dalgleish, *J. Agric. Food Chem*, 1998, **46**, 2533.
25. E. Dickinson, S. E. Rolfe, and D. G. Dalgleish, *Int. J. Biol. Macromol.*, 1990, **12**, 189.
26. E. Dickinson and Y. Matsumura, *Int. J. Biol. Macromol.*, 1991, **13**, 26.
27. E. M. Stevenson, D. S. Horne, and J. Leaver, *Food Hydrocolloids*, 1997, **11**, 3.
28. E. P. Schokker and D. G. Dalgleish, submitted for publication.
29. J. Leaver and D. G. Dalgleish, *J. Colloid Interface Sci.*, 1992, **149**, 49.
30. A. L. de Roos and P. Walstra, *Colloids Surf. B*, 1996, **6**, 201.
31. S. O. Agboola and D. G. Dalgleish, *J. Agric. Food Chem.*, 1996, **44**, 3631.
32. L. C. Ter Beek, M. Ketelaars, D. C. McCain, P. E. A. Smulders, P. Walstra, and M. A. Hemminga, *Biophys. J.*, 1996, **70**, 2396.
33. M. C. L. Maste, E. H. W. Pap, A. van Hoek, W. Norde, and A. J. W. G. Visser, *J. Colloid Interface Sci.*, 1996, **180**, 632.
34. Y. Fang and D. G. Dalgleish, *J. Colloid Interface Sci.*, 1997, **196**, 292.
35. M. Corredig and D. G. Dalgleish, *Colloids Surf. B*, 1995, **4**, 411.
36. A. R. Mackie, J. Mingins, and A. N. North, *J. Chem. Soc. Faraday Trans.*, 1991, **87**, 3043.
37. E. Dickinson, D. S. Horne, J. Phipps, and R. Richardson, *Langmuir*, 1993, **9**, 242.
38. D. G. Dalgleish, *Colloids Surf.*, 1990, **46**, 141.

39. P. J. Atkinson, E. Dickinson, D. S. Horne, and R. M. Richardson, *J. Chem. Soc. Faraday Trans.*, 1995, **91**, 2847.
40. D. G. Dalgleish and J. Leaver, in 'Food Polymers, Gels and Colloids', ed. E. Dickinson, Royal Society of Chemistry, Cambridge, 1991, p. 113.
41. D. V. Brooksbank, C. M. Davidson, D. S. Horne, and J. Leaver, *J. Chem. Soc. Faraday Trans.*, 1993, **89**, 3419.
42. S. O. Agboola and D. G. Dalgleish, *J. Sci. Food Agric.*, 1996, **72**, 448.
43. D. S. Horne, *Biopolymers*, 1984, **23**, 989.
44. F. A. M. Leermakers, P. J. Atkinson, E. Dickinson, and D. S. Horne, *J. Colloid Interface Sci.*, 1996, **178**, 681.
45. S. Damodaran and S. Xu, *J. Colloid Interface Sci.*, 1997, **178**, 426.
46. E. Dickinson, V. J. Pinfield, and D. S. Horne, *J. Colloid Interface Sci.*, 1997, **187**, 539.
47. H. Casanova and E. Dickinson, *J. Agric. Food Chem.*, 1998, **46**, 72.
48. S. O. Agboola and D. G. Dalgleish, *J. Agric. Food Chem.*, 1996, **44**, 3637.
49. E. Dickinson, V. J. Pinfield, D. S. Horne, and F. A. M. Leermakers, *J. Chem. Soc. Faraday Trans.*, 1997, **93**, 1785.

Dispersions

Dispersion Stabilization and Destabilization by Polymers

By Brian Vincent

SCHOOL OF CHEMISTRY, UNIVERSITY OF BRISTOL, CANTOCK'S CLOSE, BRISTOL BS8 1TS, UK

1 Introduction

A large number of industrial products and processes involve liquid-based dispersions (sols, emulsions or foams). Their efficacy depends on the ability of the manufacturer to control their physical properties, in particular their rheological, settling and optical properties. These, in turn, depend strongly on the state of aggregation of the constituent particles. In many cases, a totally stable (deaggregated) system is required, for example, in pharmaceuticals and cosmetics. At the other extreme, strongly aggregated particles may be required, as in many purification or separation processes, for example by settling or filtration. For efficient settling close-packed aggregates are desirable, whereas in filtration processes more open-textured aggregates (to allow adequate solvent penetration) are needed. However, in both cases the aggregation should be essentially *irreversible*. On the other hand, in many applications, an intermediate state is required, *i.e.* a state of weak, *reversible* aggregation. For example, this is important in thixotropic systems, such as paints, where the extent of aggregation, and hence also the viscosity, depend on shear rate (and time). Weak, reversible aggregation is also useful in controlling (or even preventing) the settling behaviour of concentrated dispersions, in particular in overcoming the 'caking' problem, *i.e.*, the formation of dilatant, regular, close-packed structures, which result from settling of totally stable particles. In many processed food products a controlled state of aggregation (particles, droplets, bubbles) is also required to attain the required rheological properties and texture.

All of the states of aggregation referred to may be achieved by the judicious use of polymers. The polymer molecules may be chemically bonded (grafted) to the particle surface, physisorbed from solution, or simply present in solution. The various ways in which polymers control interparticle interactions, and hence dispersion stability will be described.

Although emphasis is placed here on particulate dispersions, the same general principles also apply to emulsions and foams. For a more general introduction to the topic of the effect of polymers on interparticle interactions, the reader is referred to the book by Fleer *et al.*[1] The material presented here is intended to be at a more 'introductory' level, for those not familiar with the field.

2 Totally Stable Dispersions

In practice there are few colloidal formulations in which charge stabilization of the particles *alone* is effective in preventing aggregation. There are several reasons for this.

(a) In non-polar liquids, such as aliphatic hydrocarbons, the particle/medium interfacial charge is difficult to sustain. Even with more polar (but non-aqueous) liquids, where interfacial charge can be achieved, the charge is often difficult to control, particularly in the presence of small amounts of adventitious water, whose adsorption may lead, not only to a change in the magnitude of the charge, but possibly also to a change of sign.

(b) With aqueous dispersions of charged particles, the presence of electrolytes, in particular the divalent ions (Mg^{+2}, Ca^{+2}) present in many natural hard waters, may lead to slow coagulation, even at relatively low concentrations (*i.e.*, greater than ~10 mmol dm^{-3}).

(c) With purely charge-stabilized systems, only the *kinetics* of aggregation can be controlled (by variation of pH, ionic strength, *etc.*). If aggregation does occur then the equilibrium state would effectively be one large aggregate. This is a result of the shape of the pair potential, illustrated schematically in Figure 1. The rate of aggregation is controlled by V_{max}, but the rate of deaggregation by

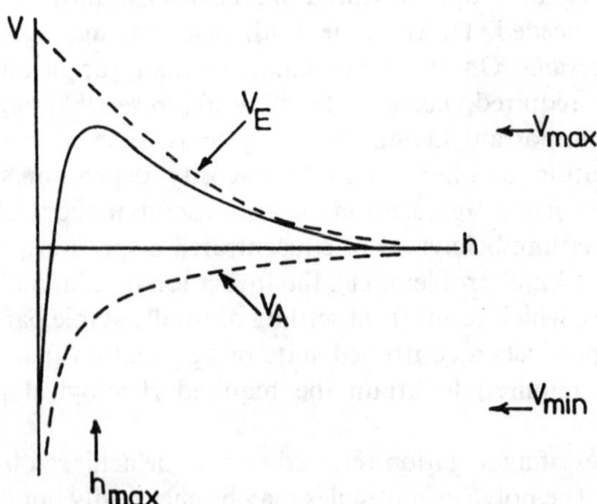

Figure 1 *Schematic representation of the interparticle pair potential V(h) between two charged colloidal particles: V_E electrostatic interaction; V_A, van der Waals attraction*

$V_{max} + V_{min}$. If the latter is greater than $\sim 10\,kT$ then aggregation is essentially irreversible.

(d) With charge-stabilized dispersions at low electrolyte concentrations, because of the long-range nature of the electrostatic term V_E (shown in Figure 1), it is difficult to prepare stable, very concentrated dispersions. Qualitatively, one can understand this from inspection of Figure 1: if the average separation of the particles is less than, say, 2 or 3 h_{max}, then the *effective* value of V_{max}, the energy barrier to coagulation, is significantly reduced. Even at lower particle volume fractions, electroviscous effects may lead to undesirably high viscosities.

(e) With commercial products, freeze–thaw cycles during storage, particularly with aqueous dispersions of charge-stabilized hydrophobic particles, can lead to coagulation. Also, so-called 'surface coagulation' of such hydrophobic particles, through their adsorption at the air–water interface, particularly as a result of shaking and bubble formation during processing and in transit, can be a problem.

Clearly, an alternative, or at least an additional, form of repulsive interaction is required in order to overcome these problems. Some form of 'structural' force may fulfil this requirement. These forces are associated with structural differences between the interfacial region of liquid in the vicinity of a particle compared with the bulk liquid. This is well known in the case of aqueous dispersions of hydrophilic particles (*e.g.*, oxides, clays). The situation is illustrated schematically in Figure 2.

In order for the two particles to come to a separation $h < 2\delta$, some perturbation of the structure in the two interfacial regions must occur. For example, 'structured' water may have to be displaced into normal bulk water. This leads to a repulsive interaction V_s. For two *hydrophobic* particles (*e.g.*, many polymer latices), the interaction is actually *attractive*; this is the origin of the so-called 'hydrophobic interaction'. The quantity δ may be of the order of the size of several or more water molecules. This is the reason, for example, why

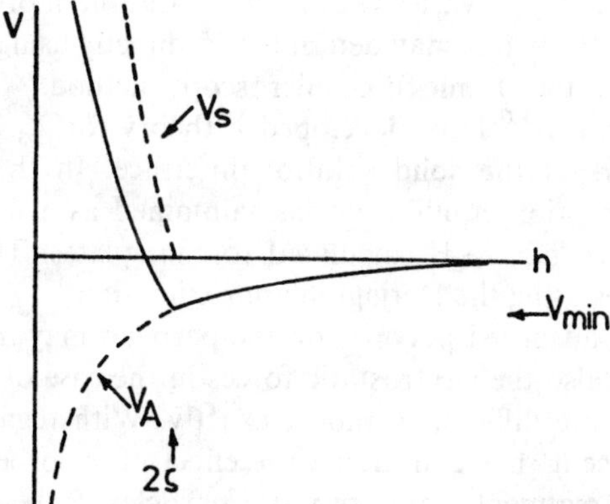

Figure 2 *The interparticle pair potential between two* neutral *particles, having a sheath (thickness δ) of 'structured' medium around each particle: V_s, structural interaction; V_A, van der Waals attraction*

hydrophilic silica particles in water are difficult to coagulate by the addition of electrolytes.

The presence of adsorbed (or grafted) polymer molecules around the particles also leads to a 'structured' layer, usually of much greater thickness (*e.g.*, δ may be up to ~ 100 nm with very high-molecular-weight polymers). The thickness of the adsorbed polymer layer can be measured by a number of techniques,[1] one of the most popular being dynamic light scattering (but only for monodispersed, spherical particles). The nature of V_s is now even more complex, since, for $h < 2\delta$, perturbation of the conformations of the adsorbed chains, as well as displacement of solvent from the interfacial region into bulk, must occur. The two respective contributions to V_s are known classically as the elastic (or volume restriction) and the mixing (or osmotic) contributions to what is generally referred to in total as the 'steric interaction'.[2] More recent theories[1] of the steric interaction have avoided this artificial split. However, it is useful to retain the division in order to understand the principal factors involved. For example, Napper[2,3] has given the following approximate expression for the steric interaction V_s between two spheres with an adsorbed polymer layer:

$$V_S = \underbrace{\frac{2\pi a kT V_2^2 \Gamma_2^2}{V_1}\left(\frac{1}{2} - \chi\right) S_{mix}}_{\text{mixing term: } V_{s,mix}} + \underbrace{2\pi a kT \Gamma_2 S_{el}}_{\text{elastic term: } V_{s,el}}. \tag{1}$$

Here a ($\gg \delta$) is the particle radius, V_1 is the solvent molecular volume, V_2 is the polymer molecular volume, Γ_2 the adsorbed amount of polymer (number of chains/area), χ is the Flory polymer–solvent interaction parameter, and S_{mix} and S_{el} are geometric functions that depend on the form of the segment concentration profile, $\rho(z)$, in the adsorbed layer normal to the interface. Analytical expressions for S_{mix} and S_{el} have been derived by Napper[2,3] for various assumed forms of $\rho(z)$. Although $\rho(z)$ has been determined experimentally using small-angle neutron scattering for individual particles,[1] one really needs to establish how $\rho(z)$ varies with their separation in order to calculate V_s accurately. Alternatively one may determine V_s directly using either the surface forces apparatus[4] or the atomic force microscopy method.[5]

Fleer and Scheutjens[1,6] have developed a theory for V_s based on a lattice model of polymers at the solid–solution interface. In their approach it is assumed that adsorption equilibrium is maintained as a function of particle separation h. Hence $\rho(z, h)$ is calculated for flat plates. The theory may be extended to spheres using the Derjaguin approximation.[1]

The presence of adsorbed polymer on the particles may modify the van der Waals forces, and also the electrostatic forces in the case of charged particles. Both these effects are difficult to model exactly. With regard to the van der Waals forces, Vincent[7] has considered the effect of adsorbed polymer layers, extending Vold's treatment[8] for composite particles. Provided that the mean segment volume fraction in the adsorbed layer is not too high, and the Hamaker constant of the polymer not too different from that of the solvent, the modification to the van der Waals forces is small.

The effect of the adsorbed polymer on the electrical double layer around a charged particle, and hence on the electrostatic interaction, is complex: both the surface charge density and the distribution of counterions between the Stern layer and the diffuse layer may well be changed. The simplest approach is to calculate the electrostatic interaction energy V_E *prior* to overlap of the adsorbed layers, *i.e.*, at $h > 2\delta$. One may then, to a first approximation, use an appropriate equation for $V_E(h)$, with the zeta potential ζ replacing the surface or Stern potential, and $h - 2\delta$ replacing h. For example, for $\kappa a > 1$, where κ^{-1} is the Debye thickness of the electrical double layer, we have

$$V_E = 2\pi\varepsilon\, a\zeta^2 \ln \{1 + \exp[-\,\kappa(h - 2\delta)]\}, \tag{2}$$

where ε is the permittivity of the medium.

In practice, homopolymers are not the most efficient type of polymer molecule to provide effective steric stabilization. The necessary requirements are as follows:

(a) high coverage (Γ needs to be of order of Γ_{plateau});
(b) efficient anchoring (χ_s for the adsorbed segments (in trains) needs to be $\gg \chi_{s,c}$, the critical value of χ_s for adsorption);
(c) extended tails (and loops, if present) (δ needs to be sufficiently large, such that at $h = 2\delta$ the van der Waals energy is $< kT$ or so);
(d) 'good' solvent environment for the segments in tails (loops) ($\chi < 0.5$).

Clearly, conditions (b) and (d), for example, cannot be fulfilled simultaneously by a homopolymer. Hence various types of copolymers have been devised to overcome this difficulty. In particular, AB block or graft copolymers have been used. In order to increase the χ_s value for the A 'back-bone', polymer interactions stronger than dispersion forces may be introduced (*e.g.*, dipole–dipole, anion–cation, or even chemical bonds).

3 Induced Strong Aggregation

Relaxation of any one of the conditions (a)–(d) above may lead to aggregation; in general, relaxation of (a) or (b) leads to strong aggregation. At low coverages ($\Gamma \ll \Gamma_{\text{plateau}}$), particles may approach to separations ($h \sim \delta$) where polymer bridges may form. This leads to an attractive interaction associated with the increase in the fraction of adsorbed segments (*i.e.*, in trains). Although there will be a repulsive term associated with solvent displacement and compression of the bridges, a relatively deep energy minimum results. Bridging flocculation is an important application in many technologies where separation is required, *e.g.*, in the removal of yeast particles from beer (where 'Isinglass', a protein extracted from the swim bladder of the sturgeon fish, is still widely used), and for bacterial 'harvesting', in processes where bacteria are used to form biopolymers, by feeding them on various chemical 'feedstocks'.

Aggregation may be induced in sterically stabilized dispersions by displacing the adsorbed polymers into bulk solution. In principle, this may be achieved by adding a solvent species that reduces χ_s effectively for the polymer + solvent mixture, such that χ_s is now less than $\chi_{s,c}$. Cohen Stuart *et al.*[10] have given an analysis of the *equilibrium* situation in this regard. However, the generally slow *rate* of such desorption processes may reduce the efficiency of this method in practice. An alternative procedure is to add a second polymeric species that has a higher χ_s value than that of the existing adsorbed polymer, which is therefore displaced. The added polymer, however, must not itself be an efficient steric stabilizer, *i.e.*, it must adsorb in a 'flat' configuration, with few segments in tails (or loops). Multiblock copolymers, in which each block has only a small number of segments, fall into this category. Hence ethylene oxide/propylene oxide multiblocks may be used in this regard to induce coalescence of oil-in-water emulsions, stabilized by proteins and other naturally occurring stabilizing polymers.

4 Weak Reversible Aggregation

Weak reversible aggregation (flocculation) may be induced by variation of δ and χ (conditions (c) and (d) above). In essence, the value of $|V_{min}|$ (Figure 2) has to be increased; by just how much is considered in the next section (§5). Clearly, a decrease in δ, for a given particle size, leads to an increase in V_{min}. One way of demonstrating the effect of variation in δ on V_{min} is to adsorb polymers of varying molecular weight onto charged stabilized particles, and then to add sufficient electrolyte to 'remove' V_E and 'reveal' V_{min}. One set of experiments which illustrates this effect was carried out by Cowell and Vincent[11] for the system of polystyrene latex particles carrying adsorbed poly(ethylene oxide) (PEO) of various molecular weights over the range 1,500 to 20,000 daltons, to which 10^{-2} mol dm^{-3} $Ba(NO_3)_2$ was subsequently added. It was shown that the lower the molecular weight of the PEO used, the lower the volume fraction of the particles at which flocculation was observed. The explanation of this result is given in section 5.

If the parameter χ is varied (by addition of non-solvent, or by a suitable variation in temperature or pressure), such that $\chi > 0.5$, then the mixing term in equation (1) becomes negative (*i.e.*, attractive). The elastic term, however, remains repulsive. The net effect is illustrated in Figure 3. (Note that the possible reduction in δ on making the solvency less is also indicated.) There is a large body of experimental data in the literature illustrating this effect, much of the earlier work being reported by Napper *et al.*[2] They showed that the critical temperature, pressure or non-solvent concentration, beyond which flocculation was observed, corresponded closely to theta-conditions ($\chi = 0.5$).

A third method of inducing weak reversible flocculation in a colloidal dispersion involves adding an excess of free (non-adsorbing) polymer to the continuous phase. This effect arises from the fact that free polymer chains are disfavoured, for entropic reasons, from approaching the particle surface. This

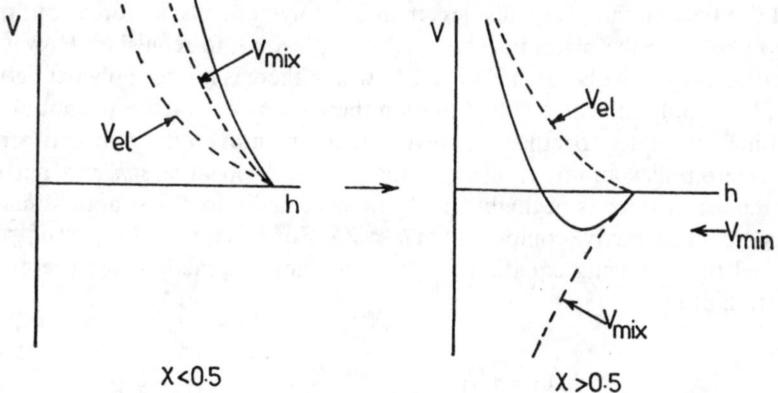

Figure 3 *Schematic representation of the introduction of V_{min} by a change in solvency: V_{el}, elastic term; V_{mix}, mixing term*

leads to the concept of a 'depletion layer' near the surface, having an effective thickness δ of the order of the diameter ($\sim 2r_g$) of the free polymer coils, at least at low polymer concentrations ($c_2 < c_2^*$, the critical coil overlap concentration). At higher polymer concentrations the value of δ tends to decrease, approaching zero in the polymer melt.[12] When two flat plates having such depletion layers approach, the change in the segment concentration profile between the plates is shown schematically in Figure 4(a); the corresponding interaction free energy change (depletion interaction energy) is shown in Figure 4(b).

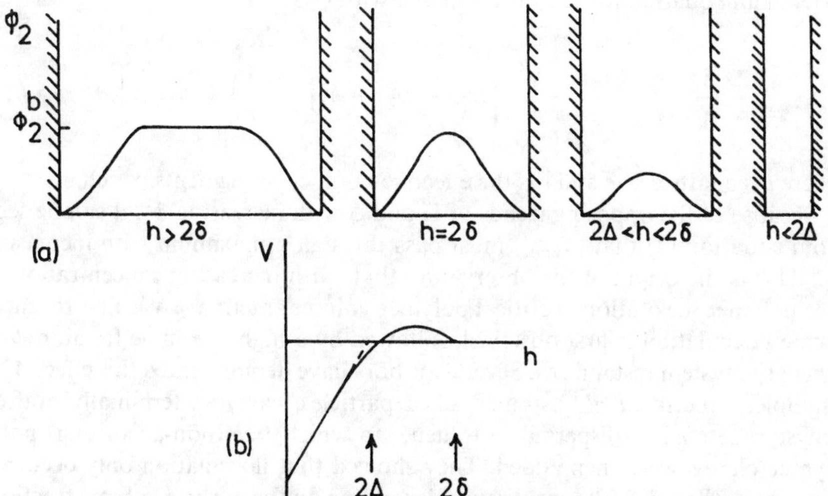

Figure 4 *Schematic representation of depletion interaction: (a) segment volume fraction profile $\phi_2(h)$ between two approaching flat plates (ϕ_2^b = bulk polymer volume fraction); (b) interaction pair potential V(h)*

It can be seen that, for h just less than 2δ, polymer coils are forced out of the region between the plates into bulk solution, leading to repulsion. However, a situation is eventually reached ($h < 2\Delta$) where there is no free polymer between the plates, only solvent. In this situation there is a net attractive (osmotic) force pushing the plates together, as solvent tends to move out from between the plates into bulk solution. It turns out that, for (not too large) *spherical* particles, the repulsion force is negligible, and one need only, to a first approximation, consider the attractive component at $h < 2\Delta$. For this region, Fleer *et al.*[12] have derived the following equation for the depletion interaction free energy as a function of h:

$$V_{\text{dep}} = \left(\frac{\mu_1 - \mu_1^0}{v_1}\right)\frac{2\pi}{3}\left(\Delta - \frac{1}{2}h\right)^2\left(3a + 2\Delta + \frac{1}{2}h\right). \tag{3}$$

Or, for $h = 0$ (*i.e.* at contact, where $V_{\text{dep}} = V_{\text{min}}$) and $2\Delta \ll 3a$, we have

$$V_{\text{min}} = \left(\frac{\mu_1 - \mu_1^0}{v_1}\right)a\Delta^2, \tag{4}$$

where μ_1^0 and μ_1 are respectively the chemical potentials of pure solvent and of solvent at polymer volume fraction ϕ_2^b, and v_1 is the molar volume of the solvent. The term $-(\mu_1 - \mu_1^0)/v_1 = \Pi$ is the osmotic pressure of the bulk polymer solution, for which either experimental data or theoretical estimates (*e.g.* Flory–Huggins) may be used. The osmotic pressure increases with increasing ϕ_2^b, but Δ decreases following the trend in δ (see above). Vincent[13] has derived an equation for the variation of δ with ϕ_2^b,

$$\frac{\delta}{r_g} - \frac{r_g}{\delta} = \frac{r_g}{l}\left(\frac{r}{\phi_2^b}\right)^{2/3}\left[\ln(1 - \phi_2^b) + \phi_2^b\left(1 - \frac{1}{r}\right) + \chi(\phi_2^b)^2\right], \tag{5}$$

where l is a lattice size and r is the effective number of segments per chain.

Because of the opposing trends of Π and Δ with increasing ϕ_2^b, it can be seen from equation (4) that $|V_{\text{min}}|$ must pass through a maximum with increasing ϕ_2^b. This is the origin of the observation that, with increasing concentration of free polymer in solution, a critical polymer volume fraction $\phi_2\dagger$ is first reached, where flocculation is first observed, followed by a higher volume fraction $\phi_2\ddagger$, where the system restabilizes. Several authors have demonstrated this effect. For example, Vincent *et al.*[14] studied silica particles, carrying terminally grafted polystyrene chains, dispersed in toluene, to which *free* (non-adsorbing) polystyrene chains were then added. They showed that flocculation only occurred for $\phi_2\dagger < \phi_2^b < \phi_2\ddagger$, the exact range depending on the particle volume fraction.

The dependence of critical flocculation conditions on particle volume fraction is a general feature of weak reversible flocculation, however induced. The thermodynamic basis for this effect is discussed below.

5 Thermodynamic Aspects of Stability

Classically, the stability of colloidal dispersions has largely been considered from the kinetic viewpoint. In more recent times greater emphasis has been placed on the thermodynamic aspects. This is particularly relevant to sterically stabilized dispersions, where, as we have seen, in many cases flocculation is weak and reversible, and associated with a shallow minimum in the interaction V_{min}. The analogy between weak flocculation and molecular phase separation may be drawn. Indeed, the equilibrium state observed with weakly flocculating particles, in which a dilute (non-aggregated) particulate phase co-exists with a more concentrated (weakly aggregated) particulate phase, is rather analogous to the vapour-state/condensed-state equilibrium observed with molecular systems.

By analogy therefore with (hard-sphere) molecules, one may write[11,15] for the free energy change ΔG_f associated with the flocculation process,

$$\Delta G_f = \Delta G_{hs} + \Delta G_i, \tag{6}$$

where ΔG_{hs} (the hard-sphere term) is the contribution to flocculation associated with the loss in entropy of the particles. It is essentially a function of the initial volume fraction ϕ of the particles. The term ΔG_{hs} ($= -T\Delta S_{hs}$) is positive and it increases as ϕ decreases; this term *opposes* flocculation. The quantity ΔG_i is the interaction term, and it is associated with the interparticle interactions through V_{min}. For systems exhibiting an attractive energy minimum, it is negative and encourages flocculation. Whether flocculation occurs therefore depends on the subtle balance of the ΔG_{hs} and ΔG_i terms. These contributions may be calculated using statistical mechanical theories, and theoretical phase diagrams may be constructed.[16–18] Such phase diagrams delineate regions of stability ($\Delta G_f > 0$) and instability ($\Delta G_f < 0$). The colloidal equivalents of the vapour, liquid and solid (at equilibrium, the colloidal crystal) phases, and their co-existence regions, may be predicted, and have been observed. However, details of this sort are beyond the scope of this introductory review article. The most important point to emphasize is that the onset of reversible flocculation in sterically stabilized dispersions depends not only on the magnitude of V_{min}, but *also* on the particle volume fraction ϕ: the higher the value of ϕ, the more intrinsically unstable is the dispersion. (Increases in ϕ may occur, inadvertently, *e.g.*, by evaporation or settling/creaming.) This concept also accounts for the results referred to in section 4 by Cowell and Vincent.[11]

References

1. G. J. Fleer, M. A. Cohen Stuart, J. M. H. M. Scheutjens, T. Cosgrove, and B. Vincent, 'Polymers at Interfaces', Chapman and Hall, London, 1993, chap. 11.
2. D. H. Napper, 'Polymeric Stabilization of Colloidal Dispersions', Academic Press, London, 1983.
3. D. H. Napper, *J. Colloid Interface Sci.*, 1977, **58**, 390.

4. J. Israelachvili and G. E. Adams, *J. Chem. Soc. Faraday Trans 1*, 1978, **74**, 975.
5. W. A. Ducker, T. J. Senden, and R. M. Pashley, *Nature* (London), 1991, **353**, 239.
6. G. J. Fleer and J. M. H. M. Scheutjens, *Adv. Colloid Interface Sci.*, 1982, **16**, 341, 361.
7. B. Vincent, *J. Colloid Interface Sci.*, 1973, **42**, 270.
8. M. Vold, *J. Colloid Interface Sci.*, 1961, **16**, 1.
9. B. Vincent, *Adv. Colloid Interface Sci.*, 1974, **4**, 193.
10. M. A. Cohen Stuart, G. J. Fleer, and J. M. H. M. Scheutjens, *J. Colloid Interface Sci.*, 1984, **97**, 526.
11. C. Cowell and B. Vincent, in 'The Effect of Polymers on Dispersion Properties', ed. Th. F. Tadros, Academic Press, London, 1982, p. 263.
12. G. J. Fleer, J. M. H. M. Scheutjens, and B. Vincent, *ACS Symp. Ser.*, 1984, **240**, 245.
13. B. Vincent, *Colloids Surf.*, 1990, **50**, 241.
14. B. Vincent, J. Edwards, S. Emmett, and A. Jones, *Colloids Surf.*, 1986, **17**, 261.
15. B. Vincent, P. F. Luckham, and F. A. Waite, *J. Colloid Interface Sci.*, 1980, **73**, 508.
16. I. Snook and W. van Megen, *Adv. Colloid Interface Sci.*, 1984, **21**, 119.
17. A. P. Gast, C. K. Hall, and W. B. Russel, *J. Colloid Interface Sci.*, 1983, **96**, 251.
18. B. Vincent, *Colloids Surf.*, 1987, **24**, 269.

Attractive Interactions and Aggregation in Food Dispersions

By C. G. de Kruif

NIZO FOOD RESEARCH, P.O. BOX 20, 6710 BA EDE, THE NETHERLANDS

1 Introduction

Interactions between food macromolecules and/or food colloids are more often attractive than repulsive. This is mainly due to the amphoteric character of food macromolecules. For instance, proteins usually have positive and negative patches. Attractive interaction on a molecular scale becomes manifest on a macroscopic scale through increased viscosity, low diffusion rates or even gelling. Although repulsive interactions may lead to the same phenomena, the repulsive forces are usually weaker and therefore the effects are extremely small. For instance, the elastic modulus of colloidal crystals is of the order of 1 Pa. Gently rocking the test tube generates stresses of 100 Pa and thus the crystal is destroyed.[1]

In sterically stabilized systems the repulsion becomes noticeable at high volume fractions only. Therefore in practical systems increased viscosity or gelation is without exception due to attractive interactions. We can recognize three types of attractive interactions.

1. London–van der Waals interactions (dispersion forces) including ion bridges, hydrogen bonding, hydrophobic interactions, and polymer bridging. Common to all these interactions is their electromagnetic character.
2. Osmotic forces, which effectively push larger particles together due to differences in (excluded) particle volume of small particles and/or polymers. The larger particles thus feel an effective attraction due to the depletion forces generated by the smaller particles.
3. Covalent interactions: food biopolymers may be crosslinked chemically. An accepted (and from a consumer standpoint preferred) route is enzymatic crosslinking, for instance the formation of ε-(γ-glutamyl)-lysine bonds by the enzyme transglutaminase, or the formation of disulfide bonds through exchange reactions which occur in proteins on heating.

Although these interactions are of a completely different origin on the molecular scale, it seems desirable to treat them within the same framework on a colloidal or macroscopic level in order to understand the physics of the observed phenomena. So, the question is: can we describe macroscopic properties without recognizing the details of the molecular interactions? It is the purpose of this contribution to present such a framework for various food systems. Without giving a complete account, it will be shown that many of the properties of food systems can be understood by characterizing the strength of the interactions between food colloids.

The strength of interaction is best characterized by the depth of the potential of mean force between two particles as a function of the separation between the particles. In Figure 1 we depict a few idealized potentials which are representative for both practical systems and theoretical models. In addition to the depth, the range is also of relevance for the macroscopic behaviour of the system. The range of interaction is defined as the distance over which two particles/macromolecules experience each other's presence. At this point it must be remarked that we consider equilibrium conditions only. We thus leave out the influence of shear-induced long-range hydrodynamic interactions. For strongly attractive systems these contributions can be neglected. Finally, in addition to the strength and range of attraction, there is a third relevant parameter, the volume fraction ϕ of particles/macromolecules in the system.

This paper is further organized as follows. First we discuss the so-called adhesive hard sphere (AHS) model. Then we discuss the consequences of making the width of the potential finite and changing the shape of the potential. Equilibrium properties can be calculated for all types of potentials. However, the calculation of transport properties is mainly limited to the AHS model. In the last section we illustrate the use of the theoretical models for describing practical systems.

2 Theoretical Background

The interaction between two spherical particles is described by a radially symmetrical pair potential as depicted in Figure 1. Given the (relative) position of all particles in space, one can calculate the (free) energy of the system. It is obvious that the position or, better, the correlation of particle positions in space depends also on the pair potential. (For more concentrated systems there are even higher-order particle position correlation functions that must be taken into account.)

For not too concentrated systems, it suffices to calculate the pair correlation function. On the other hand, one may define a radial distribution function through an effective pair potential (thus including higher-order effects). The radial distribution function is then given by

$$g(r) = \exp(-V(r))/kT \tag{1}$$

where $V(r)$ is the effective pair potential which may include higher-order effects.[2]

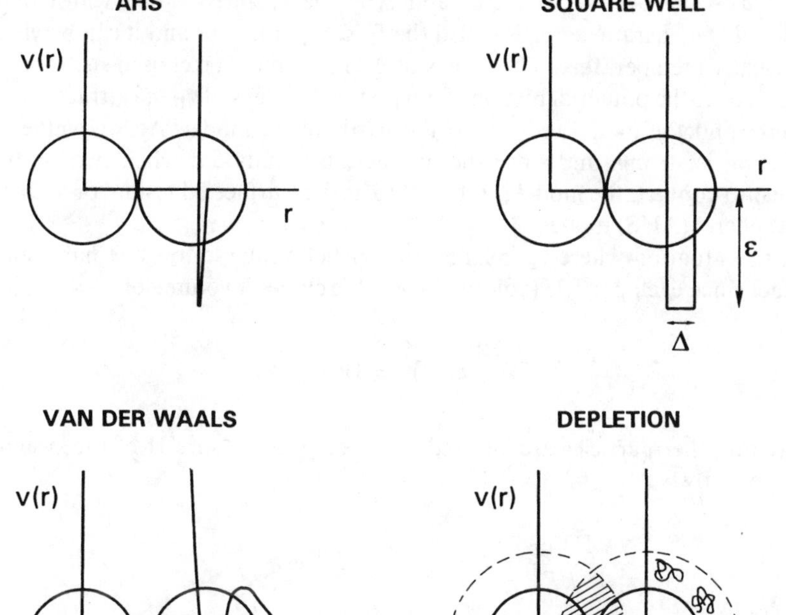

Figure 1 *Schematic representation of forms of colloidal interaction potential V(r): Baxter adhesive hard sphere (AHS) potential, where the range of the potential is infinitely short; square well potential, characterized by a finite width Δ and depth ε; DLVO-potential with primary and secondary minima and a repulsive part; depletion interaction potential according to the Vrij model. Note that these drawings are not to scale*

3 Adhesive Hard Sphere Potential

Baxter[3] solved the statistical mechanics equations by adopting the mathematically convenient potential shown in Figure 1. We will briefly introduce the AHS (or Baxter) model because several theoretical descriptions are based upon it, and because the AHS theory appears to be consistent with computer simulations.

The Baxter potential is defined by

$$\begin{aligned}
\frac{V(r)}{kT} &= \infty\,; && r < 2a \\
\frac{V(r)}{kT} &= \left(\ln\left[12\tau_{\mathrm{B}}\left(\frac{\Delta}{2a+\Delta}\right)\right]\right); && \lim_{\Delta\to 0},\ 2a \le r \le 2a+\Delta \\
\frac{V(r)}{kT} &= 0\,. && r > 2a+\Delta
\end{aligned} \tag{2}$$

Here $2a$ is the particle diameter and Δ is the infinitely small width of the potential. The parameter τ_B is called the Baxter parameter and it can be viewed as a (scaled) temperature. The values of Δ and τ_B must be chosen such that the integral over the potential remains finite. At high values of τ_B the attraction goes to zero, and the model reduces to the hard sphere model. At low values the attraction is strong, and since the interaction is only felt when two particles come into contact, the model is referred to as the 'surface adhesion' or 'adhesive hard sphere' (AHS) model.

In the situation where τ_B is large, the particles interact only at hard sphere contact since each particle (volume $= V_{HS}$) excludes a volume of

$$\frac{4}{3}\pi(2a)^3 = 8V_{HS} \tag{3}$$

where no other particles are allowed. The osmotic pressure Π of the system is then given by

$$\frac{\Pi V_{HS}}{kT} = \phi + B_2^{HS}\phi^2 + B_3^{HS}\phi^3 + \cdots, \tag{4}$$

where ϕ is the volume fraction of spheres ($\phi = \rho .\ V_{HS}$), ρ is the number density, and B_2 and B_3 are the so-called second and third osmotic virial coefficients. For hard spheres we have $B_2 = 4$ and $B_3 = 10$. For non-hard spheres B_2 takes other values; for the AHS model B_2 becomes smaller than 4. If the osmotic virial coefficients are zero, we have the equivalent of a Boyle (osmotic) gas. Then the osmotic pressure is linearly proportional to the density, which is Van't Hoff's law. The second osmotic virial coefficient may be calculated by inserting the Baxter potential in the expression for B_2':

$$B_2' = 2\pi \int_0^{\infty} \{1 - \exp[-V(r)/kT]\} r^2 \, dr. \tag{5}$$

For the Baxter potential this leads to

$$B_2 = B_2'/V_{HS} = 4 - 1/\tau_B, \tag{6}$$

where the factor 4 in equation (6) comes from the hard sphere volume and the term $1/\tau_B$ comes from the attractive interaction. So, by measuring the osmotic pressure, one may find the value of B_2 (and τ_B) as a function of, *e.g.*, temperature, salt, pI, pH, or whatever else determines the strength of the attraction.

4 The Square Well Potential

The square well potential is defined through:

$$\frac{V(r)}{kT} \begin{cases} = \infty, & r < 2a \\ = -\varepsilon, & 2a \leq r \leq 2a + \Delta \\ = 0. & r > 2a + \Delta \end{cases} \tag{7}$$

In contrast to the Baxter potential, the width Δ in equation (7) remains finite. The square well potential can be used to model the interaction of two sterically stabilized particles. The interpenetration depth of the steric layers is set equal to Δ, while the depth of the well is equated to the free energy difference between polymer layers interacting mutually or with the (poor) solvent. The second osmotic virial coefficient is simply found from

$$B_2 = 4 - \frac{\Delta}{2a}[\exp(\varepsilon) - 1]. \tag{8}$$

For two sterically interacting particles the depth of the attractive well is given by

$$\frac{\varepsilon}{kT} = L\left(\frac{\theta}{T} - 1\right), \quad (T < \theta) \tag{9}$$

with L an (enthalpic) polymer–solvent interaction parameter. Thus, by lowering the solvent temperature below the θ-temperature of the polymer chains on the surface of the particle, one induces net attractive interactions between the particles. For small values of $\Delta/2a$, the square well result is equated to the Baxter result at the level of B_2.

5 Depletion Interaction Potential

The depletion interaction potential is depicted in Figure 1 and defined in the Vrij[4] description by

$$\frac{V(r)}{kT} \begin{cases} = \infty, & r < 2a \\ = -Vol_{\mathrm{overl}} \times \Pi_{\mathrm{pol}}, & 2a \leq r \leq 2a + 2R_g \\ = 0. & r > 2a + 2R_g \end{cases} \tag{10}$$

Here Vol_{overl} is the geometrically determined volume of overlap of the depletion layers around two spheres, where R_g is the radius of gyration of the polymer. Simply said, the (unbalanced) osmotic pressure of the polymer or (equivalently) the smaller particles pushes the two colloidal particles together.

The phenomenon of depletion interactions is very common in food dispersions. It occurs when food thickeners are added, or when an excess of, *e.g.*, caseinate is added to an emulsion. The range of the potential is determined by the size of the polymer, and it may thus be of the same range as the particle size in contrast to the AHS and square well models.

6 Other Interaction Potentials

There are several other phenomena that may give rise to an attractive interaction between particles, *e.g.*, the secondary van der Waals minimum, weak bridging flocculation or weak ion binding, hydrophobic forces, *etc.* In general one may characterize the strength and range of the attraction by a few parameters pertinent to the system at hand, and then follow similar routes as described herewith.

It was shown by Regnault and Ravey[5] that the shape of the potential is irrelevant as long as its range is short. The reason for this is probably that one experimentally probes the integral over the potential rather than the potential itself. It is probably for this reason that the AHS model gives such good results. Menon *et al.*[6] refined and extended the Baxter model to finite widths of the potential, basically confirming the above conclusion. For a longer-ranged potential, say $\Delta/2a > 0.1$, the problem is more involved. However, one can use computers to solve the integral equations numerically; see, for instance, the work of Gillan[7] and ten Grotenhuis.[8] Usually the results change somewhat quantitatively, but not qualitatively, with respect to the AHS case.

7 Phase Behaviour of AHS

If the pair interaction strength increases, the colloidal dispersion may show all kinds of phase transitions just as are found with simple gases and liquids on lowering the temperature. Short-range attractions lead to a so-called liquid–solid or gas–solid coexistence while with long-range interactions the system shows transitions of the gas–liquid type and remains more 'fluid'. In Figure 2 we

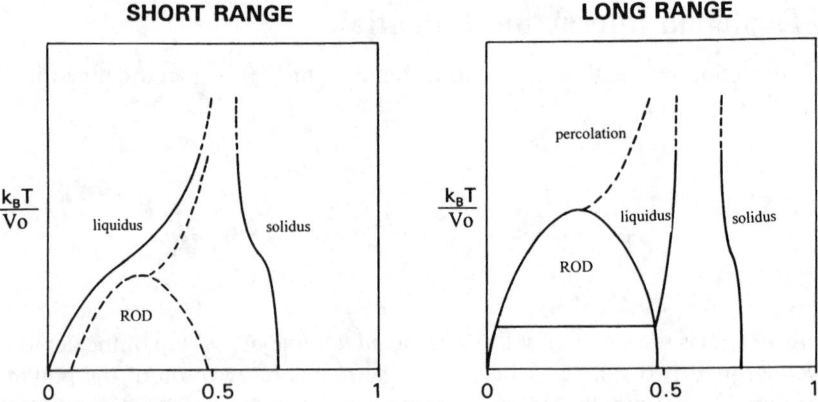

Figure 2 *Phase diagram for colloids with attractive interactions (not to scale). Short-ranged interactions: the dashed (parabolic) liquidus may be hidden under the (colloidal) gas–solid coexistence lines. The percolation line is represented by the upward dashed line. Long-ranged interactions: these give a (colloidal) gas–liquid region of demixing and, at sufficiently strong attraction, a (colloidal) gas–solid demixing region. The dashed line represents the percolation line. ROD is the region of demixing*

have sketched the phase diagram for the two cases. A more detailed discussion is given by Poon *et al.*[9] An interesting feature of the AHS model is that one can calculate the percolation transition. The percolation line is defined by the condition that the so-called pair connectedness function diverges, meaning that there is a 'continuous' path of colloidal particles from one side of the system to the other. It must be realized that this path is not stationary but changes in time. The percolation line may be correlated with the flocculation and gelation of the system:[10]

$$\tau_{\text{perc}} = \frac{19\phi^2 - 2\phi + 1}{12(1-\phi)^2}. \tag{11}$$

This percolation line is represented by the dashed upward line in the diagram of Figure 2.

8 Attractive Potentials and Food Science

There are numerous examples in the literature where use is made of the AHS model. Since we do not intend to give a review, we have chosen a few examples which may be found in practical food systems. These examples may illustrate the applicability and usefulness of the AHS model for understanding the behaviour of practical systems.

Small Globular Proteins

It has long been known that proteins can be crystallized by changing the solvent conditions, notably by the addition of polyvalent salts, *e.g.*, ammonium sulfate is known to induce crystallization.[11] The addition of so-called crystallizing agents seems to affect the protein interactions. George and Wilson[12] measured the osmotic pressure of various globular proteins under conditions where these proteins are known to crystallize. They found that the second osmotic virial coefficient was in a window between 0 and $-6V_{\text{HS}}$ where V_{HS} is the equivalent hard sphere volume of the protein. It was further found that even lower values of B_2 lead to amorphous sediments.

A particularly interesting example is the crystallization behaviour of lysozyme.[13–18] This protein from hen-egg white readily crystallizes under the right conditions. However, under the liquid–solid coexistence phase line (the solubility line) lies a metastable liquid–liquid demixing region (ROD) (see Figure 2). So, by cooling a lysozyme solution, one may observe a liquid–liquid phase separation instead of the thermodynamically more stable solid phase. The reason for this is that the formation of crystal nuclei requires an activation free energy which thus allows the formation of a metastable phase.

It must be noted that an attractive potential is not a prerequisite for crystallization. Colloidal crystals of highly charged particles are easily formed due to the long-range repulsive potential ($B_2 \gg 4$). Even for a suspension of monodisperse hard spheres (where $B_2 = 4$) crystallization occurs at $\phi = 0.5$ only due to excluded (repulsive) volume effects.

Casein Micelles in Milk

Casein micelles are proteinaceous particles containing tiny calcium phosphate clusters. The casein micelles are 100 nm in radius with a log-normal size distribution ($\sigma = 0.5$). This association colloid is extremely stable. It withstands boiling, addition of salts, and even drying, without a severe loss in properties. At room temperature the casein particles can be characterized as hard spheres. The micelles have an outer protein layer of κ-casein (and β-casein?). This steric stabilization layer or 'brush' thus provides the colloidal stability of the casein micelles. If the brush is destroyed, the micelles lose their stability and they flocculate. Casein micelles are destabilized during cheese-making and yogurt-making. In cheese-making an enzyme, chymosin, hydrolyses the brush layer, and if the chain density is reduced by 80–90% the particles irreversibly flocculate. Before that, however, there are reversible attractive interactions. The interactions can then be described using the AHS model. Two casein micelles are modelled to interact through a hard core repulsion preceded by an attractive (square) well of depth ε and width Δ chosen to be 0.3 nm. Thus, the only freely adjustable parameter is ε, which is correlated with the number of

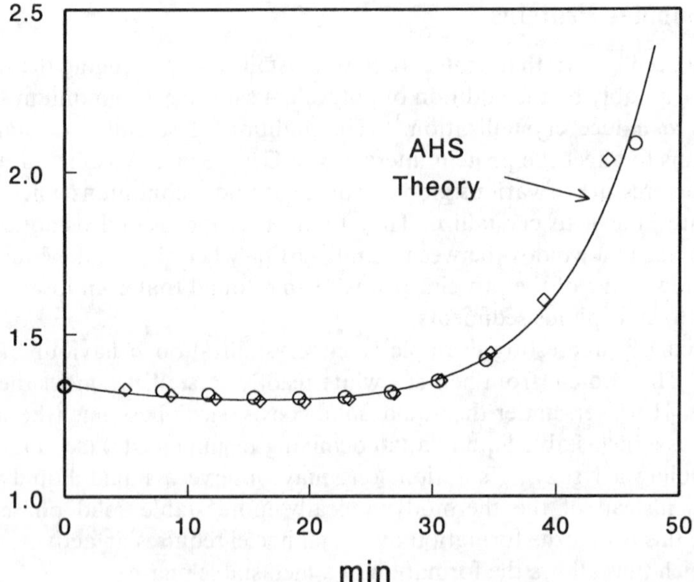

Figure 3 *Relative viscosity of renneted casein micelles as present in skim milk as a function of time. The drawn line represents the theoretical calculation in which the micelles are considered as adhesive hard spheres*

chains still on the surface of the micelle. Using simple first-order enzyme kinetics, the well depth is given by de Kruif:[19]

$$\frac{\varepsilon}{kT} = hk\frac{[E]}{[E_0]}t. \tag{12}$$

Here, h is a proportionality factor (≈ 2), k is the enzyme reaction constant, $[E]$ is the enzyme concentration relative to a reference concentration, and $[E_0] = 0.01$ wt% for a standardized enzyme preparation. As a result of 'cutting the κ-casein hairs' the micelles become slightly smaller, as observed by dynamic light scattering and viscosity measurements.[20] Using the AHS model we can account consistently for both equilibrium and transport properties of the milk. In Figure 3 we give the viscosity as a function of renneting time. The drawn line is that calculated using the AHS model.

Yogurt

In yogurt-making the pH of the milk is gradually lowered through the conversion of milk sugar (lactose) into lactic acid by lactic acid bacteria. If the pH reaches a critical value pC (which depends among other factors on the pre-treatment of the milk), the κ-casein brush collapses on the surface and steric stability is lost. As a result the caseins flocculate. Although the process is rather abrupt, there is a pH region before the point of flocculation, pC, where the micelles can be modelled as AHS. The potential well depth is given by de Kruif[21] as

$$\frac{\varepsilon}{kT} = \frac{1}{\mathrm{p}C - \mathrm{pH}}. \tag{13}$$

In Figure 4 we have given the apparent particle radius as measured by dynamic light scattering as a function of pH. The collapse of the κ-casein brush thus occurs at pC = 4.9 for skim milk. This value is higher for heated yogurt milk. This is probably due to the deposition of whey protein on the micelles. This shifts pC to higher values, but the AHS model can still be applied to describe the system.[22]

If the value of pC coincides with the collapse of the protein brush, then it is clear that partially renneting the brush will increase pC to higher values. In Figure 5 we have given the scaling relation between brush density (proportional to σ) and the average charge density ($1/\alpha$) where α is the average distance between charges. The charge density $1/\alpha$ is directly related to the degree of dissociation of the carboxylic acid groups and can therefore be calculated from the pH of the milk.[23] From Figure 5 it is clear that the micelles lose their stability, *i.e.*, that the κ-casein brush collapses due to renneting (= low brush density) and due to lowering of the pH (= low charge density on the κ-casein 'hairs'). But it is not the electrostatic repulsion between two different micelles which is diminished. The lower charge density of the chains is supposed to reduce their solvency as a polyelectrolyte, and that is why the 'hairs' collapse.

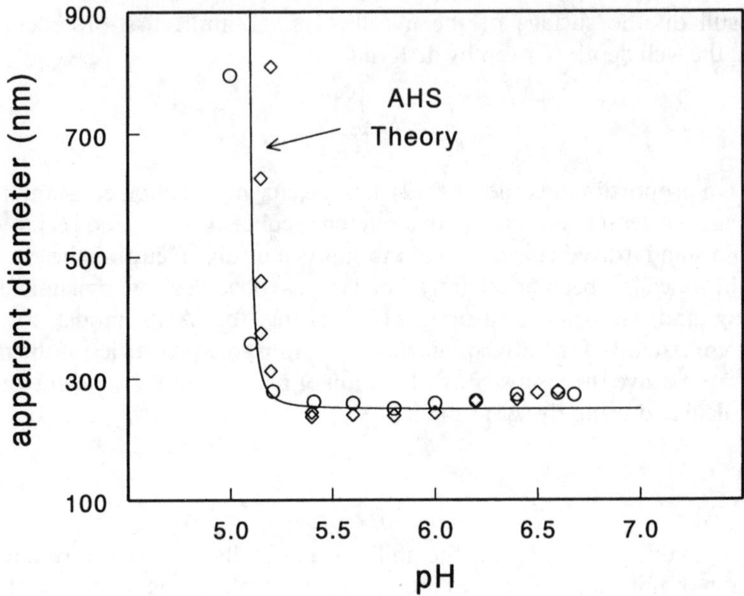

Figure 4 *Apparent particle diameter of the casein micelles as determined with dynamic light scattering as a function of pH*

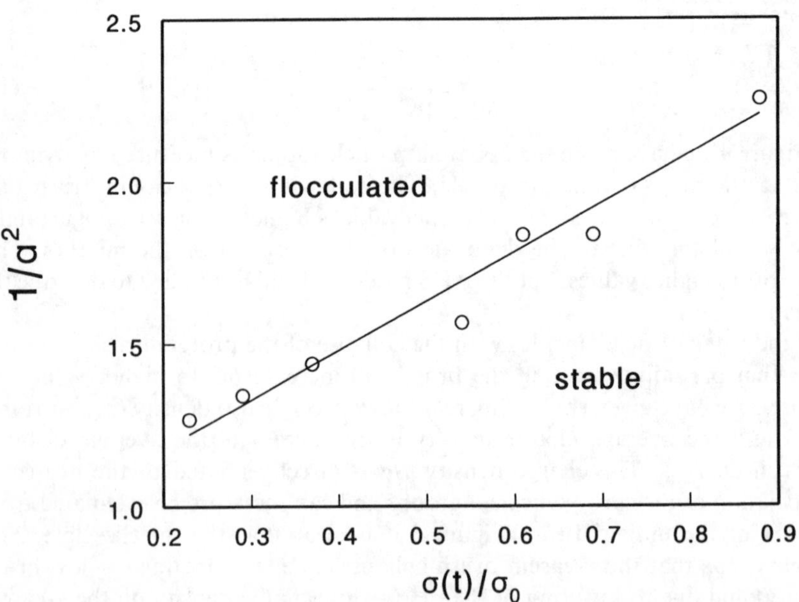

Figure 5 *Stability threshold of casein micelles. Above the line the casein brush is collapsed and therefore the system flocculates. The scaling relation is between charge density squared ($1/\alpha^2$) and chain density σ*

Caseins + Polysaccharides

It is very common to add so-called thickening/stabilizing polysaccharides to milk (derived) products. Notably the addition of carrageenans (κ, ι, λ), xanthan, guar and several others is customary.

Starting with the work of Snoeren,[24] it has usually been assumed that the polysaccharides form bridging functions between the micelles due to their adsorption. However, in practice, bridging flocculation seems to be more the exception than the rule. The dominant mechanism is the depletion interaction. The phenomenon of depletion interaction was first described by Oosawa and Asakura[25] for a polymer between two flat walls. It was Vrij[4] who developed a simple, but physically appealing, model for the interaction between 'hard' particles as induced by a non-adsorbing polymer. In Figure 1 a sketch of the depletion mechanism is given. Around each sphere is a so-called depletion layer into which the centre of gravity of a polymer cannot enter. This leads to a reduced (polymer) osmotic pressure (Π) in the depletion layer. In the Vrij model the system gains free energy if two depletion volumes overlap ($Vol_{\rm overl}$), since more volume becomes available for the polymer. The gain in free energy is $\Delta G \propto \Delta\Pi\ Vol_{\rm overl}$. Therefore two particles in a (dilute) polymer solution experience an effective attractive potential as depicted in Figure 1. The result of such an attractive interaction is that the system may phase separate into a colloid-rich/polymer-poor phase and a colloid-poor/polymer-rich phase. In the Vrij model this phase line can be calculated, given the colloid particle size and the volume fraction. For the polymer one needs to know its radius of gyration, its molecular mass, and its weight concentration.

We recently isolated an exocellular polysaccharide (EPS) as produced by lactic acid bacteria.[26] This EPS (coded EPS-B40) has R_g = 86 nm and $M_n = 1.5 \times 10^6$; and it is rather monodisperse (M_w/M_n = 1.13). By adding EPS-B40 to casein micelles we determined the phase boundary[26] as depicted in Figure 6. The drawn line is the prediction of the Vrij theory with no adjustable parameters.

Polydispersity in particle size seems to influence the crystallization behaviour. From the phase diagram as measured by Bourriot *et al.*[27] we conclude that the polydispersity of the polymer shifts the phase boundary to higher polymer concentration. Bourriot *et al.*[27] required about 6 times more of their non-adsorbing polysaccharide (guar) as compared to our monodisperse EPS-B40.

Emulsions

Crude oil-in-water emulsions are easily made by vigorously mixing the oil in water using a suitable surfactant, *e.g.*, sodium dodecyl sulfate (SDS), or a protein like sodium caseinate. Although polydisperse, the emulsion droplets as such are perfectly stable in time and do not coalesce or gel. The emulsion droplets in cream or milk are reasonably stable in their native form, and the viscosity of cream is close to what may be expected for a dispersion of hard spheres.[28]

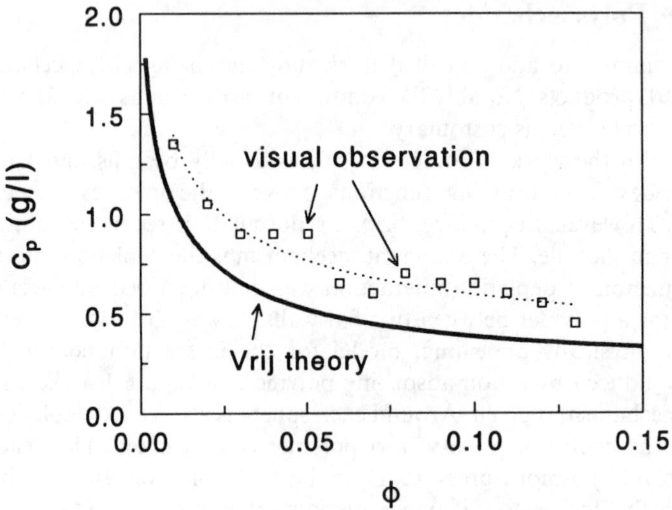

Figure 6 *Phase diagram of casein micelles + added (random coil) polysaccharide (EPS B40). At high EPS concentration c_p the system phase separates into two liquid phases (layers) of which the lower is rich in casein micelles and the upper is rich in EPS; ϕ is volume fraction of casein micelles. The drawn line is a theoretical prediction of the Vrij model*

Under certain conditions the interaction between emulsion droplets becomes attractive. Here we would like to discuss three cases. We derive some examples from the work of Bibette and coworkers, the reason being that Bibette has developed[29] a method to fractionate crude emulsions into highly monodisperse emulsions in which the complicating influence of polydispersity is eliminated. Bibette *et al.*[30] used an emulsion of silicone oil stabilized by SDS. On addition of extra NaCl the emulsion droplets exhibit an attractive interaction which also depends on temperature. With increasing strength of the attraction the system was found to show liquid–solid coexistence, as is expected for short-range attractions. If the attractions are made stronger, the system flocculates (without breaking the emulsion) into a rigid but stable gel network of emulsion droplets. In the emulsion droplet fractionation process as developed by Bibette *et al.*,[30] direct use is made of depletion interactions. An excess of SDS is added to a crude emulsion. The SDS forms surfactant micelles. As a result of the difference in size between surfactant micelles and emulsion droplets a phase transition of the gas–solid type is induced.[30] Since the depletion force is proportional to the radius of the emulsion droplets, the large droplets are phase-separated first. By repeating this process monodisperse emulsions of different sizes can be obtained.

The same phenomenon of depletion flocculation was observed by Dickinson and co-workers[31–33] in emulsions stabilized by sodium caseinate. An excess of sodium caseinate also forms micelles, 15–20 nm diameter, that induces flocculation and enhanced creaming of the emulsions. Since the dispersions used by Dickinson are practical, and thus polydisperse, it is not possible to get an ordered state, and therefore a 'glass state' results.

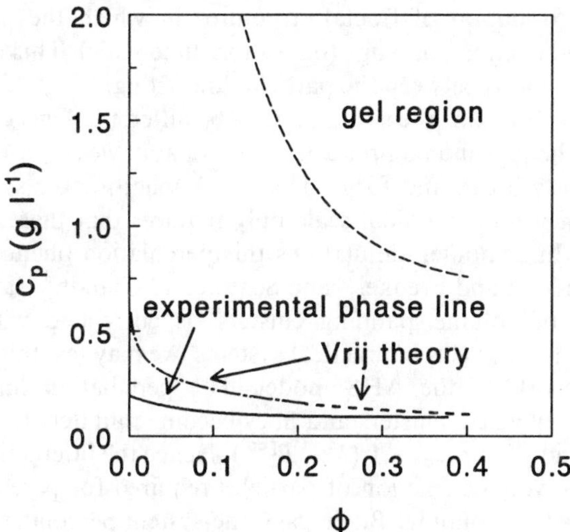

Figure 7 *Phase diagram of an emulsion of volume fraction ϕ with added (random coil) polysaccharide (EPS B40). At high EPS concentration c_p the system phase separates into two liquid layers of which the lower is rich in EPS and the upper is rich in emulsion droplets. The drawn line is a theoretical prediction of the Vrij model. Above the upper dashed line the system is trapped into a gel or glass state*

The last example is an emulsion to which a polysaccharide has been added. A highly practical example is salad dressing, which is a concentrated emulsion with xanthan added. Parker *et al.* showed[34] that this system is inherently unstable due to the depletion forces of the xanthan. However, the system is 'frozen' into a gel network, and is thus kinetically stabilized, *i.e.*, the shelf-life is long enough. A particularly interesting phenomenon first addressed by Parker[34] is the delayed creaming in such dispersions. It appears that gel networks formed on preparing emulsion/polysaccharide systems after a certain time (hours/weeks) do show creaming. This phenomenon of delayed creaming is only partially understood, and it seems to be related to gravity forces.

In Figure 7 we show the phase behaviour of an emulsion to which the polysaccharide EPS-B40 is added. At low concentrations we observe a phase separation, while at higher EPS concentrations we observe gelling.

Flocculation, Floc Time and Gelation

When casein micelles are renneted they are slowly destabilized. After a certain time—in cheese practice this is about 20 minutes—flocs or specks are visible on the wall of the test tube.[35] Similarly cooling a polymer-stabilized latex or silica dispersion in a marginal quality solvent leads to flocculation at a certain temperature. Heating β-lactoglobulin solutions also leads to formation of large flocs and clusters.[36] Usually after this initial flocculation a gel is formed. For low particle concentrations (*i.e.*, $\phi < 0.1$) this behaviour can easily be understood by assuming the formation of fractal structures in which the particles are strongly bound and cannot rearrange (on a short time scale). This implies that the attractive interactions between the particles are strong.

At higher volume fractions the situation may be different. A gel can again be formed, but now the pair interaction can be reasonably weak, so that particle bonds spontaneously break and form. Thus on a macroscopic scale a gelled network exists which on a colloidal scale only requires that there is a certain bond probability. In computer simulations this percolation phenomenon was studied by Kranendonk and Frenkel[37] and Seaton and Glandt.[38] If percolation (*i.e.* the presence of volume-spanning clusters) is correlated with the first formation of large flocs or gels in practical systems, we may test this hypothesis within the framework of the AHS model. The percolation line (*i.e.*, the asymptotic boundary where clusters and flocs become infinitely large) is given by Chiew and Glandt[10] as $\tau_{perc} = 1/(4 - B_2^{perc})$. Hence the interaction strength is correlated to the volume fraction of particles required for percolation. For instance, for $B_2 = 0$ (a colloidal Boyle gas), the system percolates at $\phi_{perc} = 0.25$.

We have tested[39] whether this model could be applied to renneting of milk. Above it was shown that the pair interaction of casein micelles is a unique function of the degree of enzymatic cleavage of the outer κ-casein hairs. We also showed[39] that the enzymatic reaction was independent (within experimental error) of the volume fraction of casein micelles.

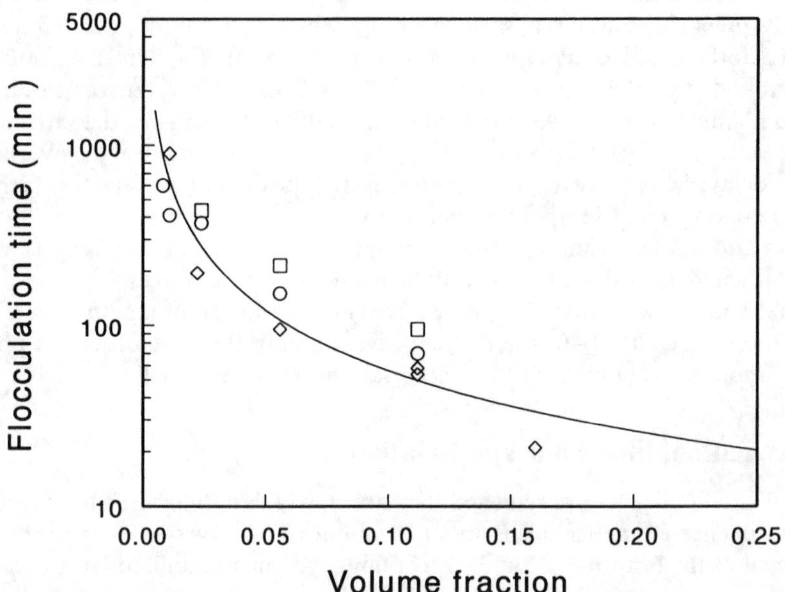

Figure 8 *Flocculation time of casein micelles as a function of volume fraction at a constant chymosin/casein ratio. The drawn line represents the time when the system passes the percolation line*

On renneting concentrated skim milk at a constant casein/enzyme ratio, it is observed that the floc time is progressively lowered with increasing volume fraction of casein micelles. By applying the AHS model to the casein micelles, we can calculate the point in time where the system passes the percolation line after the addition of rennet. In Figure 8 we plot the experimentally observed floc time against the calculated floc time (*i.e.*, at percolation). The consistency between experiment and theory is rather good, indicating that the concept of adhesive hard spheres is useful for understanding practical food systems.

9 Conclusions

Attractive interactions in colloidal systems may be of various origins which are reflected in the range and strength of the attraction. Modelling these interactions with an appropriate (model) potential allows the application of statistical mechanics theories to describe properties such as osmotic pressure, light scattering, flocculation and phase behaviour. If a mapping can be made with the AHS model which was initially developed by Baxter[3] for atomic systems, one can even calculate the first-order ϕ dependence of transport properties such as viscosity and diffusivity. We have shown that the AHS model provides a general basis for describing the physical stability of practical food systems.

References

1. P. M. Chaikin, J. M. di Meglio, W. D. Dozier, H. M. Lindsay, and D. A. Weitz, in 'Physics of Complex Fluids', ed. S. A. Safran and N. A. Clark, Wiley, New York, 1987, p. 65.
2. J. K. G. Dhont, 'An Introduction to Dynamics of Colloids', Elsevier, Amsterdam, 1996.
3. R. J. Baxter, *J. Chem. Phys.*, 1968, **49**, 2770.
4. A. Vrij, *Appl. Chem.*, 1976, **48**, 471.
5. C. Regnaut and J. C. Ravey, *J. Chem. Phys.*, 1989, **91**, 1211.
6. S. V. G. Menon, C. Manohar, and K. Srinivasa Rao, *J. Chem. Phys.*, 1991, **95**, 9186.
7. M. G. Gillan, *Mol. Phys.*, 1979, **38**, 1781.
8. E. ten Grotenhuis (computer program, NIZO, personal communication).
9. W. C. K. Poon, P. N. Pusey, and H. N. W. L. Lekkerkerker, *Physics World*, April 1996, 27.
10. Y. C. Chiew and E. D. Glandt, *J. Phys. A., Math. Gen.*, 1983, **16**, 2599.
11. R. Aschaffenburg and J. Drewry, *Biochem. J.*, 1956, **65**, 273.
12. A. George and W. W. Wilson, *Acta Cryst.*, 1994, **D50**, 361.
13. N. Asherie, A. Lomakin, and G. B. Benedek, *Phys. Rev. Lett.*, 1996, **23**, 4832.
14. M. Malfois and F. Bonneté, *J. Chem. Phys.*, 1996, **105**, 3290.
15. D. Rosenbaum, P. C. Zamora, and C. F. Zukoski, *Phys. Rev. Lett.*, 1996, **76**, 150.
16. M. Muschol and F. Rosenberger, *J. Chem. Phys.*, 1995, **103**, 10424.
17. R. Hoskins, I. D. Robb, P. A. Williams, and P. Warre, *J. Chem. Soc., Faraday Trans.*, 1996, **92**, 4515.
18. W. C. K. Poon, *Phys. Rev. E*, 1997, **55**, 3762.

19. C. G. de Kruif, Th. J. M. Jeurnink, and P. Zoon, *Neth. Milk Dairy J.*, 1992, **46**, 123.
20. C. G. de Kruif, *Langmuir*, 1992, **8**, 2932.
21. C. G. de Kruif, *J. Colloid Interface Sci.*, 1997, **185**, 19.
22. A. Parker, personal communication.
23. C. G. de Kruif and E. B. Zhulina, *Colloids Surf. A.*, 1996, **117**, 151.
24. T. H. M. Snoeren, T. A. J. Payens, J. Jeunink, and P. Both, *Milchwissenschaft*, 1975, **30**, 393.
25. S. Asakura and F. Oosawa, *J. Chem. Phys.*, 1954, **22**, 1255.
26. R. Tuinier and C. G. de Kruif, submitted for publication.
27. S. Bourriot, C. Garnier, and J.-L. Doublier, *Les Cahiers de Rhéologie*, 1997, **15**, 284.
28. L. W. Phipps, *J. Dairy Res.*, 1969, **36**, 417.
29. J. Bibette, T. G. Mason, H. Gang, D. A. Weitz, and P. Poulin, *Langmuir*, 1993, **9**, 3352.
30. J. Bibette, D. Roux, and B. Pouligny, *J. Phys. II (France)*, 1992, **2**, 401.
31. E. Dickinson, M. Golding, and M. J. W. Povey, *J. Colloid Interface Sci.*, 1997, **185**, 515.
32. E. Dickinson and M. Golding, *Food Hydrocolloids*, 1997, **11**, 13.
33. E. Dickinson, *J. Dairy Sci.*, 1997, **80**, 2607.
34. A. Parker, P. A. Gunning, K. Ng, and M. M. Robbins, *Food Hydrocolloids*, 1995, **9**, 333.
35. N. J. Berridge, *Analyst (London)*, 1952, **77**, 57.
36. M. Verheul, Ph.D. Thesis, University of Twente, Enschede, 1998.
37. W. Kranendonk and D. Frenkel, *Mol. Phys.*, 1988, **64**, 403.
38. N. A. Seaton and E. D. Glandt, *J. Chem. Phys.*, 1987, **87**, 1785.
39. C. G. de Kruif, in 'Food Colloids and Polymers', ed. E. Dickinson and P. Walstra, Royal Society of Chemistry, Cambridge, 1993, p. 54.

Food Dispersion Stability

By Rajendra P. Borwankar, Bruce Campbell, Christian Oleksiak, Theodore Gurkov,[1] Darsh T. Wasan,[2] and Wen Xu[2]

KRAFT FOODS, 801 WAUKEGAN ROAD, GLENVIEW, IL 60025, USA
[1]UNIVERSITY OF SOFIA, 1126 SOFIA, BULGARIA
[2]ILLINOIS INSTITUTE OF TECHNOLOGY, CHICAGO, IL 60616, USA

1 Introduction

As a world's leading food processor, Kraft Foods is deeply interested in gaining a fundamental understanding of the stability of its food products, many of which are food emulsions and/or foams. Emulsions and foams are thermodynamically unstable and will separate over time. The primary modes of destabilization of emulsions involve creaming, flocculation and coalescence. These processes occur concurrently and tend to build on each other. In the case of foams, coalescence and Ostwald ripening are the primary culprits. While we admit that Ostwald ripening has not been seriously investigated as much as other mechanisms, we believe it is not very prevalent in food systems.

In the case of emulsions, Hartland and Gakis[1] and later Lobo *et al.*[2] developed models of stability of batch emulsions which tend to cream. Van den Tempel[3] and Borwankar *et al.*[4] have accounted for simultaneous flocculation and coalescence using simple models incorporating 'average' flocculation and coalescence rate constants and other simplifying assumptions. We later developed[5] a more detailed model by extending Smoulchowski theory to account for coalescence within aggregates. At the core of all these attempts to describe more rigorously the stability of emulsions is the coalescence rate constant. Over the past three or four decades, it has been generally accepted that coalescence in foams and emulsions is controlled by the thinning and rupture of thin liquid films between bubbles or droplets. Hence, we have focused the majority of our effort on the role of thin films in the stability of food emulsions and foams.

This paper summarizes the salient findings of our most recent work in the area of food dispersion stability. Stratification and location of emulsifier play a major role in governing the drainage and stability of thin films, especially in the case of low-molecular-weight surfactants. Particularly interesting are the results on non-equilibrium systems which are able to rationalize the otherwise contradictory results between theory and practice. Proteins, the major workhorse in the stabilization of food emulsions and foams, make a fascinating, albeit complicated, study. Through the study of thin films of proteins we are now beginning to unravel the details of protein stabilization of emulsions and foams. While much of our attention has been focused on the study of single thin liquid films and their role in governing food dispersion stability, we have also recently started investigating emulsion systems in a macroscopic way using Kossell diffraction, which has allowed us to study the role of fat particle structure in the stabilization of foamed emulsions. In this paper we first present the microscopic world of thin liquid films and later move on to reporting the investigations using Kossell diffraction. We hope to show that we have pushed the limits of the application of the techniques for thin film studies and Kossell diffraction from model systems to real systems. Results on real food systems will be presented.

2 Thin Film Studies

The first thin film studies revolved around the use of hydrodynamic modelling to determine the lifetimes of thin liquid films as they drain and eventually rupture. These models highlight the roles of interfacial tension gradients (Gibbs–Marangoni effect) and interfacial viscoelasticities (see, for example, references 6 and 7). But, in relation to food dispersions, these models have three major shortcomings. Firstly, the interfacial tension gradient mechanism fails to explain the stability of emulsions which are prepared by incorporating an emulsifier in the droplet phase, since according to this mechanism locating the emulsifier in the droplet phase should be detrimental to the interfacial tension gradients which are needed to be sustained for emulsion stability. Secondly, the mechanism fails to explain the stability behaviour of emulsions prepared at surfactant concentrations above the critical micelle concentration (CMC), where interfacial tension gradients are diminished. And, thirdly, these models predict lifetimes of thin films to be of the order of seconds or at most minutes, whereas food dispersions are required to be, and are, stable for days, weeks or months. Later, a mechanism of stratification or micellar stabilization was observed[8] which addresses the second point. Eventually, this mechanism was extended[9] to show that thermodynamically metastable films can result under certain conditions, thereby addressing the third point.

In our investigations of systems of interest to food industrial practice, we have discovered new mechanisms relevant to low-molecular-weight surfactants under non-equilibrium conditions which provide satisfactory explanations of practical observations. We have also now firmly established that structure formation, closely related to stratification, is widely prevalent in food systems and is likely to be responsible for their long-term stability.

Emulsion Stabilization under Non-equilibrium Conditions

Low-molecular-weight surfactants which are soluble both in water and in oil are frequently used as emulsifiers. Normally, they are initially dissolved in one of the phases only. Just after formulation of the emulsion, there is a time period when intensive surfactant redistribution and mass transfer through the interfaces takes place. On the other hand, during this period the adsorption of amphiphilic molecules at the newly created oil–water surfaces may still be incomplete. When two droplets with such unsaturated and highly mobile oil–water interfaces collide, it is clear that, due to the lack of sufficient surface elasticity and viscosity, the thin liquid film between them will drain very fast, and rupture will eventually occur, unless other factors favouring stabilization are present. Since some time has to be allowed until a tightly packed rigid adsorption layer of surfactant develops, whereupon direct repulsive forces (steric, electrostatic, structural) come into play and so provide long-term stability, our attention is directed to the initial transit period which may be crucial in determining whether the emulsion will survive. Our model studies with thin liquid films made in a glass capillary have shown that new stabilizing effects are operative in non-equilibrium conditions, when the surfactant is being transferred from one of the bulk phases to the other. We have looked at both cases, one where the surfactant is initially dissolved in the droplet phase and the other where it is initially dissolved in the thin film phase.

At pointed out by Hartland,[10] when a surface-active solute diffuses across the interface from the continuous phase (film) to the dispersed phase (droplet), this should lead to stabilization. Because the volume of the film is small, the solute concentration there falls more rapidly than in the bulk of the continuous phase, and hence the interfacial tension is greater around the film centre and correspondingly smaller at the periphery and in the meniscus. A tensile restoring force emerges, which is opposite in direction to the surfactant density gradient[11] (*i.e.*, from higher to lower surface pressure). It sets the fluid interface in motion, and liquid is dragged towards the film centre. This is a manifestation of the well-known Marangoni effect,[10,11] and is expected to provide high stability. Such a qualitative picture has been confirmed by recent experiments,[12] although it turns out that the real processes are far more complex.

In the paper of Velev *et al.*,[12] a fascinating cyclic phenomenon was encountered during observations of aqueous emulsion films between oil phases in the presence of non-ionic surfactant (soluble both in water and in oil). When the surfactant is initially dissolved in the film phase, and diffuses across the interface towards the oil, its transfer induces spontaneous growth of a lens-like region (dimple) around the film centre (Figure 1). The thickness along the periphery remains more or less constant. Upon reaching a certain size, the dimple flows out into the meniscus, and a new one starts to form. The process is cyclic and usually goes on for many hours. Its driving force was proven to be the surfactant mass transfer.[12] A clear distinction must be made between this cyclic dimpling and the so-called hydrodynamic dimple. The latter appears only when the film is thinning, and is not connected in any respect with mass transfer through the

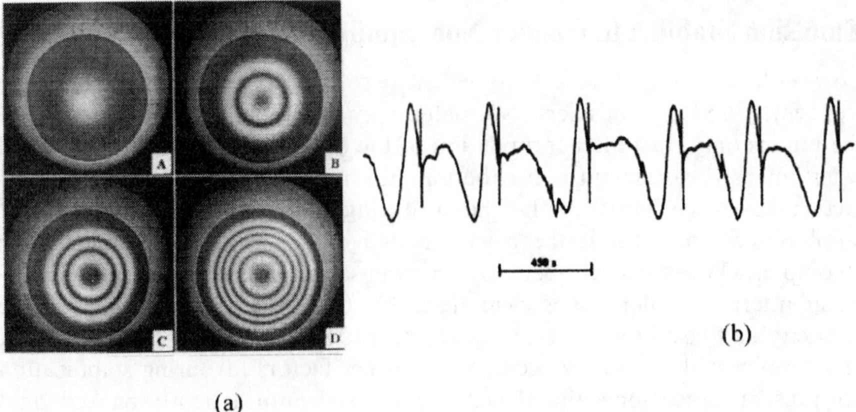

Figure 1 *Oscillating dimple in an emulsion film (diameter $\approx$ 330 μm). (a) Interferometric image in reflected light of wavelength λ. Black and white colours correspond to multiples of $\lambda/(4\mathrm{n})$, where* n *is the refractive index in the film; (b) intensity as a function of time of light reflected from a small fixed area of the film*
(Reproduced with permission from ref. 12)

phase boundaries. The film surfaces are deformed under the action of the viscous friction. At a certain stage of thinning the hydrodynamic dimple irreversibly disappears and the film becomes plane-parallel.

A comprehensive theoretical description of the process of dimple growth has been proposed by Danov *et al.*[13] These authors have proven that the molecular surface diffusion and the surface viscosity are insignificant factors in the phenomenon. Moreover, the diffusion flux of surfactant from the film phase towards the interface turns out to be negligible as well. The main flux is by surface convection, *i.e.*, by the inward flow along the oil–water boundary, directed from the meniscus to the film, which provides a continuous supply of new surfactant (see Figure 2). A major role can be attributed to the Marangoni effect, connected with interfacial tension gradients. As a fluid element moves to the film centre, along the interface, it continuously loses surfactant because the latter goes to the oil phase. Therefore, the surface concentration decreases and the interfacial tension σ rises with diminishing values of the radial coordinate r. The induced gradients of σ set the fluid surface into motion. Thus, liquid is dragged into the film, feeding the dimple, as shown in Figure 2.

The theoretical calculations of dimple shapes are in good agreement with experimental dimple profiles measured interferometrically.[13] Quasi-stationary shapes are possible for not very big diffusion fluxes (j_0). Otherwise, large negative pressures develop near the film periphery and eventually lead to expulsion of the dimple. The analysis of the flow properties reveals the appearance of vortices inside the film; they show up after a certain period of dimple growth. In this system the stabilizing effect, which keeps the film very thick for a long time, can be attributed to hydrodynamic reasons, in particular, to net repulsion when the pressure is integrated along the liquid–liquid interfaces.

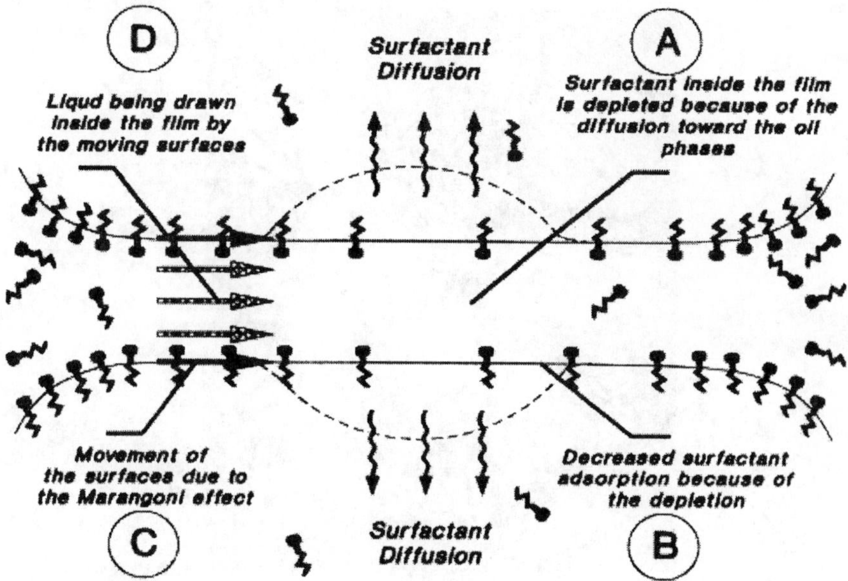

Figure 2 *Schematic illustration of how the surfactant fluxes lead to the phenomenon of the oscillating dimple*
(Reproduced with permission from ref. 13)

The complementary case, when surfactant is initially dissolved in the oil (droplets) and afterwards diffuses across the phase boundaries towards the aqueous film, is even more interesting.[14] Very good stability was observed, contrary to the expectations based on the conventional theory, which maintain that locating emulsifier in the droplet phase is detrimental to formation of interfacial tension gradients which must be sustained to confer emulsion stability.[6] One should also point out that this stability cannot be attributed to electrostatic or any other conventional type of repulsion. Experiments were carried out[14] with aqueous films between xylene phases, in the presence of non-ionic surfactant (Tween 20) which was initially dissolved in the oil. The water contained 0.1 mol l^{-1} NaCl. The films remained thick (above 100 nm) and absolutely stable, until the equilibrium distribution of surfactant was reached (which took up to 48 hours). Then, the films thinned down and ruptured. In this system one observes a specific dynamic pattern, consisting of intensive liquid circulation through channels and exchange of mass in the lateral direction between the film and the Plateau border (see Figure 3).

In order to clarify the physical mechanism of the phenomenon, the surfactant concentration in the oil phase was varied.[14] It turns out that the effects emerge in a threshold manner: below a certain 'critical' concentration, there is no stabilization at all. Experiments were performed also with films whose phases contained given amounts of surfactant (C_0) before being put in contact with the oil. The stabilizing effect and the general pattern of the process were found to be the same for $C_0 = 0, 1, 10, 100$ and even $500 \times$ CMC. The value C_0 represents a

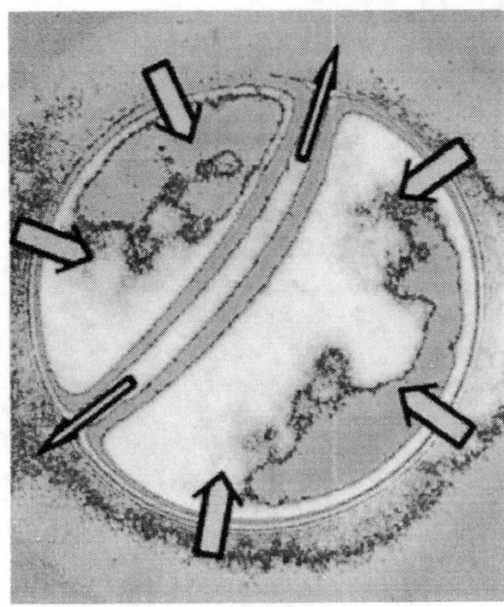

Figure 3 *Aqueous film stabilized by 5 × 10^{-4} M Tween 20. The surfactant is initially dissolved in the surrounding xylene phase and diffuses to the film across the interfaces. The arrows indicate liquid fluxes brought about by gradients in osmotic pressure of surfactant micelles*
(Reproduced with permission from ref. 14)

background concentration, evenly distributed both in the film and in the meniscus during the experiment (Figure 4). Such a high surfactant content does not influence the film stability, which indicates that only the *excess* concentration in the film with respect to the meniscus is important.

These findings may be rationalized by the following explanation (see Figure 4). As the surfactant diffuses through the interfaces, its concentration in the aqueous film gradually increases. Two types of solute species exist in the water phase: single molecules (monomers) and micellar aggregates. The diffusion of the monomers is fast, and they are uniformly distributed in the aqueous phase. The local monomer concentration in the film cannot increase above the CMC because the individual molecules, after crossing the oil–water boundary, quickly form micelles. As the monomer concentration in water is constant, the diffusion flux coming from the oil cannot stop until the equilibrium distribution is reached. The thin and broad film gets enriched with micelles, whose concentration in the voluminous Plateau border remains low (the Brownian diffusion of the big aggregates is slow). The increased osmotic pressure of *micelles* in the film engenders convective fluxes of liquid; material is sucked from the meniscus, and after saturation with surfactant it rushes out through the channels of higher thickness (Figure 3).

As discussed by Ford *et al.*,[15] the primary role of emulsifiers is stabilizing emulsions that arise upon droplet rupture during high energy emulsification in

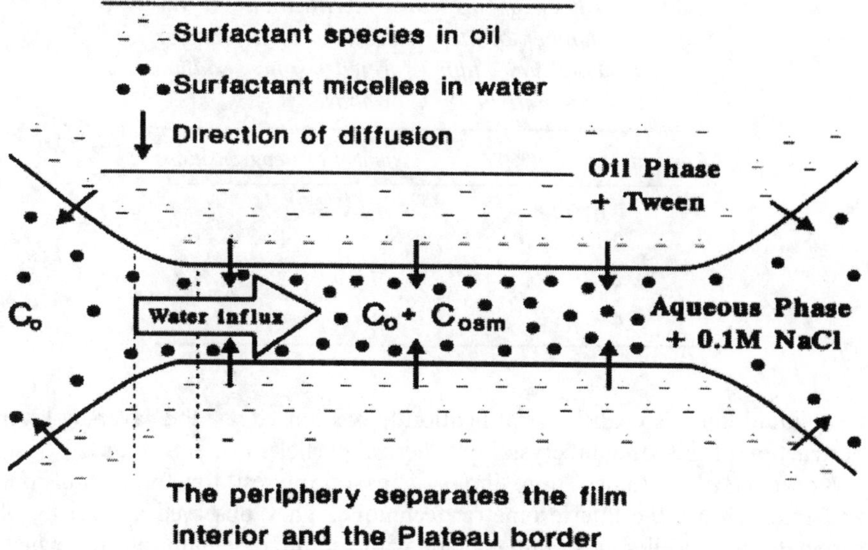

Figure 4 *Sketch of an aqueous film explaining the origin of osmotic pressure differences and liquid fluxes which are due to the non-equilibrium distribution of surfactant micelles*

conventional emulsion processing equipment. 'Fast' emulsifiers (*e.g.*, LEO-10) adsorb rapidly so that the interfacial tension of interest in droplet breakage is quite close to the equilibrium interfacial tension, resulting in what Armbruster calls mechanically limited dispersion.[16] Then it was not very apparent to us how such systems may be stabilized under the intense environment during emulsification (high compressive forces during collisions). Perhaps the mechanisms discussed above are relevant. For a 'slow' emulsifier (Armbruster's example is egg yolk), the operative interfacial tension during breakage is much higher than the equilibrium interfacial tension, resulting in mechanically limited dispersion. In addition, the resulting emulsion is not stabilized by the above mechanisms because no significant equilibration of egg yolk occurs, since its proteins are not soluble in the oil phase, and recoalescence is not completely prevented.

Stratification in Food Systems

The conventional (DLVO) theory for the stability of dispersions takes into account only two kinds of molecular interactions: electrostatic and van der Waals. An important type of non-DLVO surface force is the structural interaction, which manifests itself when the continuous phase contains small colloidal particles such as, for example, micelles or latex spheres. In general, whenever a surface bounds a liquid phase, ordering is induced among the particles close to the wall. In the case of a film, the structured regions near the two opposing surfaces overlap; this gives rise to an oscillatory disjoining pressure and interaction energy.[9] This effect causes a step-wise thinning of the

Table 1 *Effect of protein concentration on the number of step transitions of a foam or emulsion film formed from sodium caseinate at 40 °C or above*

Concentration (wt%)	*Number of step transitions*
0.01	0
0.1	1
0.5	2
2	3
4	4

thin liquid film (so-called 'stratification'), recognized as the layer-by-layer destruction of the colloidal crystal of spherical particles inside the film.

Koczo *et al.*[17] studied thin film systems of interest for food emulsions and foams using the interferometric technique. They observed a layering of caseinate sub-micelles in thin films made with sodium caseinate solution which was similar to the stratification observed with surfactant micelles and latex particles. As shown in Table 1, the number of step transitions was found to increase with increasing concentration of sodium caseinate, and for the extent of stratification it did not matter whether the film was a foam film or an emulsion film (with oil phase of *n*-hexadecane). The calculated effective volume fraction of the sub-micelles in a 2 wt% solution of caseinate is 20%. The three-step transitions observed at this concentration are in good agreement with previous work[9] in which three-step transitions were observed for non-ionic and anionic micellar systems in the range of 10–20% effective volume fraction.

The layering of sodium caseinate sub-micelles in foam and emulsion films results in increased drainage time as the layers are removed one-by-one via a series of step transitions. Furthermore, under certain conditions (low temperature, small film size), the film transition can be completely inhibited so that drainage stops with the film still containing one or more layers of micelles. Such films are rather thick and so can be very stable. Thus, the layering of sub-micelles can prevent two oil droplets or air bubbles from approaching together. This is proposed as a new mechanism of stabilization for these systems.

Such layering need not be confined specifically to sodium caseinate. Indeed, Koczo *et al.* proposed[17] that layering could occur in other systems, especially those containing globular proteins. One of the systems of interest to us is similar to a frozen whipped topping formulation. Frozen whipped toppings are prepared by making an oil-in-water emulsion followed by cooling to crystallize the fat. The emulsion is then aerated and whipped to create a whipped topping that can be frozen. Typically, sodium caseinate and lipid emulsifiers are used along with gums. Our investigations have revealed that polysorbate micelles, and even macromolecules of the food hydrocolloids studied (xanthan and guar gum), can induce layering in thin liquid films. And it was found that, when sodium caseinate, lipid emulsifiers and gums are all present simultaneously, as

Table 2 *Effect of various food ingredients on the number of film transitions of a foam film*

Solution	*Temperature (°C)*	*Number of step transitions*
0.01% Polysorbate 60	25–80	0
0.1% Polysorbate 60	65, 80	1
0.5% Polysorbate 60	45	0 (film size < 0.3 mm)
0.5% Polysorbate 60	45	2 (film size > 0.3 mm)
0.5% Polysorbate 60	65, 80	2
0.5% Polysorbate 60 + 2% sodium caseinate	25, 65	3
0.5% Polysorbate 60 + 2% sodium caseinate	80	4
Polysorbate 60, sodium caseinate, xanthan, guar, sugars, corn syrup	65–80	6–9
0.1% xanthan gum		a
0.05% guar gum		a

[a] Step transitions observed, but number not recorded; step height equals size of random coil of gum molecule, indicating microlayering of gum molecules.

in a real whipped topping system, up to 9 film transitions could be observed (see Table 2).[18]

Model Investigations on Emulsion Systems Stabilized by Proteins

The amphiphilic nature of proteins makes them suitable for stabilizing oil-in-water emulsions, especially in food applications. Here we discuss some recent results from model studies with two types of proteins: globular (BSA, bovine serum albumin; BLG, β-lactoglobulin) and disordered (β-casein). Their molecular structures have been characterized in bulk aqueous solutions, *e.g.*, circular dichroism and hydrodynamics, but not yet fully determined for molecules adsorbed on liquid interfaces.[19]

Marinova *et al.*[20] have explored the properties of thin aqueous films (diameter 200–400 μm) using a Scheludko cell. This approach has been shown to provide insights regarding the interaction of two approaching protein-stabilized interfaces under a variety of conditions of ionic strength, pH, specific ion effects, and ageing time. Specifically, the technique has been used to measure the interaction energies between pairs of interfaces through their contact angles as well as their film thickness. With the Scheludko cell, electrostatic interactions have been shown to be operative at pH values away from the isoelectric point of the protein. It has been found[20] that, without added salt, the films containing 0.015 wt% BSA at pH = 6.6 (isoelectric point pI = 4.8) remain very thick and stable. Even a minor amount of inorganic electrolyte (as little as 10^{-3} M NaCl) is sufficient to screen the electrostatic repulsion, and this results in the formation of very thin Newton black films. Another piece of evidence[20] for the importance of the molecular charge is provided by the fact that the contact angle varies with pH and passes through a maximum around the isoelectric point, p*I*.

Electrostatic interactions are also observed in thin films stabilized by BLG and β-casein. BLG behaves similarly to BSA although the BLG interfacial behaviour appears to be more pH sensitive near its p*I* (p*I* = 5.4). The BLG emulsion films containing 0.15 M NaCl at pH $\approx$ 6.2 were found to be highly unstable, with rupture occurring almost immediately after the appearance of black spots. Thin films stabilized by β-casein are also sensitive to ionic strength (thick films are obtained without added salt, while the presence of 0.15 M NaCl allows Newton black films to form[20,21]). This electrostatically dominated behaviour is qualitatively similar to prior observations by other investigators, and it effectively validates the Scheludko cell approach for the use of protein-stabilized thin films as models for droplet–droplet interactions in emulsions.

New findings obtained from the above mentioned technique include the establishment of the involvement of protein aggregates in the thinning process. The precise behaviour of these aggregates during film thinning is dependent on protein type and solution conditions. For example, BLG aggregates at pH $\approx$ 6.2 seem to be essentially stiff and they cannot be broken down under the action of the capillary pressure inside the film. However, the gradual disintegration of large protein clusters and an increase in the contact angle accompanied the formation of highly stable BLG Newton black films at pH $\approx$ 7.0. The process is illustrated in Figure 5, where a sequence of photographs shows the time course of protein aggregate disappearance. Finally, after nearly five minutes, the whole film thins to a Newton black film. In the case of β-casein, it is well known that aggregation occurs in solution at room temperature. Any protein lumps caught in black emulsion films were observed to be readily disintegrated by the capillary pressure, which indicates that the aggregation is reversible and that the interfaces are mobile.

Of the proteins investigated, β-casein has some unique properties which can impact on interfacial behaviour, *e.g.*, its high affinity for Ca^{2+} ions. When the film thickness is measured interferometrically,[20,21] a decrease from ~20 to ~10 nm is observed when the Ca^{2+} concentration is raised above 10 mM (Figure 6). The effect may be attributed to compression of the protein adsorption layers (which has been confirmed to happen at a single interface),[20] or to partial desorption. At the same time, the energy of adhesion between the two interfaces of a Newton black film increases enormously.[21] The cross-binding (at 0.01 wt% casein and 20 mM $CaCl_2$) is so strong that the film surface cannot be detached on applying the maximum capillary pressure accessible in the conventional capillary cell.[21]

Figure 5 *Interferometric images taken at six consecutive moments during thinning of an aqueous film of β-lactoglobulin in the presence of 0.15 M NaCl and 0.001 M phosphate buffer at pH 6.9. Progressive destruction of protein lumps is observed and finally a Newton black film forms. The distance between marks corresponds to 30.25 μm*

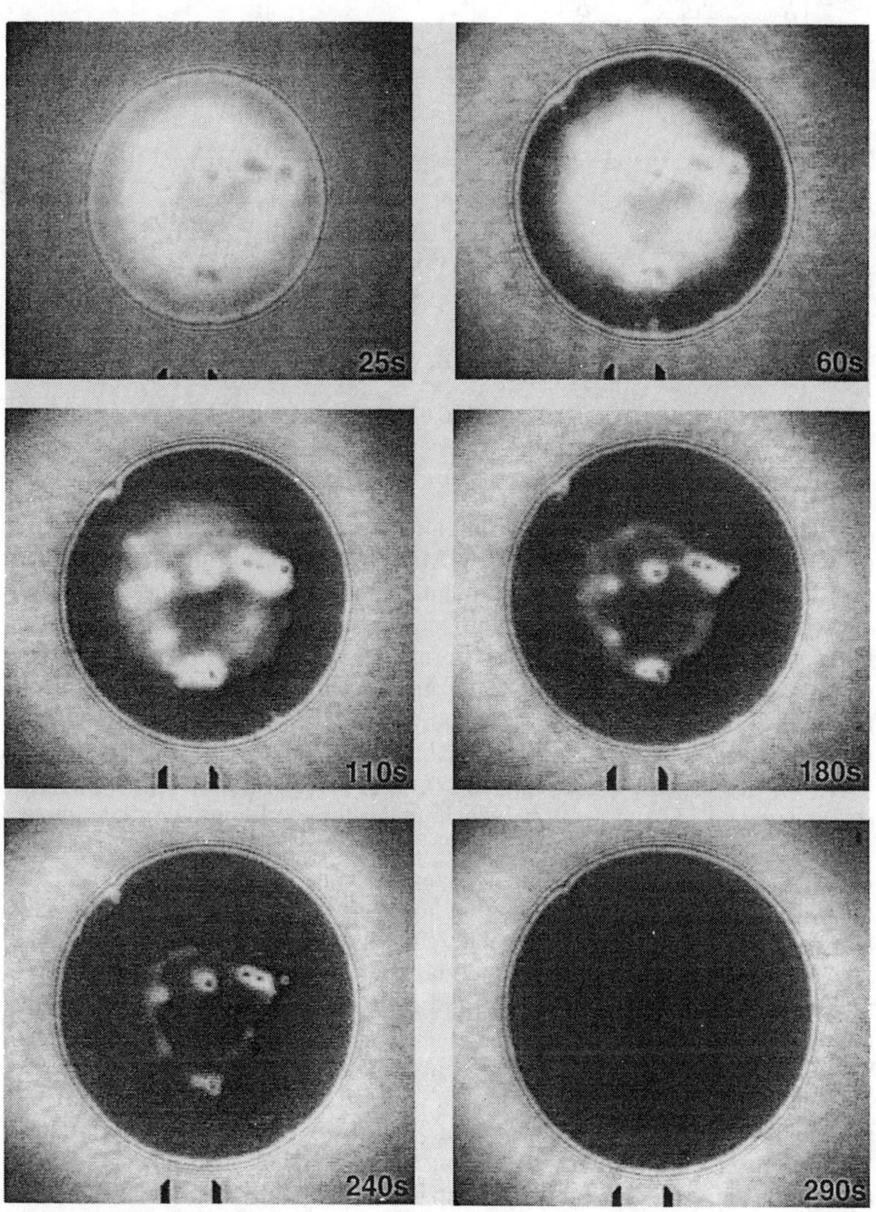

Figure 5

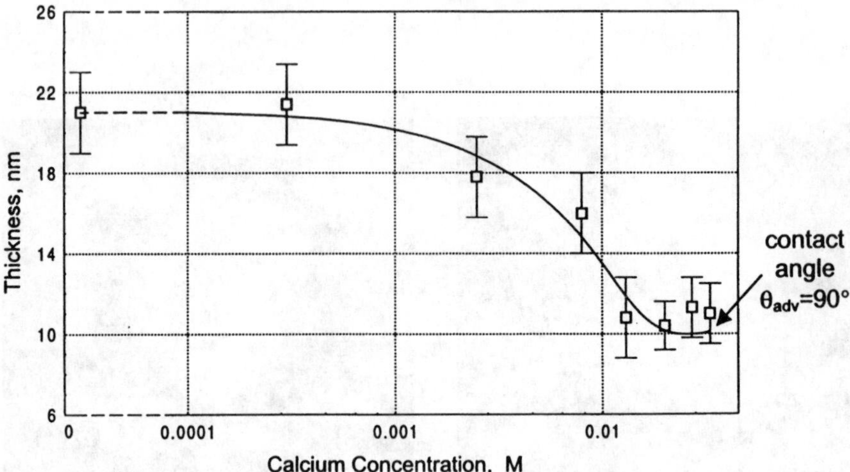

Figure 6 *Thickness of emulsion films with 0.01% (wt) β-casein in the presence of Ca^{2+} ions. Ionic strength* I = *0.15 M (adjusted by NaCl), pH* = *5.0*

Utilizing the thin film approach, interfacial ageing effects of adsorbed proteins have been characterized by Marinova *et al.*[20] Thin film ageing is investigated by comparing fresh thin film formation (studied immediately after introducing and exposing the oil surfaces to the protein solution) with aged film formation (in which the film is formed 30 minutes after exposing the oil surface to the protein solution). Thin film systems stabilized by BSA in the presence of 0.15 M NaCl were observed to exhibit pronounced ageing effects. Freshly made films typically form very thin, plane-parallel Newton black films. Films formed after 30 minutes ageing contain protein aggregates of irregular shape. Unlike BLG and β-casein, which, under the conditions described above, aggregate both in the bulk and on the film surface, BSA aggregates do not occur in the bulk as confirmed by dynamic light scattering. Therefore, the presence of these aggregates is due to aggregation *at the surfaces*. This aggregation is seen to be reversible. Initially the films formed from aged systems thin down and entrap liquid and protein lumps. Subsequently, excess material is gradually squeezed out, and the particles are dispersed under the action of the capillary pressure.

A theoretical explanation of this phenomenon has recently been proposed.[22] This is consistent with the action of surface diffusion as the underlying mechanism for protein disaggregation, as shown by the good fit obtained between theory and experiment. From the best fit one finds the maximum adsorption along the rim where the lump is attached to the interface. The results suggest the existence of protein multilayers at the liquid surface.

Another manifestation of the ageing effect is the hysteresis of the contact angle and its time dependence. The time dependence of the contact angle θ was measured for BSA-stabilized films for more than one hour after loading the two phases in the cell.[20] At fixed capillary pressures, the value of θ was found to keep on increasing for approximately one hour in an open film (for example, from $\theta = 0.57°$ up to $\theta = 0.8°$ at pH = 6.4). After that, a more or less constant angle is reached. In addition, well pronounced hysteresis of θ develops in these films upon film separation. Hysteresis in these systems is observed experimentally by pushing liquid into the meniscus, which in turn reduces the capillary pressure until the contact line starts shrinking. In this manner one determines the advancing angle, θ_{adv}, which characterizes the force required to detach the adhered film surfaces. The initial value of θ_{adv} with 0.015 wt% BSA is estimated[20] to be *ca.* 2°. After ~30 minutes, a substantial increase of θ_{adv} is such that its measurement becomes impossible, since the Newton fringes around the film periphery cannot be discerned. These films are 'solid-like', *i.e.*, the two interfaces are stuck firmly and irreversibly. In rheological terms, one would say that a 'yield stress' should be applied in order to disjoin the surfaces. The corresponding energy of adhesion thus increases sharply with ageing. Some numerical calculations for the interfacial shape in the transition zone film/meniscus are underway, which will give us the opportunity to evaluate the line tension in films stabilized by proteins. Let us mention here that the notion of 'equilibrium angle' is meaningful only as long as the surfaces retain their fluid nature. Only then can the film diameter adjust for the respective values of the contact angles and capillary pressures in the cell.

3 Structure Formation and Stability of Food Emulsions and Foams

The research discussed above is based on studying two isolated droplets or bubbles and the intervening film between them. It is anticipated that the results from such systems can be applied to understand the behaviour of emulsions and foams as a whole. Recently, we have also pursued a macroscopic approach where the structure of the emulsion and/or foam can be probed directly. Kossell diffraction or 'back' light scattering has been known for many decades. Earlier information obtained by the technique was qualitative. We have recently developed an experimental technique based on this phenomenon that allows the obtaining of quantitative information about the structure of food systems in a non-invasive way without elaborate need for sample preparation. This technique involves shining a monochromatic laser beam on a sample in a glass cell and measuring the back light scattering using a vertically polarized CCD digital camera. Digital signal processing allows for the diffraction pattern to be analyzed to give the particle dispersion structure. The technique was developed and tested on model systems[23] before we ventured into food systems of interest. Again, we studied the frozen whipped topping system.

We have measured the effects of sodium caseinate on fat particle packing structure.[24] The results of back light scattering experiments on a 5.4% fat emulsion system containing two different levels of sodium caseinate are shown in Figure 7. The higher peak of the sample containing twice as high a sodium caseinate level as the other sample indicates that increasing sodium caseinate

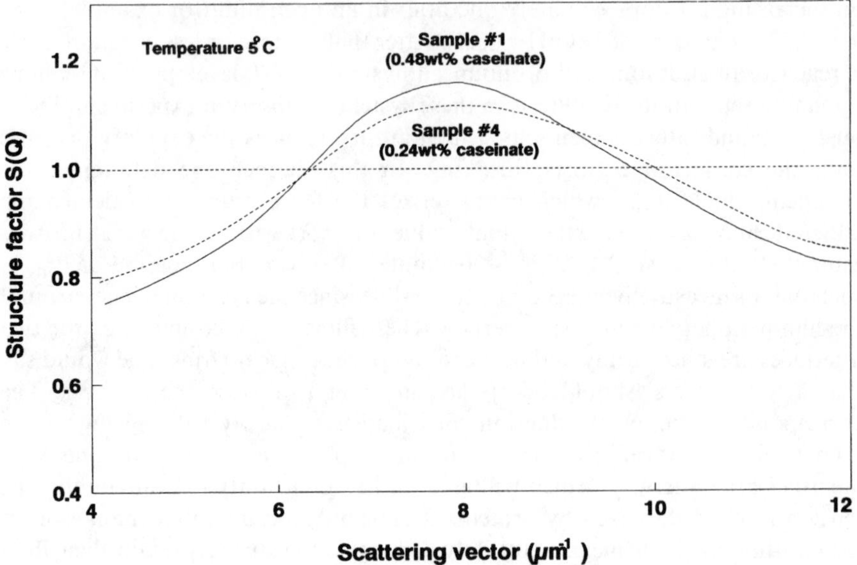

Figure 7 *The effect of caseinate on the fat particle structure inside food emulsions (fat concentration = 5.14%)*

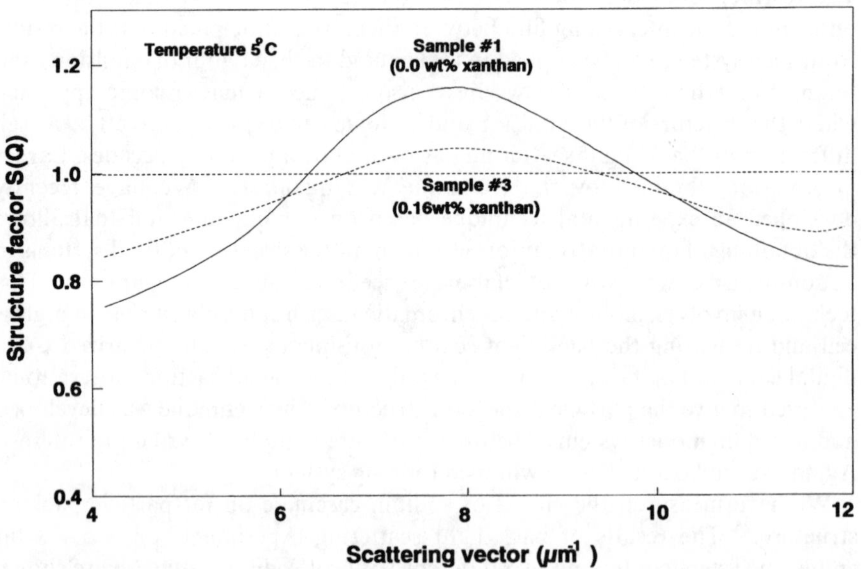

Figure 8 *The effect of xanthan on the fat particle structure inside food emulsions (fat concentration = 5.14%)*

concentration facilitates fat particle structure formation. This is consistent with the observed stabilization of emulsions by sodium caseinate sub-micelles.[17] Theoretical calculations show[24] that the structural energy barrier can be much larger than 3 kT when sodium caseinate sub-micelle concentration reaches 20%.

The effect of xanthan gum on food emulsion stability has also been studied. As shown in Figure 8, the peak height of the sample without xanthan was much higher than that of the sample with xanthan indicating destabilization of the emulsion by xanthan gum. Based on the observation that flocculation of emulsions by xanthan gum can occur at concentrations as low as 10–20 ppm, Koczo *et al.* have argued[25] that the destabilization by xanthan molecules here is by a different mechanism than the depletion flocculation mechanism suggested by Cao *et al.*[26] Rather, they showed the destabilization to be similar to the phase separation of geometrically incompatible polymers demonstrated by Flory.[27] Here, the xanthan molecules (anisotropic phase of elongated xanthan molecules with high aspect ratio) are incompatible with the emulsion droplets (isotropic phase).

The back light scattering method has also been applied to demonstrate fat particle structuring occurring during the whipping process when such an emulsion is whipped in a two-stage process. As measured by the normalized structure factor value, the fat particle structure was improved after each successive whipping stage. A well developed fat particle structure between the air cells is crucial to the stability of the whipped topping.

References

1. S. Hartland and N. Gakis, *Proc. R. Soc. London A*, 1979, **369**, 137.
2. L. Lobo, I. Ivanov, and D. Wasan, *AIChE J.*, 1993, **39**, 322.
3. M. van den Tempel, *Rec. Trav. Chim.*, 1953, **72**, 419, 433.
4. R. P. Borwankar, L. A. Lobo, and D. T. Wasan, *Colloids Surf.*, 1992, **69**, 135.
5. K. D. Danov, I. B. Ivanov, T. D. Gurkov, and R. P. Borwankar, *J. Colloid Interface Sci.*, 1994, **167**, 8.
6. I. B. Ivanov, *Pure & Applied Chem.*, 1980, **52**, 1241.
7. Z. Zapryanov, A. K. Malhotra, N. Adrengi, and D. T. Wasan, *Int. J. Multiphase Flow*, 1983, **9**, 105.
8. A. D. Nikolov, D. T. Wasan, P. A. Kralchevski, and I. B. Ivanov, in 'Ordering and Organization in Ionic Solutions', ed. N. Ike and I. Sogami, World Scientific Publishing, Singapore, 1988.
9. A. D. Nikolov and D. T. Wasan, *Colloids Surf. A*, 1997, **128**, 243.
10. S. Hartland, in 'Thin Liquid Films', ed. I. B. Ivanov, Marcel Dekker, New York, 1988, p. 663.
11. D. A. Edwards, H. Brenner, and D. T. Wasan, 'Interfacial Transport Processes and Rheology', Butterworth-Heinemann, Boston, 1991.
12. O. D. Velev, T. D. Gurkov, and R. P. Borwankar, *J. Colloid Interface Sci.*, 1993, **159**, 497.
13. K. D. Danov, T. D. Gurkov, T. Dimitrova, I. B. Ivanov, and D. Smith, *J. Colloid Interface Sci.*, 1997, **188**, 313.

14. O. D. Velev, T. D. Gurkov, I. B. Ivanov, and R. P. Borwankar, *Phys. Rev. Lett.*, 1995, **75**, 264.
15. L. D. Ford, R. Borwankar, R. W. Martin, Jr., and D. Holcomb, in 'Food Emulsions', 3rd edn, ed. S. E. Friberg and K. Larsson, Marcel Dekker, New York, 1997.
16. H. Armbruster, 'Untersuchungen zum kontinuierlichen Emulgierprozess in Kolloidmuhlen unter Beruksichtigung spezifischer Emulgatoreigenscheften und Stromungsverhatnissen im Disperfierspalt', Dissertation, University of Karlsruhe, Karlsruhe, Germany, 1990.
17. K. Koczo, A. D. Nikolov, D. T. Wasan, R. P. Borwankar, and A. Gonsalves, *J. Colloid Interface Sci.*, 1996, **178**, 694.
18. K. Koczo, D. T. Wasan, and R. P. Borwankar, Paper presented at the AIChE Annual Meeting, San Francisco, November, 1994.
19. E. Dickinson, *J. Chem. Soc. Faraday Trans.*, 1992, **88**, 2973.
20. K. G. Marinova, T. D. Gurkov, O. D. Velev, I. B. Ivanov, B. Campbell, and R. P. Borwankar, *Colloids Surf. A*, 1997, **123–124**, 155.
21. O. D. Velev, B. Campbell, and R. P. Borwankar, manuscript in preparation.
22. T. D. Gurkov, K. G. Marinova, A. Zdravkov, C. Oleksiak, and B. Campbell, *Prog. Colloid Interface Sci.*, submitted.
23. W. Xu, A. Nikolov, D. T. Wasan, A. Gonsalves, and R. P. Borwankar, *J. Colloid Interface Sci.*, 1997, **191**, 471.
24. W. Xu, Ph.D. Thesis, Illinois Institute of Technology, Chicago, 1997.
25. K. Koczo, D. T. Wasan, R. P. Borwankar, and A. Gonsalves, *Food Hydrocolloids*, 1998, **12**, 43.
26. Y. Cao, E. Dickinson, and D. J. Wedlock, *Food Hydrocolloids*, 1990, **4**, 185; 1991, **5**, 443.
27. P. J. Flory, *Macromolecules*, 1978, **11**, 1138, 1141.

Emulsifying Properties of β-Casein and its Hydrolysates in Relation to their Molecular Properties

By P. E. A. Smulders, P. W. J. R. Caessens, and P. Walstra

DIVISION OF FOOD SCIENCE, DEPARTMENT OF FOOD TECHNOLOGY AND NUTRITIONAL SCIENCES, WAGENINGEN AGRICULTURAL UNIVERSITY, P.O. BOX 8129, 6700 EV WAGENINGEN, THE NETHERLANDS

1 Introduction

Proteins generally have excellent emulsifying properties and they are therefore often used in food emulsions. The emulsion forming and stabilizing properties depend on protein molecular properties such as molar mass, conformation, and charge, and on physico-chemical conditions such as pH and ionic strength. Despite numerous studies, the relation between these properties remains uncertain.

When studying emulsions, a distinction has to be made between formation and stabilization. During emulsion formation, droplets are deformed and broken up, proteins adsorb at the newly created oil–water interface, and droplets collide and possibly recoalesce.[1,2] The final average droplet size of an emulsion is the result of a steady state between break-up and recoalescence of droplets.

The most common types of physical instabilities for food emulsions are creaming, aggregation, and (partial) coalescence.[3] Coalescence of droplets is often the most important type of instability, since this process eventually leads to complete separation of the oil and water phase. Coalescence takes place when the film between two droplets ruptures, and it is more likely to occur for large, creamed, or aggregated droplets.[3]

Bovine caseins are often used in food emulsions. One of the major caseins is β-casein, a protein with little secondary structure,[4,5] no intramolecular cross-links,[6] and approximately random coil properties. The N-terminal domain (residues 1–43) of the protein is highly hydrophilic and contains five phosphoseryl residues, while the remainder of the protein contains many hydrophobic residues.[6] At neutral pH most of the protein net charge is carried by the N-terminal domain, and the net charge of the remainder of the protein is nearly zero.[7] The uneven distribution of residues gives β-casein its strongly amphiphilic, detergent-like properties.[6] At temperatures below 4 °C, β-casein exists as a monomer,[8] whereas at higher temperatures micelles are formed above a critical concentration.[9] The micellization is mainly the result of hydrophobic interactions involving the C-terminal domain of the protein[10] and it is opposed by electrostatic interactions, as shown by the effects of temperature, ionic strength,[9] and pH.[11] The conformation of β-casein at the oil–water interface appears to be similar to its conformation in micelles.[12] The flexible conformation of β-casein allows adsorption in a 'brush-like' conformation with a dense layer of hydrophobic parts near the oil–water interface and the charged N-terminal domain protruding far into the aqueous phase.[13]

Due to these distinct and well-known features, β-casein and its peptides provide a useful tool for structure–function research. Caessens *et al.*[14] found large differences in the emulsion forming and stabilizing properties of well-defined β-casein peptides. In this investigation the properties of some of these peptides are studied in more detail in order to elucidate the relation between emulsion and molecular properties of proteins and peptides. To that end, peptide properties were compared with properties of β-casein under various physico-chemical conditions.

2 Materials and Methods

Materials

Bovine β-casein (90% β-casein based on dry weight, 95% β-casein based on total nitrogen, w/w) was obtained from Eurial (Rennes, France). Amphiphilic β-casein peptides (RET1: f 1/29–105/107) and hydrophobic β-casein peptides (PEL1: f 106/108/114–209) were prepared as described by Caessens *et al.*[14]

Soya oil (Reddy) and tricaprilin (Sigma) were made surfactant-free by use of silicagel 60 (Merck). All other chemicals were of analytical grade and were purchased from Merck.

Methods

Protein and peptides were dissolved in 0.02 M HCl-imidazole buffer at pH 6.7 or 9 containing 0.02% sodium azide and given amounts of sodium chloride, yielding a total ionic strength of 0.075 M unless stated otherwise. Soya oil-in-water emulsions (oil volume fraction: $\varphi = 0.2$) were prepared using a laboratory-scale high-pressure homogenizer (Delta Instruments, Drachten, the Netherlands), operating at room temperature at a pressure of 5 MPa. Homogenization was repeated until a constant droplet-size distribution was obtained. Emulsions were prepared at least in duplicate.

The droplet-size distribution of emulsions was determined using a spectroturbidimetric method with a modified spectrophotometer (Zeiss PMQ II) as described previously.[14] The resulting surface–average diameter d_{32} was used to calculate the specific surface area A of the emulsion using the equation:

$$A = \frac{6\varphi}{d_{32}}. \tag{1}$$

Before measurement the emulsions were diluted with a 0.2% sodium dodecyl sulfate solution in order to disperse and stabilize the droplets and to dissociate any droplet aggregates. The state of aggregation of droplets was checked by light microscopy.

The protein surface excess concentration was determined using a depletion method. Oil droplets were separated from the emulsions by centrifugation (20 min at 13000 g in Eppendorf Z231 M). The protein concentrations of the resulting subnatant and of the original solution were measured using BCA protein reagent (Pierce Ltd.), calibrated for each specific protein or peptide. The concentration difference divided by the specific surface area yields the surface excess.

The recoalescence rate of emulsion droplets during homogenization was determined as described by Smulders and Walstra.[16] Emulsions with soya oil or tricaprilin were prepared separately, mixed in equal amounts, and subsequently re-homogenized. During homogenization, samples were taken and diluted in sucrose solutions with a refractive index equal to the refractive index of an equivolumic mixture of soya oil and tricaprilin. The turbidity of these mixtures was measured with the previously described turbidimeter at a wavelength of 590 nm. The difference between the refractive index of the oil (n_o) and that of the aqueous phase (n_a) is related to the turbidity (T) by:[17]

$$F = n_a\sqrt{\frac{T}{\varphi}} \propto |n_o - n_a|. \tag{2}$$

This value of F decreases when mixing of the oils due to droplet recoalescence occurs. Hence, the decrease in the ratio F/F_{max} during homogenization (where F_{max} is the initial turbidity) is a measure of the recoalescence rate.

The coalescence stability of emulsions was studied by following the change in average droplet size during storage at room temperature. Some of the emulsions were allowed to cream, while others were slowly rotated under exclusion of air. The change in droplet size was followed by measurement of the turbidity at 380 and 1700 nm. The turbidity at one wavelength was used to estimate the droplet size using the approximate relation:[18]

$$\frac{T}{\varphi} = \frac{\text{constant}}{d_{32}}. \tag{3}$$

The emulsions were diluted prior to measurement as described above. Results for measurements at 380 and 1700 nm agreed well with each other.

3 Results and Discussion

Droplet Size of Emulsions

The average droplet size of emulsions prepared with amphiphilic peptides (f 1/29–105/107) was smaller than those of emulsions prepared with intact β-casein (Figure 1a), yet comparable to the average droplet size of emulsions obtained with some globular proteins (the latter results to be published). The recoalescence rate of β-casein emulsions was found to be higher than the recoalescence rate of emulsions with amphiphilic peptides (Figure 2a). Recoalescence of droplets leads to an increase in the average droplet size of emulsions. For emulsions with β-casein or amphiphilic peptides, the difference in recoalescence rate was used to estimate the increase in average droplet size by assuming that the extent of break-up droplets was comparable for both emulsions. The calculated increase in droplet size was too small to explain fully the difference in droplet sizes, and therefore also variation in droplet break-up is to be expected.

Droplet break-up is possibly affected by the presence of β-casein micelles, which exist at all the used concentrations.[9] A possible consequence of the presence of β-casein micelles is that the micellar solution may be less capable of lowering the interfacial tension during droplet deformation and thus may inhibit droplet break-up.[1] β-Casein micelles are covered with a 'hairy' layer of highly charged, hydrophilic N-terminal tails.[20] Caessens *et al.*[14] showed that peptides consisting of parts of this N-terminal domain only slightly decrease the surface tension of air–water interfaces. It could therefore be imagined that the emulsifying properties of β-casein are affected by its micelle forming behaviour. Micelles are not likely to be present in amphiphilic peptide solutions, due to removal of a large part of the C-terminal

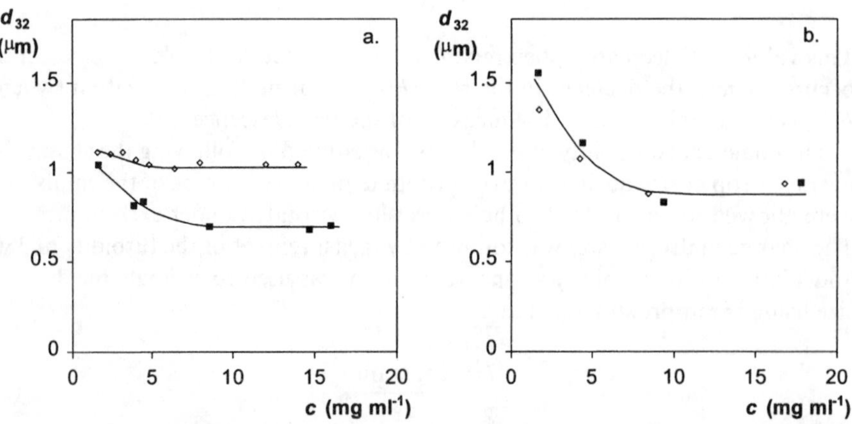

Figure 1 *Average droplet size* d_{32} *as function of emulsifier concentration* c *for emulsions made with (a) β-casein (◇) or amphiphilic peptide (■) at pH 6.7 and (b) β-casein (◇) or hydrophobic peptide (■) at pH 9*

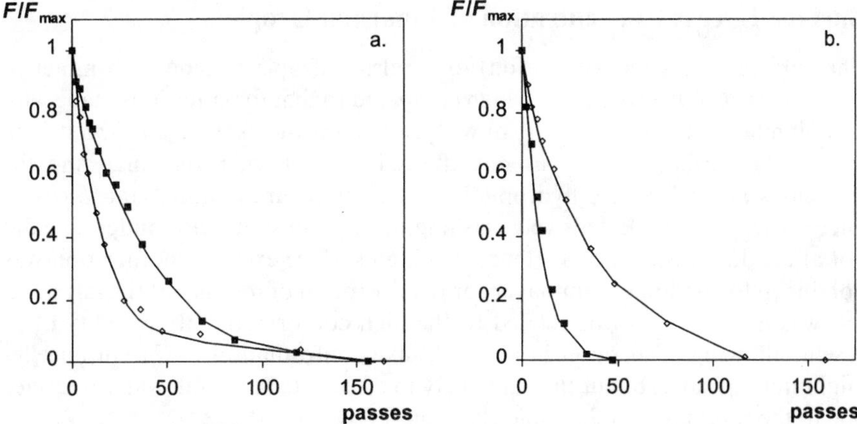

Figure 2 *Relative recoalescence extent,* F/F_{max}, *of emulsion droplets* ($c = 2$ *mg ml*$^{-1}$) *made with (a) β-casein (◇) or amphiphilic peptide (■) at pH 6.7 and (b) β-casein (◇) or hydrophobic peptide (■) at pH 9 as a function of number of passes through the homogenizer*

domain,[10] which might explain the smaller droplet size of emulsions with these peptides.

The emulsifying properties of β-casein at pH 9 (Figure 1b), where electrostatic interactions impede micellization,[10] would appear to confirm this hypothesis. The average droplet size of β-casein emulsions at this pH was smaller than that of emulsions formed at pH 6.7. This difference could again only be partially explained by a decrease in recoalescence rate at pH 9 (Figure 2b). The differences in the average droplet size were, however, small, and therefore the effect of micelles has to be further studied.

The hydrophobic peptides (f 106/108/114–209) of β-casein, which are poorly soluble at pH 6.7, are mainly present in solution as solid particles. Emulsions made with these peptides were, due to bridging flocculation,[19] strongly aggregated, and they immediately separated into a cream and a clear aqueous layer. Relatively large droplets with a high surface excess were formed (for example, for $c = 100$ mg ml^{-1}, we have $d_{32} = 2.5$ μm and $\Gamma = 16$ mg m^{-2}); these droplets were stable against coalescence. The properties of the hydrophobic peptides appear to be mainly determined by the size and number of solid particles present in solution. The peptides were therefore further studied at pH 9, where their solubility is high.

The average droplet size of emulsions made with hydrophobic peptides at pH 9 (Figure 1b) was, despite the higher recoalescence rate (Figure 2b), found to be equal to the droplet size of emulsions made with intact β-casein. This suggests that the ability of hydrophobic peptides to lower the interfacial tension during droplet deformation, and hence to break-up the droplets, was enhanced, while their ability to stabilize against recoalescence during emulsification was not so good.

Surface Excess Concentration of Emulsion Droplets

The surface excess (concentration) of emulsion droplets made with intact β-casein at pH 6.7 was found to be high compared to that for emulsions made with amphiphilic peptides (Figure 3a) or with globular proteins (typically about 2 mg m^{-2}). The surface excess was not affected by preparing and analyzing the emulsions at 4 °C, where hydrophobic interactions are minimal and β-casein micelles are absent.[8] Extensively washing of the emulsions with buffer also did not affect the surface excess. Hence, the high surface excess concentration was not due to formation of multilayers or to adsorption of micelles at the interface, but was more likely to be related to the high conformational flexibility of β-casein, allowing adsorption in a 'brush'-like conformation. β-Casein micelles might initially adsorb, but they are likely to dissociate at the oil–water interface due to the reversibility of the micellization.[8]

The surface excess of emulsion droplets made with amphiphilic peptides was much smaller than that of the β-casein emulsions (Figure 3a). Nylander and Wahlgren[21] also found a lower surface excess for a slightly different β-casein peptide (f 1–93). They suggested that peptides without the large C-terminal domain to orient the N-terminal tail adopt a flat, less extended conformation at the interface, which explains the lower surface excess.

The surface excess of the β-casein emulsions was somewhat smaller at pH 9 than at pH 6.7 (Figure 3), presumably due to an expansion of the conformation of adsorbed proteins caused by increased electrostatic repulsion. The surface excess of emulsions made with hydrophobic peptides (Figure 3b) was high and comparable to the surface excess of β-casein emulsions at this pH. The hydrophobic peptides apparently formed, like β-casein, a densely packed layer at the interface.

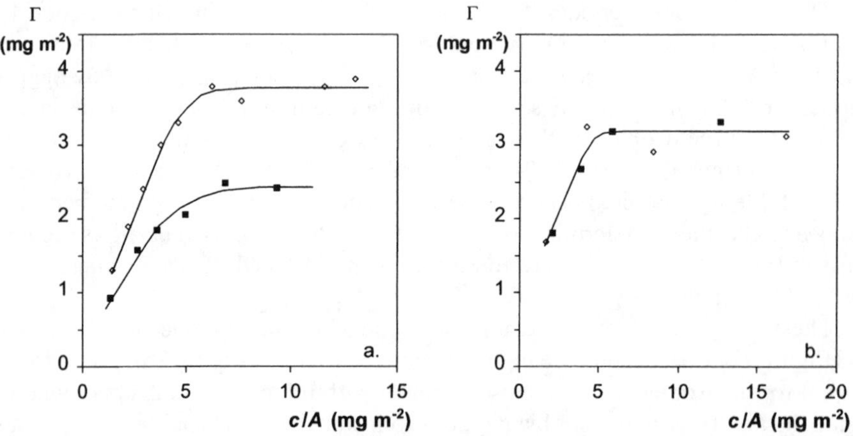

Figure 3 *Surface excess (concentration)* Γ *of emulsions* versus *concentration* c *over specific surface area* A *with (a) β-casein (◇) or amphiphilic peptide (■) at pH 6.7 and (b) β-casein (◇) or hydrophobic peptide (■) at pH 9*

Washing of emulsions made with amphiphilic or hydrophobic peptides was found to lead to a decrease of the surface excess. Multilayer formation of hydrophobic peptides at the interface appears unlikely, since the surface excess was not different when the emulsions were prepared and analyzed at 4 °C. The decrease in surface excess was therefore presumably caused by desorption of peptides from the oil–water interface. Most proteins desorb very reluctantly from an interface; however, desorption of peptides is much more likely to occur due to their reduced molar mass.[3]

Coalescence Stability

Protein-stabilized emulsions are generally stable against coalescence as long as the droplets are small and the surface excess is sufficiently high.[3] This applies here to emulsions stabilized with β-casein, which only initially showed some slight coalescence at very low protein concentrations ($c = 0.5\,\text{mg ml}^{-1}$). No coalescence was detected for emulsions made with higher protein concentrations.

Peptide-stabilized emulsions were all observed to be less stable against coalescence than β-casein emulsions (Figure 4). The emulsions creamed only very slowly, and therefore no difference was found in the stability of emulsions stored rotating or non-rotating. The hydrophobic peptides possessed relatively good emulsion stabilizing properties, presumably due to steric repulsion associated with the relatively high surface excess of the emulsion droplets, but also due to electrostatic repulsion arising from the greater charge of the peptides at pH 9.

The amphiphilic peptides, which possessed good emulsion forming properties, were clearly less suitable as emulsion stabilizers. Emulsions with a low concentration of amphiphilic peptides rapidly coalesced during the first few

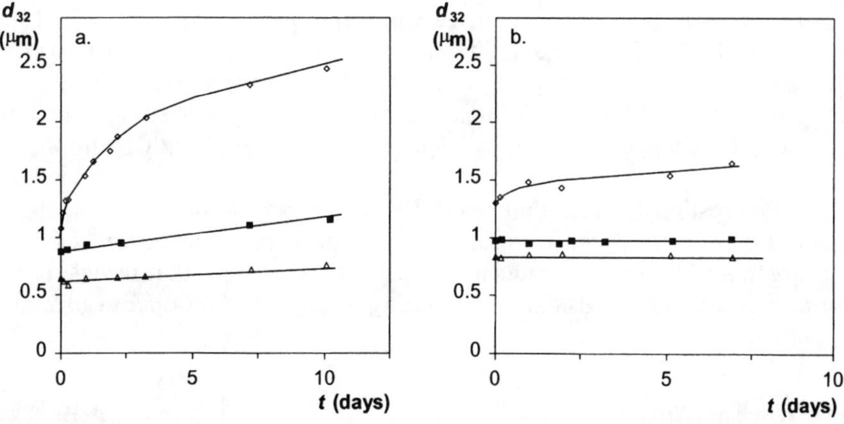

Figure 4 *Average droplet size* d_{32} *of emulsions containing (a) amphiphilic peptides at pH 6.7 and (b) hydrophobic peptides at pH 9* versus *storage time* t. *Concentrations: 2 mg ml*$^{-1}$ *(◇), 4 mg ml*$^{-1}$ *(■), and 8 mg ml*$^{-1}$ *(△)*

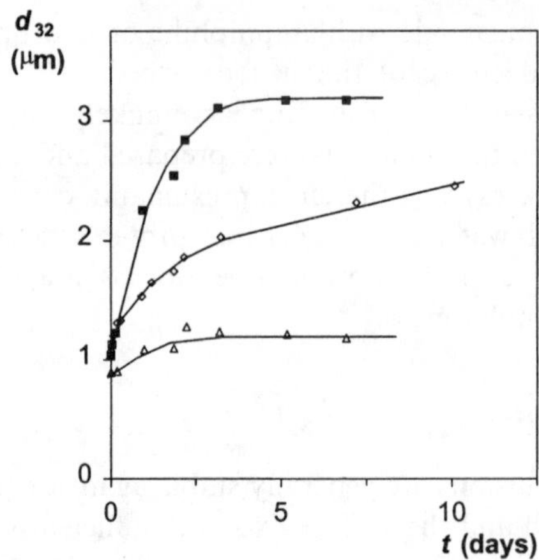

Figure 5 *Average droplet size* d_{32} *of emulsions containing 2 mg ml*$^{-1}$ *amphiphilic peptides* versus *storage time* t. *Physico-chemical conditions: pH 6.7 and ionic strength* I = *0.075 M* (◇); *pH 6.7 and* I = *0.150 M* (■); *and pH 9 and* I = *0.075 M* (△)

days after emulsification. After several days, the coalescence rate decreased and became comparable to the coalescence rate of emulsions with higher peptide concentrations. The decrease in coalescence rate was probably caused by an increase in surface excess due to a decrease of the specific surface area.

The charged groups of the N-terminal domain play an important role in stabilizing β-casein emulsions.[22] This stabilizing role will tend to be even greater for emulsions made with amphiphilic peptides, since the low surface excess may only provide limited stabilization by steric repulsion. Changes in the electrostatic interactions have indeed a large effect on the emulsion stability (Figure 5). An increased ionic strength diminishes the electrostatic interaction range, yielding less stable emulsions, whereas the stability is increased by an increase in peptide charge (and hence an increase in surface potential) as a result of the higher pH. The coalescence stability at higher peptide concentrations was less affected by change in ionic strength, presumably due to the presence of more charged N-terminal tails at the interface. An increase in ionic strength had no effect on the coalescence stability of emulsions prepared with hydrophobic peptides.

It is interesting to note that, contrary to the behaviour of the emulsion stabilizing properties, the emulsion forming properties are not affected by increase in ionic strength (results not shown). This confirms that recoalescence during emulsification and coalescence during storage of emulsions are governed by different mechanisms.[15]

4 Conclusions

The emulsion forming properties of amphiphilic and hydrophobic peptides were found to be better than or comparable to the emulsion forming properties of β-casein. The differences in emulsion stabilizing properties were found to be much larger, suggesting that they are much more dependent on the molecular properties of peptides. The coalescence stability of emulsions appears to be related to the ability of peptides to form a dense layer at the interface and thus to the relative hydrophobicity of the peptides. Electrostatic interactions, and hence peptide net charge, becomes a more important factor when steric repulsion is limited due to a low surface excess.

References

1. P. Walstra, *Chem. Eng. Sci.*, 1993, **48**, 333.
2. P. Walstra and P. E. A. Smulders, in 'Modern Aspects of Emulsion Science', ed. B. P. Binks, Royal Society of Chemistry, Cambridge, 1998, p. 56.
3. P. Walstra, in 'Encyclopedia of Emulsion Technology 4', ed. P. Becher, Marcel Dekker, New York, 1996, p. 1.
4. L. K. Creamer, T. Richardson, and D. A. D. Parry, *Arch. Biochem. Biophys.*, 1981, **211**, 689.
5. E. R. B. Graham, G. N. Malcolm, and H. A. McKenzie, *Int. J. Biol. Macromol.*, 1984, **6**, 155.
6. H. E. Swaisgood, 'Developments in Dairy Chemistry—1', ed. P. F. Fox, Applied Science, London, 1982, p. 1.
7. L. K. Creamer, *Biochim. Biophys. Acta*, 1972, **271**, 252.
8. T. A. J. Payens and B. W. van Markwijk, *Biochim. Biophys. Acta*, 1963, **71**, 517.
9. D. G. Schmidt and T. A. J. Payens, *J. Colloid Interface Sci.*, 1972, **39**, 655.
10. G. P. Berry and L. K. Creamer, *Biochemistry*, 1975, **14**, 3542.
11. R. Niki and S. Arima, *Agric. Biol. Chem.*, 1969, **33**, 826.
12. L. C. ter Beek, M. Ketelaars, D. C. McCain, P. E. A. Smulders, P. Walstra, and M. A. Hemminga, *Biophys. J.*, 1996, **70**, 2396.
13. D. G. Dalgleish and J. Leaver, *J. Colloid Interface Sci.*, 1991, **141**, 288.
14. P. W. J. R. Caessens, H. Gruppen, S. Visser, G. A. van Aken, and A. G. J. Voragen, *J. Agric. Food Chem.*, 1997, **45**, 2935.
15. P. Walstra, *J. Colloid Interface Sci.*, 1968, **27**, 493.
16. P. E. A. Smulders and P. Walstra, 'Recoalescence of Emulsion Droplets Stabilized by β-Casein or by β-Lactoglobulin during Emulsion Formation', Poster presented at 'Food Colloids: Proteins, Lipids and Polysaccharides' symposium, Ystad, 1996.
17. P. Walstra, *Neth. Milk Dairy J.*, 1965, **19**, 1.
18. P. Walstra, *Neth. Milk Dairy J.*, 1969, **23**, 238.
19. P. Walstra, in 'Food Emulsions and Foams', ed. E. Dickinson, Royal Society of Chemistry, London, 1987, p. 242.
20. A. L. Andrews, D. Atkinson, M. T. A. Evans, E. G. Finger, J. P. Green, M. C. Phillips, and R. N. Robertson, *Biopolymers*, 1979, **18**, 1105.
21. T. Nylander and N. M. Wahlgren, *J. Colloid Interface Sci.*, 1994, **162**, 151.
22. F. A. Husband, P. J. Wilde, A. R. Mackie, and M. J. Garrood, *J. Colloid Interface Sci.*, 1997, **195**, 77.

Influence of Emulsifier and Pore Size on Membrane Emulsification

By Volker Schröder and Helmar Schubert

INSTITUT FÜR LEBENSMITTELVERFAHRENSTECHNIK, UNIVERSITÄT KARLSRUHE, KAISERSTR. 12, D-76128 KARLSRUHE, GERMANY

1 Introduction

To produce emulsions of defined properties numerous emulsifying technologies are presently available.[1] Rotor/stator systems or high-pressure homogenizers are most commonly used. In the dispersing zone of this type of apparatus, large droplets of a premix are deformed and disrupted in laminar or turbulent flow, or by cavitation.[2] The use of membranes for emulsification purposes is a relatively new technology.[3–6]

Droplet formation in a membrane emulsification process is different from existing methods, since small droplets of the disperse phase are formed directly in the dispersing zone, *i.e.* at the membrane surface. The dispersed phase is pressed through the membrane pores into the continuous phase flowing inside the membrane tube. As droplets of the disperse phase form, emulsifier molecules occupy the surface of the growing droplets. Adsorption of surfactants at an interface results in a decrease of interfacial tension. The adsorption kinetics of different emulsifiers can be characterized by measurement of the dynamic interfacial tension. Droplets of a critical diameter detach from the membrane surface under the influence of drag forces caused by the flowing continuous phase. By applying membranes with a certain pore-size distribution, emulsions with small droplets and narrow droplet-size distributions can be produced. Membrane emulsification allows the use of shear-sensitive ingredients due to the small shear stresses generated in the process. It is possible to produce oil-in-water, water-in-oil, or even multiple emulsions, depending on the properties of the membrane used (hydrophilic or lipophilic).[4]

2 Production of Emulsions Using Membranes

Using membranes for emulsification, the droplets of the dispersed phase are produced directly in the dispersing zone, *i.e.* at the membrane surface. Figure 1

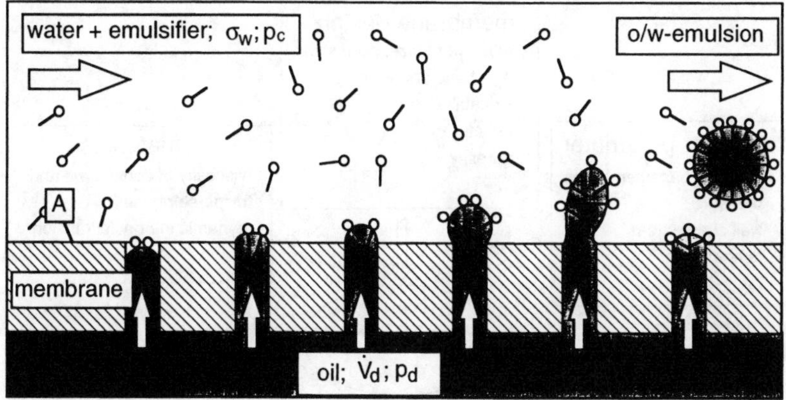

Figure 1 *Membrane emulsification process for the production of (oil-in-water) o/w emulsions.* σ_w = *wall shear stress of the continuous phase;* p_c = *pressure of the continuous phase;* $\dot{V}_d$ = *volume flow-rate of the dispersed phase;* p_d = *pressure of the dispersed phase; A = membrane surface area*

shows the membrane emulsification process for the case of oil-in-water emulsions.

When the dispersed phase (oil) is pressed through the pores of a microporous membrane, droplets will form when the dispersed phase reaches the membrane surface. The continuous phase (water and emulsifier) flows along the membrane surface. Droplets of the dispersed phase are detached from the pores at a certain diameter. The droplet surface is occupied and stabilized by emulsifier molecules during droplet formation. Apart from the mean droplet diameter and the droplet-size distribution, the emulsification result is expressed in terms of the dispersed phase flux,

$$J_\mathrm{d} = \frac{\dot{V}_\mathrm{d}}{A}, \tag{1}$$

where $\dot{V}_\mathrm{d}$ is the volume flow-rate of the disperse phase and A is the area of the membrane surface. This definition of the dispersed phase flux allows comparison of the results of different types or sizes of membranes. Dispersed phase flux is an essential parameter allowing evaluation of the process of membrane emulsification.

The emulsification result depends on a number of parameters, which can be divided into three groups: process parameters, membrane design and materials (Figure 2). The most important process parameters are the trans-membrane pressure and the wall shear stress of the continuous phase.[7] The wall shear stress can be calculated from the pressure loss of the flowing continuous phase, *i.e.* pressure difference between ends of the membrane module. Trans-membrane pressure can be calculated from the pressure difference between the continuous and dispersed phase sides (see Figure 1):

$$\Delta p_\mathrm{TM} = p_\mathrm{d} - p_\mathrm{c}. \tag{2}$$

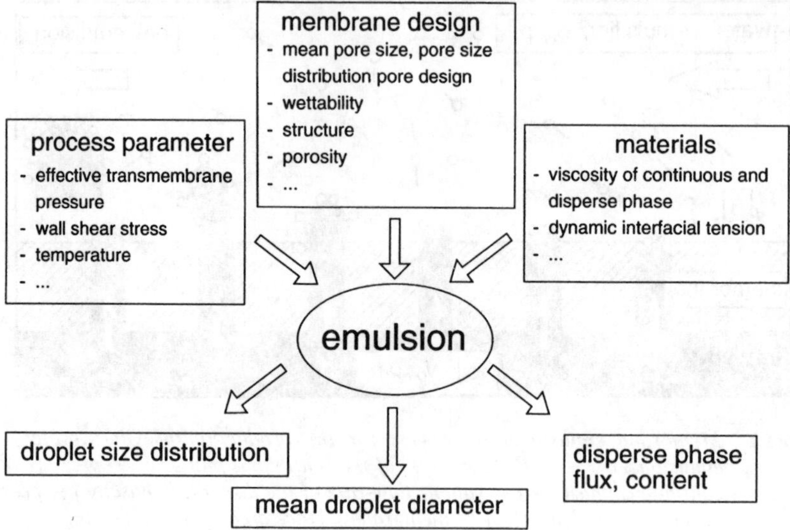

Figure 2 *Parameters influencing the emulsification result*

Further, the effective trans-membrane pressure Δp_e can be defined as

$$\Delta p_e = \Delta p_{TM} - p_{cap,d} \tag{3}$$

where $p_{cap,d}$ is the capillary pressure of the dispersed phase. The capillary pressure can be calculated according to the Laplace equation,

$$p_{cap,d} = \frac{4 \cdot \gamma}{d_p} \cdot \cos \delta, \tag{4}$$

where γ is the interfacial tension, d_p is the pore size, and δ is the contact angle. Membranes used for emulsification should not be wetted by the dispersed phase. Otherwise individual droplet formation is undermined resulting in broad droplet-size distributions. The contact angle δ describes the wettability of the membrane by the liquid phases.

The influence of the mean pore size on the emulsification result is discussed in detail in section 5. Porosity of the membrane has an influence on both droplet size and dispersed phase flux. The higher the porosity, the higher the flux through membrane pores, and the higher the probability of droplet coalescence at the membrane surface. Dynamic interfacial tension has a strong influence on droplet formation. This will also be discussed in section 5.

3 Droplet Formation and Detachment

When the dispersed phase pressed through the pores first reaches the membrane surface, droplet formation starts with half-sphere shaped droplets. Droplet size increases until, at a critical volume, the droplet becomes unstable and starts to

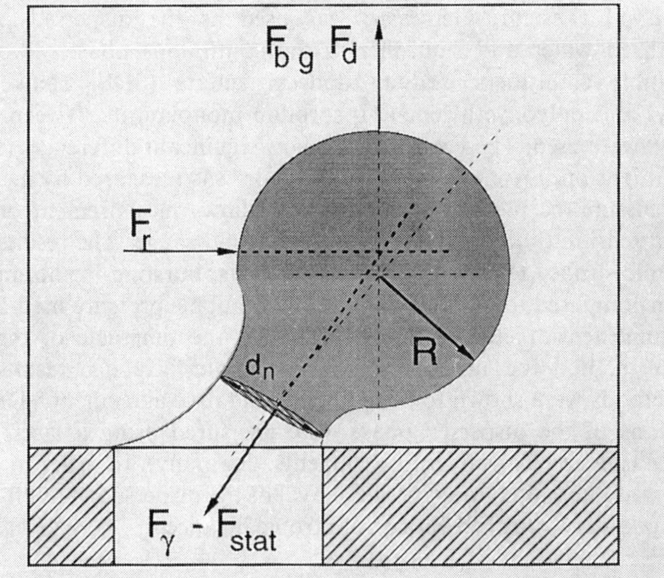

Figure 3 *Forces acting at a single droplet.* F_γ = *interfacial tension force;* F_{stat} = *dynamic effect of the pressure difference between the phases;* F_r = *flow resistance force;* F_d = *dynamic lift force;* F_{bg} = *buoyancy and gravitation force;* d_n = *neck diameter;* R = *droplet radius*[8]

rise from the membrane surface. The diameter of the droplet neck decreases rapidly until the droplet finally detaches from the pore. Droplet formation then starts again at the pore.

Figure 3 schematically shows the balance of forces acting on a single droplet. Droplet formation is mainly controlled by the interfacial tension force, the dynamic effect of the pressure difference between the phases, the flow resistance force, and the dynamic lift force.[8] Apart from the interfacial tension force, the other above mentioned forces tend to induce droplet detachment. Additional forces like the linear momentum force or the buoyancy and gravitation force can be neglected according to our calculations.[8] The interfacial tension force and the dynamic effect of the pressure difference between the phases are proportional to the interfacial tension and the neck diameter. The flow resistance force and the dynamic lift force depend on the wall shear stress of the continuous phase. Both forces are a function of droplet size too.

The influence of pore size and interfacial tension on the emulsification result was investigated while varying the effective trans-membrane pressure and the wall shear stress of the continuous phase. The results are discussed in section 5.

4 Materials and Methods

Oil-in-water emulsions were produced using microporous ceramic membranes (Membraflow Filtersysteme, Germany) with mean pore sizes of 0.1–3 μm.

Vegetable oil (Lesieur, Germany) was used as the dispersed phase and demineralized water and emulsifier as the continuous phase. Two different water-soluble emulsifiers, sodium dodecyl sulfate (SDS, Henkel KGaA, Germany) and polyoxyethylene (20) sorbitan monolaurate (Tween 20, Roth, Germany) were used. These surfactants show significant differences in dynamic interfacial tension. Dynamic interfacial tension was measured by means of the bursting membrane method.[9] This method allows measurement of dynamic interfacial tension (liquid–liquid) in the subsecond range. The results obtained for dynamic surface tension (liquid–gas) with the bursting membrane method have been compared to those of the maximum bubble pressure method and the drop volume tensiometer by Stang.[10] Though the timescale of the bursting membrane method does not represent the exact interfacial age, results obtained for all methods were shown to be in good qualitative agreement.[8] Droplet-size distributions of the dispersed phase were measured using a laser scattering system (MastersizerX, Malvern Instruments, Germany). In order to determine the dispersed phase flux, the rate of removal of the dispersed phase from the oil storage tank was controlled with an electronic balance.

5 Results and Discussion

Influence of Mean Pore Size

The influence of the mean pore size on the mean droplet diameter (Sauter average) is shown in Figure 4. Droplet size increases with increasing pore size. This is due to the fact that the volume and the diameter of droplets at the beginning of the formation, *i.e.* for hemispherical droplets, are already larger, due to the larger pore diameter. Furthermore, the interfacial tension force is proportional to the pore perimeter as long as the droplet does not start to form a neck. Therefore the interfacial tension force is higher for a higher mean pore diameter of the membrane. Consequently, the flow resistance force and the dynamic lift force have to be higher, too, in order to detach the droplet. This results in larger droplet diameters. Coalescence of droplets dispersed in the flowing continuous phase and at the membrane surface can be neglected due to the low concentration of droplets in the continuous phase and the fast stabilization of the newly formed interfaces by SDS. Besides the effect of the pore size on the droplet diameter, the effective trans-membrane pressure also affects the resulting droplet diameter. The larger the effective trans-membrane pressure, the larger the droplets.

The dispersed phase flux is strongly influenced by the mean pore size of the membrane and by the effective trans-membrane pressure (Figure 5). The dispersed phase flux obeys Darcy's law:

$$J_{\mathrm{d}} = \frac{B_0}{\eta_{\mathrm{d}}} \cdot \frac{\Delta p_{\mathrm{e}}}{\Delta L}. \tag{5}$$

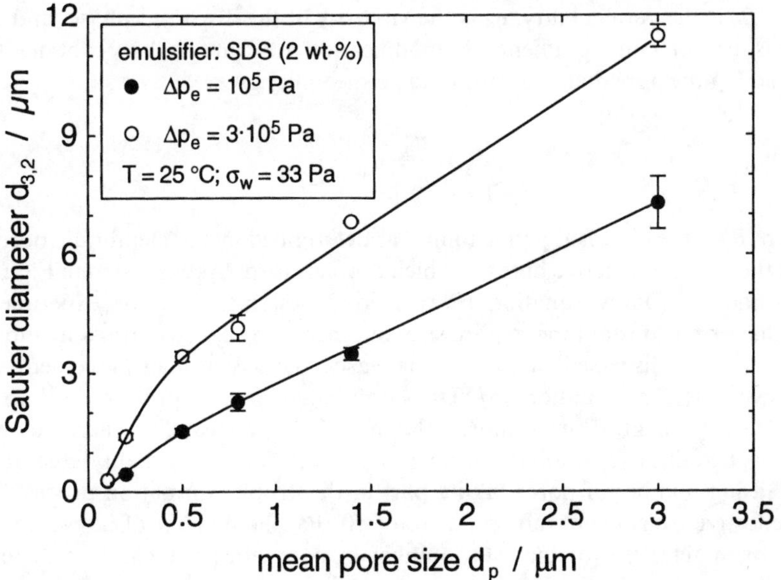

Figure 4 *Influence of the mean pore size* d_p *and effective trans-membrane pressure* Δp_e *on the mean droplet diameter (Sauter diameter* d_{32}*); wall shear stress* $\sigma_w = 33$ *Pa;* T = *25°C*

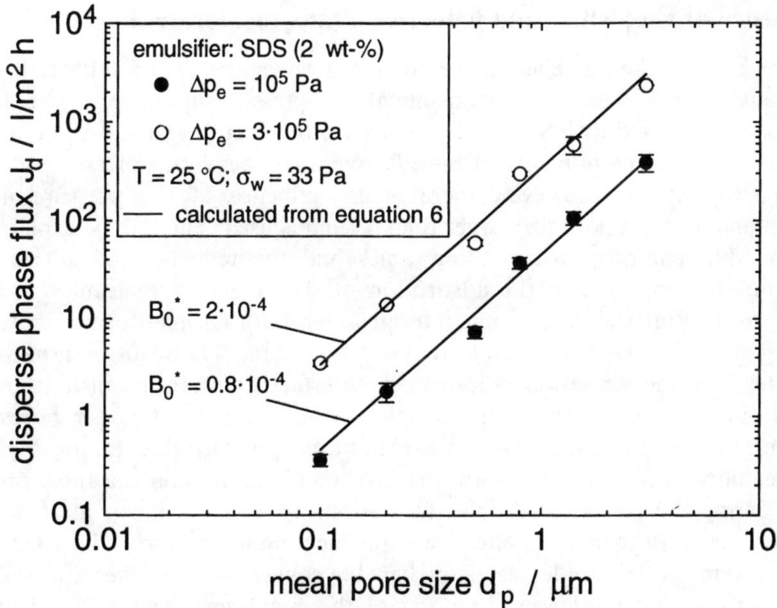

Figure 5 *Influence of mean pore size* d_p *and effective trans-membrane pressure* Δp_e *on the dispersed phase flux* J_d*; wall shear stress* $\sigma_w = 33$ *Pa;* T = *25°C*

Here B_0 is the permeability, η_d is the viscosity of the dispersed phase, and $\Delta p_e/\Delta L$ is the pressure gradient. A modification of Darcy's Law obtained by extracting the pore diameter from the permeability gives

$$J_d = \frac{B_0^*}{\eta_d} \cdot d_p^2 \cdot \frac{\Delta p_e}{\Delta L}, \tag{6}$$

where B_0^* is the modified permeability, as determined by the membrane porosity and the number of active pores at which droplets form. As expressed in Figure 5, the modified Darcy equation gives a good description of the experimental results. For a constant mean pore size, but increasing effective trans-membrane pressure, the dispersed phase flux increases. This is also in agreement with Darcy's law, *i.e.* equation (5). Dispersed phase flux is proportional to the pressure gradient. Furthermore, the modified permeability increases with increasing effective trans-membrane pressure. This is probably due to an increasing number of pores taking part in the droplet formation process. For an effective trans-membrane pressure of 10^5 Pa, the number of active pores is approximately 10% of the total number of pores. In principle, it is therefore possible to increase the number of pores taking part in the droplet formation process and thus to enhance the dispersed phase flux by increasing the effective trans-membrane pressure. But this may also result in a larger average droplet size due to faster droplet formation times and a higher rate of coalescence of droplets at the membrane pores.

Influence of Emulsifier and Effective Trans-membrane Pressure

Figure 6 shows the influence of the emulsifier on the emulsification result for different values of the effective trans-membrane pressure and the pore size. For emulsions stabilized with SDS and produced with a membrane of 0.1 μm mean pore size, there is no influence of the effective trans-membrane pressure on the mean droplet size. However, there is an influence of the pressure if a membrane of mean pore size larger than 0.1 μm is used. This can be explained by the different droplet formation times and the reduction of interfacial tension with time due to the adsorption of the emulsifier molecules at the newly formed interfaces (dynamic interfacial tension). Droplet formation time is long at small pores compared to that at larger pores. The adsorption of SDS is faster than the formation of a droplet. Interfacial tension is constant at its equilibrium value. However, the droplet formation time at larger pores is shorter for the same effective trans-membrane pressure due to the higher permeability of the membrane. In this case interfacial tension is a function of time during the period of droplet formation. This results in an increase of droplet size with increasing effective trans-membrane pressure. The effect of the pressure on the droplet size is higher, the slower the emulsifier absorbs at the interface, and the larger the pores of the membrane. Tween 20 adsorbs more slowly at newly formed interfaces.[8] For the 0.1 μm membrane there is no influence of the effective trans-membrane pressure on droplet size. Droplet

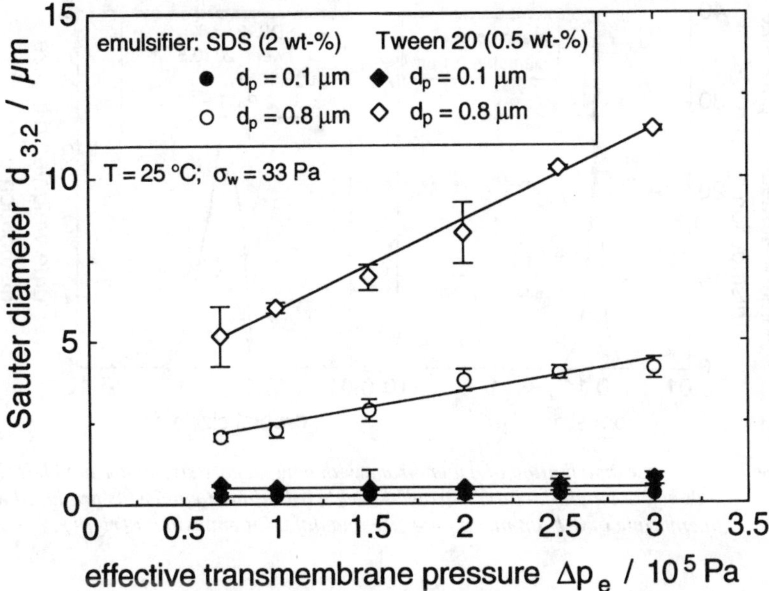

Figure 6 *Influence of emulsifier and effective trans-membrane pressure* Δp_e *on the mean droplet size (Sauter diameter* d_{32}*) for different mean pore sizes* d_p*; wall shear stress* σ_w *33 Pa;* T = 25 °C

formation again is slow compared to the reduction of dynamic interfacial tension. Droplet diameters for Tween 20 are about twice the size of the droplets stabilized with SDS. This is in agreement with the ratio of equilibrium interfacial tensions.[8]

For the 0.8 µm membrane, again there is an increase of the droplet diameter with increasing effective trans-membrane pressure. Due to the slower change in dynamic interfacial tension of Tween 20, droplets stabilized with Tween 20 are larger. Droplets have to grow until detaching forces like flow resistance force and dynamic lift force reach values high enough to detach the droplets from the pores. This results in a higher slope of the curve for Tween 20 than that for SDS for a constant membrane pore size. Furthermore, coalescence of the droplets at pores due to poorer stabilizing properties is also possible. Droplet-size distributions for a constant effective trans-membrane pressure and mean pore size can be taken from Figure 7.

The influence of the dynamic interfacial tension is emphasized if the wall shear stress of the continuous phase is varied instead of the effective trans-membrane pressure (Figure 8). Besides SDS, Tween 20 was used at two different concentrations (0.05 and 0.5 wt%). This offers the possibility to investigate the same emulsifier for different values of dynamic interfacial tension. The tension is reduced more slowly if a Tween 20 concentration of 0.05 wt% is used instead of 0.5 wt%. These concentrations of Tween 20 are both above the critical micelle concentration. The oil content of the emulsions was *ca.* 1 vol%. For each of the

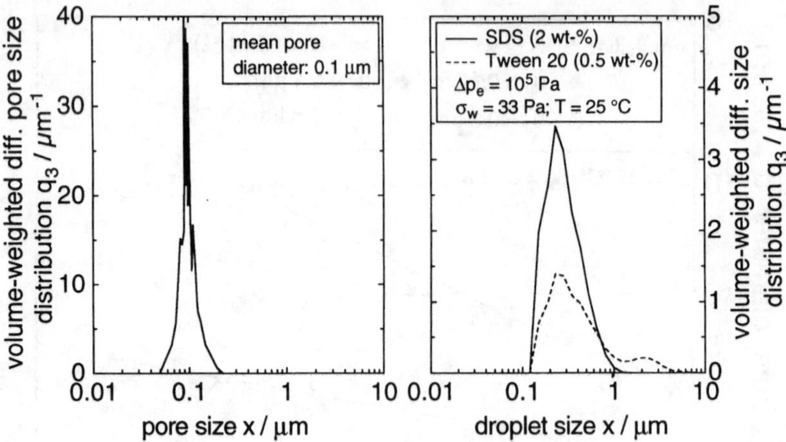

Figure 7 *Pore-size distribution of a membrane with a mean pore size of 0.1 μm (left); data taken from manufacturer. Droplet-size distribution of emulsions produced with a membrane of 0.1 μm mean pore size and different emulsifiers (right)*

systems (water, emulsifier, oil) there is a strong influence of wall shear stress of the continuous phase on the droplet size. The smaller the wall shear stress, the larger the droplets. Above a critical value of the wall shear stress, the influence on the droplet size disappears. For the fast adsorbing emulsifier (SDS), the slope of the curve is smaller and the critical wall shear stress is shifted to smaller

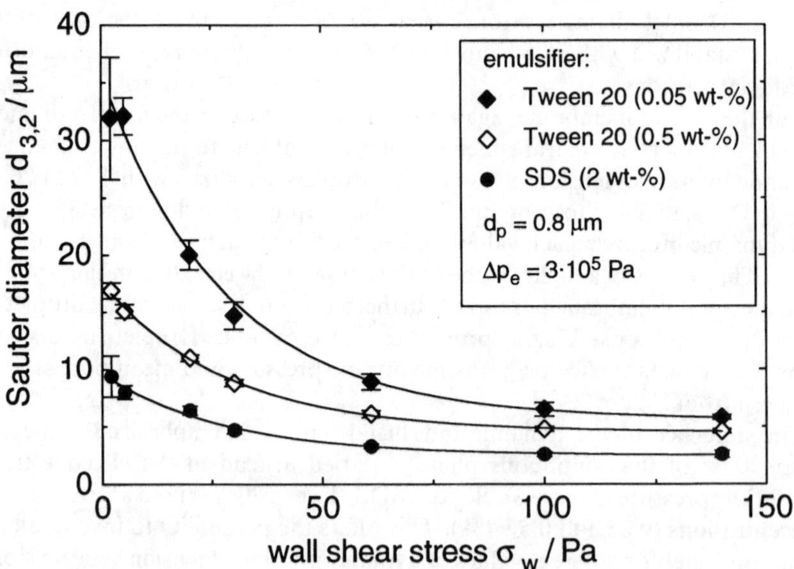

Figure 8 *Influence of emulsifier and wall shear stress σ_w of the continuous phase on the mean droplet size (Sauter diameter* d_{32}*). $\Delta p_e = 3 \times 10^5$ Pa; $d_p = 0.8$ μm;* T = *25°C*

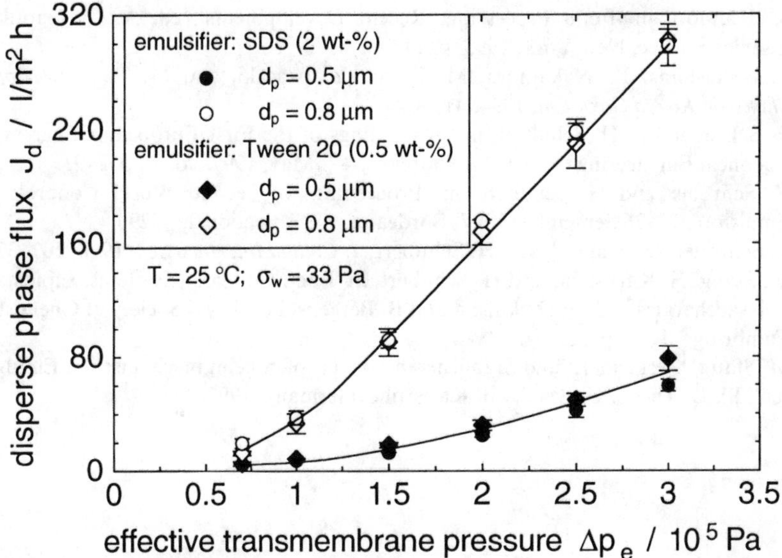

Figure 9 *Influence of emulsifier, dynamic interfacial tension, and effective trans-membrane pressure Δp_e on the dispersed phase flux J_d*

values compared to the slowly adsorbing Tween 20. Formation of new interface is faster than the adsorption of the emulsifier molecules in the case of slowly adsorbing emulsifiers. The forces for detaching the droplets from the pores are small if the wall shear stress is low. Due to slow adsorption and thus a higher value of the interfacial tension force, droplets have to grow bigger to be detached from the pore. Again, coalescence of forming droplets may occur at the membrane pores if the stabilization kinetics of the emulsifier is not fast enough.

The dispersed phase flux is not influenced by emulsifier properties (Figure 9). It is suggested that number of pores taking part in the emulsification process is therefore independent of the type of emulsifier, but is dependent on the trans-membrane pressure.

Acknowledgement

This work was supported by the DFG (Deutsche Forschungsgemeinschaft, Bonn, Germany).

References

1. H. Karbstein and H. Schubert, *Chem. Eng. Process.*, 1995, **34**, 205.
2. P. Walstra and I. Smulders, in 'Food Colloids—Proteins, Lipids and Polysaccharides', ed. E. Dickinson and B. Bergenståhl, Royal Society of Chemistry, Cambridge, 1997, p. 367.
3. S. Matsumoto, *J. Texture Stud.*, 1986, **17**, 141.

4. K. Kandori, in 'Food Processing: Recent Developments', ed. A. G. Gaonkar, Elsevier Science, New York, 1995, p. 113.
5. T. Nakashima, K. Nakamura, M. Kochi, Y. Iwasaki, and M. Tomita, *Nippon Shokuhin Kogyo Gakkashi*, 1994, **41**, 70.
6. V. Schröder and H. Schubert, in 'Proceedings of the First European Congress on Chemical Engineering (ECCE 1)', Florence, 4–7 May, 1997, vol. 4, p. 2491.
7. V. Schröder and H. Schubert, in 'Proceedings of Second World Congress on Emulsion', 23–26 September 1997, Bordeaux, vol. 1, section 2, p. 290.
8. V. Schröder, O. Behrend, and H. Schubert, *J. Colloid Interface Sci.*, 1998, **202**, 334.
9. M. Stang, H. Karbstein, and H. Schubert, in 'Food Colloids—Proteins, Lipids and Polysaccharides', ed. E. Dickinson and B. Bergenståhl, Royal Society of Chemistry, Cambridge, 1997, p. 382.
10. M. Stang, 'Zerkleinern und Stabilisieren von Tropfen beim mechanischen Emulgieren', Ph.D. Thesis, University of Karlsruhe, Germany, 1997.

Contribution of Oil Phase Viscosity to the Stability of Oil-in-Water Emulsions

By Peter J. Wilde, Michel Cornec†, Fiona A. Husband, and David C. Clark‡

FOOD BIOPHYSICS DEPARTMENT, INSTITUTE OF FOOD RESEARCH, NORWICH LABORATORY, NORWICH RESEARCH PARK, COLNEY LANE, NORWICH NR4 7UA, UK

1 Introduction

Many processed foods are complex dispersions such as foams and emulsions. The stability of such dispersions is of paramount importance as it affects storage, shelf-life, and consumer acceptability. Food emulsions can be stabilized against coalescence by proteins, low-molecular-weight emulsifiers, or mixtures of the two. The mechanism of stabilization is different in each case: emulsifiers act via a Gibbs–Marangoni mechanism[1] involving minimal interactions of surface adsorbed molecules and lateral diffusion at the interface, whereas proteins form a condensed viscoelastic film of immobile, interacting molecules at the interface.[2] These two mechanisms are mutually incompatible and instability often arises when a mixed component interface is created.[3–5] Fundamental knowledge of adsorbed layers and their behaviour has been derived from experiments involving competitive adsorption in mixed protein emulsifier systems at the oil–water interface.[6–14] Most of this work has been performed at the hydrocarbon oil–water interface which is different in terms of surface free energy and viscosity to the triglyceride oil–water interfaces encountered in real food emulsions.[15–17] The effect of oil phase viscosity on the droplet-size distribution has been studied,[18–20] although no clear model has fully described the behaviour. Therefore the aim of this study was to differentiate between the respective influences of the composition and the viscosity of the oil phase in the formation and the stability of emulsions. Some emulsions formed from mixtures of β-lactoglobulin and Span 80 were characterized at both the alkane–water and soya oil–water interfaces.

†Current address: Agricultural and Biological Department, Purdue University, West Lafayette, IN 47907-1146, USA

‡Current address: DMV International, NCB laan 80, P.O. Box 13, 5460 BA Veghel, Netherlands

2 Material and Methods

The β-lactoglobulin (lot 91H7005), n-tetradecane (>99 wt%) and Florisil were obtained from Sigma Chemical Company (Poole, Dorset). Sorbitan monoleate (Span 80) was obtained from Fluka Chemicals (Gillingham, Dorset). Paraffin oil was purchased from BDH Laboratory (Poole, Dorset) and commercial soya oil from a local retailer. The fluorescent probe dodecylamino fluorescein (DDAF) was obtained from Molecular Probe Inc. (Eugene, OH). Both oils were made free of surface-active contaminants by twice passing through a Florisil column as described by Gaonkar.[21]

Oil-in-water emulsions were made using a low-volume single-pass valve homogeniser (Avestin); five passes were made at a pressure of 32 MPa at room temperature.[13] Oil phases (20 wt% of total emulsion) were n-tetradecane or soya oil containing various amount of Span 80, dissolved at room temperature. The aqueous phase consisted of β-lactoglobulin (1.5 mg ml^{-1}) dissolved in 10 mM sodium phosphate buffer, pH 7, using surface chemically pure water. All experiments were performed in triplicate at room temperature (22 °C). The droplet-size distribution was determined by light diffraction using a Malvern Mastersizer. The specific surface area (SSA) (or surface area per unit mass of emulsion) of the freshly made emulsion was calculated by the Mastersizer from the full distribution. The viscosity of oil was measured by stress viscometry using a Bohlin controlled stress rheometer.[3] The geometry used was one of three, depending on sample viscosity: double-gap DG 40/50, cone-plate 4/40, or concentric cylinder C25. The pure oils and oils containing Span 80 behaved in a Newtonian manner. The shear-rate was varied between 0.5 and 200 s^{-1} depending on the viscosity of the sample and the geometry of the measuring head. The resulting viscosity value was taken as the mean viscosity over the shear-rate range used.

Emulsion stability to coalescence was assessed from the change in droplet-size distribution upon accelerated creaming. The SSA was measured immediately after the emulsion preparation. Accelerated coalescence was obtained by centrifuging the emulsion at 13,000 g for 15 minutes. The cream layer was redispersed in the aqueous phase by gentle agitation and the SSA of the reformed emulsion was measured again by the Malvern Mastersizer. The extent of coalescence was determined from the difference in observed SSA before and after centrifugation.

The mobility of the adsorbed layer of thin films was investigated by measurement of lateral diffusion of the fluorophore DDAF, using the fluorescence recovery after photobleaching (FRAP) technique. An aqueous thin film was formed between two oil droplets held in the aqueous sample medium. One droplet was held stationary at the base of a glass cell by a hydrophobic spot formed with octadecyltrichlorosilane; the other droplet, suspended from a modified syringe tip, was lowered onto the stationary droplet. This allowed a thin film to form, and the drainage rate and mobility behaviour were observed through a Nikon inverted microscope. Once the thin film has drained down to its equilibrium thickness, the surface lateral diffusion was measured, using the FRAP technique.[12–14]

3 Results

The emulsification behaviour of Span 80 in the presence and absence of β-lactoglobulin, with either alkane or triglyceride oil, is shown in Figure 1. In the presence of β-lactoglobulin, addition of more than 100 mM Span 80 was found to increase the specific surface area (SSA). However, it appears that β-lactoglobulin and Span 80 act antagonistically since the data for the mixed Span 80 + β-lactoglobulin emulsion are shifted to higher Span 80 concentrations compared to the data for Span 80 alone. This system has been described in detail previously.[3,5] Considering the same system at the soya–water interface, it can be observed that the increase in the SSA is much less pronounced. At high concentrations of Span 80 (>400 mM) a decrease in the SSA was observed in all systems. However, this decrease in the SSA occurs at higher Span 80 concentrations with the soya oil.

The viscosity of solutions of Span 80 dissolved in *n*-tetradecane and in soya oil are presented in Figure 2. Soya oil is much more viscous than *n*-tetradecane. The viscosity of the oils increases at concentrations above 415 mM Span 80. This increase is more pronounced in the case of *n*-tetradecane. Indeed, the *n*-tetradecane undergoes a 100-fold increase in viscosity between 200 and 1200 mM Span 80, whereas the soya oil viscosity only increases by a factor of 5.

The effect of oil phase viscosity on emulsifying properties was investigated by increasing the viscosity of *n*-tetradecane with paraffin oil. Figure 3 clearly shows

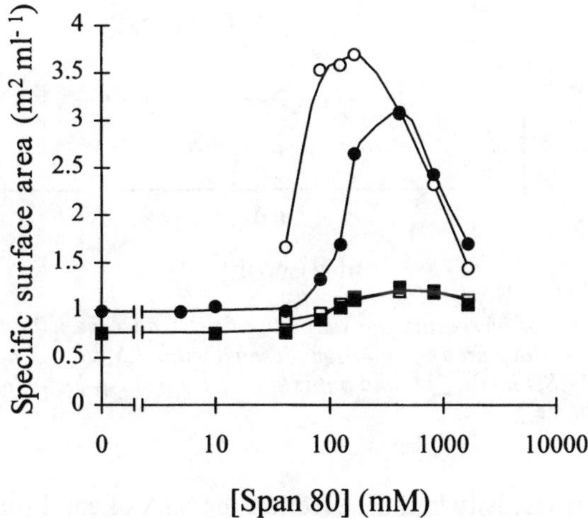

Figure 1 *The influence of Span 80 concentration on the emulsifying capacity of emulsions using* n*-tetradecane (○, ●) or soya oil (□, ■) as the oil phase. The specific surface area is plotted as a function of the concentration of added Span 80: Span 80 alone (○, □); Span 80 + 1.5 g l^{-1} β-lactoglobulin (●, ■)*

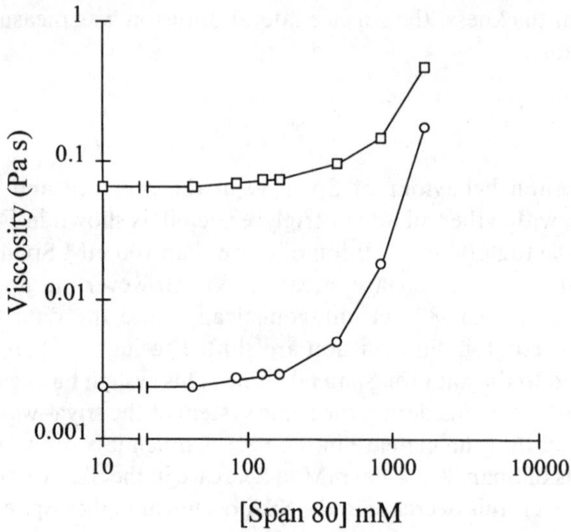

Figure 2 *The effect of Span 80 on the bulk viscosity of* n-*tetradecane* (○) *and soya oil* (□)

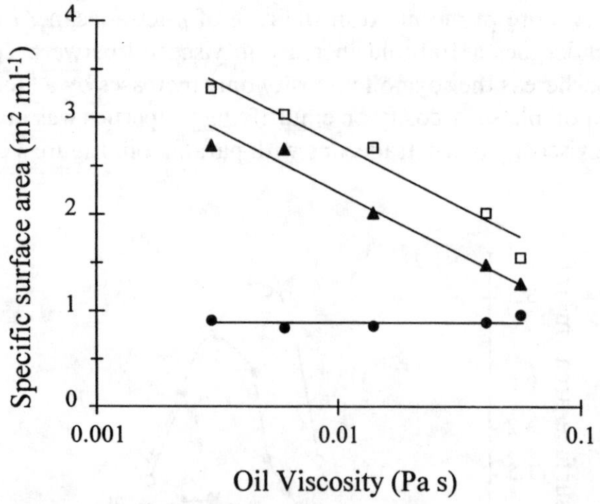

Figure 3 *The effect of increasing the viscosity of* n-*tetradecane with paraffin oil on the specific surface area of emulsions formed with: 1.5 g l^{-1} β-lactoglobulin* (●), *166 mM Span 80* (□), *and a mixture of 1.5 g l^{-1} β-lactoglobulin + 166 mM Span 80* (▲)

that the oil phase viscosity had no effect on the SSA of emulsions formed from β-lactoglobulin alone. In the case of Span 80, and the mixed β-lactoglobulin + Span 80 emulsions, the SSA was found to decrease as the oil phase viscosity increased. The interfacial tension was found to increase from 51.4 ± 0.4 mN m^{-1} to 52.5 ± 0.4 mN m^{-1} at the highest concentration of paraffin oil.

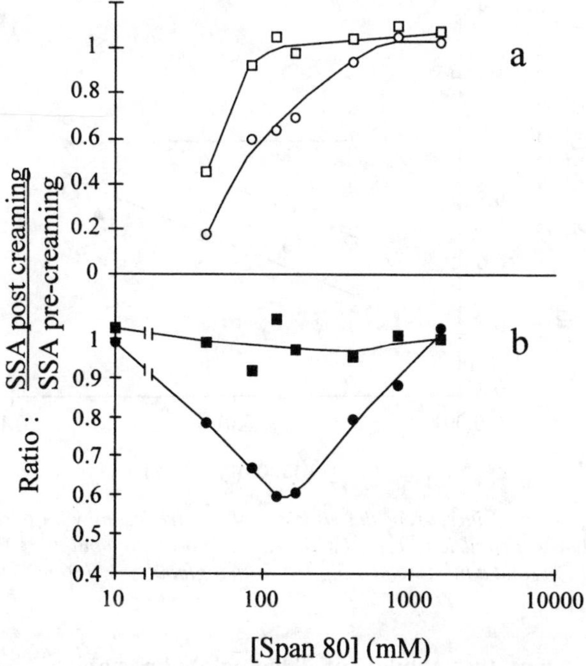

Figure 4 *Effect of Span 80 concentration on the stability of emulsions made with* n-*tetradecane (○, ●) or soya oil (□, ■) as the oil phase. The stability is expressed as the ratio of the specific surface area (SSA) of the emulsion after centrifugation at 13,000* g *for 15 minutes to the SSA of the fresh emulsion prior to centrifugation for (a) Span 80 alone and (b) in the presence of 1.5 g l^{-1} β-lactoglobulin*

Emulsion stability was determined by monitoring the extent of droplet coalescence induced by accelerated creaming. Figure 4 shows the effect of Span 80 concentration on emulsion stability of various emulsion systems. Considering first the control emulsions containing no protein (Figure 4a), emulsion stability increases with increasing Span 80 concentration. However, a lower Span 80 concentration is required to increase the stability of the soya oil emulsion to a value of 1. It can also be observed (Figure 4b) that addition of Span 80 does not destabilize β-lactoglobulin emulsions formed with soya oil. In contrast, addition of Span 80 to the β-lactoglobulin/*n*-tetradecane emulsion induces destabilization over a concentration range of 40–166 mM. However, the stability against coalescence was found to be increased by addition of higher levels of Span 80 and was restored for a Span 80 concentration of 1660 mM. This result may be related to the increase in the bulk oil viscosity due to the high amount of dissolved Span 80. To test this hypothesis, the *n*-tetradecane viscosity was increased by adding paraffin oil. Figure 5 shows the effect of oil phase viscosity on the stability of emulsions stabilized by fixed concentrations of β-lactoglobulin, Span 80 or both. The results reveal that the oil phase viscosity

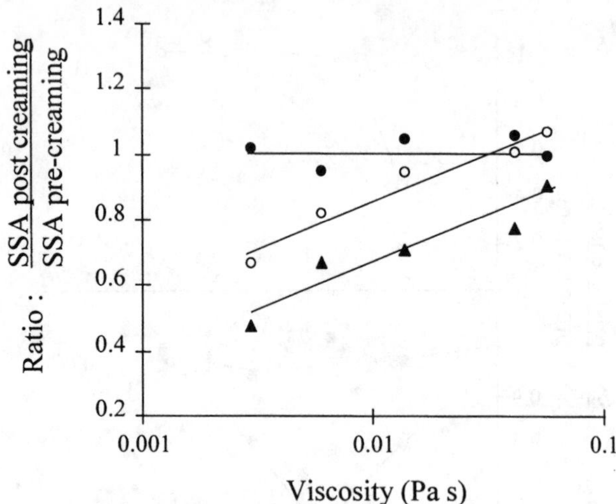

Figure 5 *The effect of increasing the viscosity of* n-*tetradecane with paraffin oil on the stability of emulsions formed with: 1.5 g l^{-1} β-lactoglobulin (●), 166 mM Span 80 (○), and a mixture of 1.5 g l^{-1} β-lactoglobulin + 166 mM Span 80 (▲)*

does not influence the stability of β-lactoglobulin stabilized emulsions. In contrast, the stability of emulsions formed with Span 80 alone and in the presence of β-lactoglobulin are lower at low oil viscosity, but increase as the oil phase viscosity increases.

Thin films formed from a solution of 4.15 mM Span 80 were found to show drainage characteristics typical for emulsifier stabilized films, commonly referred to as mobile or asymmetric drainage,[2,9] which allows a rapid liquid drainage from the film. The effect of the bulk oil viscosity on the drainage properties of Span 80 stabilized films was studied and it was observed that liquid drainage from Span 80 thin films becomes slower as the viscosity is increased. The lateral surface diffusion coefficient (D) of the fluorescent probe (DDAF) in the adsorbed layer of thin film stabilized by Span 80 was measured as a function of the viscosity using the FRAP technique. Results are shown in Figure 6. A decrease in the magnitude of the lateral diffusion coefficient of DDAF was observed with increasing oil phase viscosity.

4 Discussion

Previous work has shown[3,5,19,20] that the oil phase viscosity influences the emulsification process. Here, the results also suggest that the relationship between oil phase viscosity and emulsification depends on the adsorption and interfacial behaviour of the emulsifiers. In addition the viscosity of the oil also influences emulsion coalescence stability. Therefore we will discuss the emulsification and emulsion stability processes separately.

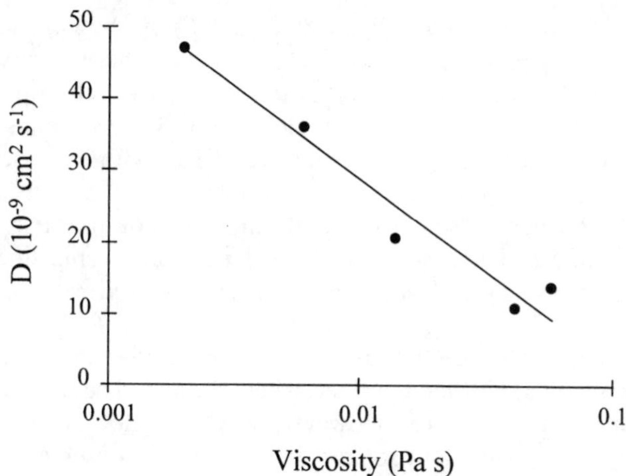

Figure 6 *The effect of increasing the viscosity of* n-tetradecane *with paraffin oil on the surface lateral diffusion coefficient (*D*) of the fluorescent probe molecule DDAF in the adsorbed layer of Span 80 stabilized thin films*

Emulsifying Properties

The effect of Span 80 on emulsification properties of β-lactoglobulin stabilized emulsions is shown in Figure 1. The results have been discussed in detail previously.[3] Essentially the increase in SSA above 100 mM Span 80 is due to a reduction in the interfacial tension[5] leading to a smaller droplet size, which has been observed with other emulsifier systems.[16,17] The decrease in SSA at higher Span 80 concentrations is a direct result of an increase in oil phase viscosity[3] which has been observed by other workers.[19,20] Considering the dynamic forces required to break up a deformable oil droplet during homogenization, large changes in liquid velocity are required over a length scale similar to that of the droplet. As the deformation of the droplet is determined by the size, viscosity and interfacial tension of the droplet, the timescale required to achieve break-up is proportional to both the droplet radius and the dispersed phase viscosity, and inversely proportional to the interfacial tension.[20] The soya oil has generally poorer emulsification properties than *n*-tetradecane (Figure 1), despite having a lower interfacial tension than *n*-tetradecane.[18] The other main contributory factor to droplet breakup is the dispersed phase viscosity[19,20] (Figure 2); therefore the higher viscosity of the soya oil reduces emulsification efficiency. This is supported by the fact that, at high Span 80 concentrations (>1 M), the oil phase viscosity of the soya oil and *n*-tetradecane become similar (Figure 2). This trend is duplicated in Figure 1, where the SSA values of the oils also become similar. Unfortunately, the relationship between oil viscosity and emulsification does not hold for emulsions stabilized by β-lactoglobulin (Figure 3). The two emulsions in the presence of Span 80 obey the relationship between viscosity and emulsification, but the protein-stabilized emulsion shows no effect of oil-phase viscosity. Increasing the viscosity of the oil reduces the rate of adsorption of the oil-soluble Span 80 (results not shown), and it could therefore result in a reduction in the SSA. However, the viscosity of the oil should not affect the rate of adsorption of the protein from the aqueous phase, which is still present in the mixed β-lactoglobulin + Span 80 emulsion (Figure 3). This emulsion system still shows a decrease in SSA with increased oil phase viscosity; therefore the differences in emulsification behaviour between β-lactoglobulin and Span 80 stabilized emulsions does not appear to be related directly to the interfacial tension. Also the reduction in emulsification with increased oil-phase viscosity has previously been observed with water-soluble emulsifiers.[20,22]

The other main difference between the systems is that the emulsions with Span 80 possess a dynamic fluid interface, where the adsorbed molecules are free to diffuse laterally at the surface.[5] However, the β-lactoglobulin adsorbed layer is rigid, highly viscoelastic, and so free diffusion is not observed.[2,5,13] Recent work by Williams *et al.*[22] has demonstrated that a significant attribute of an emulsion droplet for emulsification is its interfacial viscoelasticity. Droplets with a highly elastic interfacial layer have very different break-up properties from those with a less elastic interface. This effectively separates the dispersed and continuous phases, so that the relative viscosities do not come directly into contact, as they would with a fluid interface, where the dynamics and flow of the surface layer is strongly influenced by the flow of the adjacent phases (Gibbs–Marangoni mechanism). Therefore, the interfacial viscoelastic properties of emulsion droplets play a significant role in the relationship between oil-phase viscosity and emulsification behaviour.

Emulsion Stability

Emulsion stability against coalescence has been studied for both oil systems by submitting the emulsion to a centrifugal field. Figure 4a shows that, in the absence of protein, low emulsion stability was observed for emulsions stabilized by low concentrations of Span 80. This is not surprising, since the steric stabilization properties of Span 80 for oil-in-water emulsions is poor, as demonstrated previously using orthokinetic stability measurements.[5] The increase in emulsion stability at higher Span 80 concentrations (Figure 4a and b) is unexpected, as this effect was not observed during previous coalescence measurements[3] which monitored an oil droplet coalescing with a planar interface. Therefore the effects of oil-phase viscosity were investigated, and it was observed that increasing the oil-phase viscosity leads to an increase in stability of the Span 80 stabilized emulsions (Figure 5). The fact that the stability of the soya oil emulsions is restored at lower Span 80 concentrations than for *n*-tetradecane emulsions (Figure 4a) suggests that this is an effect of the oil-phase viscosity also (Figure 2). The mechanism behind the increase in stability is not clear. The increase in oil-phase viscosity, as discussed in the previous section, clearly affects the deformation of the droplets, and the accelerated creaming method will certainly involve droplet deformation. The observation that thin films stabilized by Span 80 display slower drainage and surface diffusion behaviour (Figure 6) when the oil-phase viscosity is increased possibly suggests that this is a dynamic problem. If the rate of deformation were to be slowed, then the point at which coalescence occurs may also be delayed, resulting in an apparent increase in stability.

The effect of surface viscosity, though, is clearer. Presence of β-lactoglobulin alone results in a stable emulsion, irrespective of the oil-phase viscosity (Figures 4b and 5) or the method of measurement.[3–5] The highly elastic protein network formed on the droplet's surface is efficient at stabilizing emulsion droplets through steric stabilization and so reducing droplet deformation. The effects of Span 80 on the stability of the emulsions (Figure 4b) are due to the disruption of the protein adsorbed layer, resulting in a much weaker, less elastic network, and hence poor emulsion stability towards coalescence.[3–5]

Therefore, one can say that reducing the droplet deformation may influence emulsion stability to coalescence. However, the viscoelastic properties of the β-lactoglobulin-stabilized interface appear to have more influence on the stability of the droplet than an increase in the oil-phase viscosity. This is because the reduction in droplet deformation caused by increasing the oil-phase viscosity only appears to affect the stability of emulsions which undergo accelerated creaming by centrifugation, whereas the β-lactoglobulin emulsions are stable to coalescence independent of the method of measurement.

5 Conclusions

The effect of an oil-soluble surfactant (Span 80) on emulsifying capacity of emulsions was studied at both the *n*-tetradecane–water and soya oil–water interfaces. Span 80 was found to be a much better emulsifier of *n*-tetradecane than soya oil, due to the higher viscosity of the soya oil reducing the deformability of the oil droplet. Oil-phase viscosity does not affect β-lactoglobulin stabilized emulsions due to the highly elastic adsorbed layer, which itself controls droplet deformability. Increasing oil-phase viscosity improves the resistance of oil droplets against coalescence by accelerated creaming, probably by decreasing the rate of drainage from thin liquid films or by reducing the rate of drop deformation.

Acknowledgements

This research was in part supported by a Human Capital and Mobility grant (ERB CHRXT930322) to M.C. from the European Union. The authors gratefully acknowledge financial support from BBSRC via the Core Strategic Grant of the Institute.

References

1. W. E. Ewers and K. L. Sutherland, *Aust. J. Sci. Res. Ser. A*, 1952, **5**, 697.
2. D. C. Clark, A. R. Mackie, P. J. Wilde, and D. R. Wilson, *Faraday Discuss.*, 1994, **98**, 253.

3. M. Cornec, P. J. Wilde, P. A. Gunning, F. A. Husband, M. L. Parker, and D. C. Clark, *J. Food Sci.*, 1998, **63**, 39.
4. J. Chen, E. Dickinson, and G. Iveson, *Food Struct.*, 1993, **12**, 135.
5. M. Cornec, A. R. Mackie, P. J. Wilde, and D. C. Clark, *Colloids Surf. A*, 1996, **114**, 237.
6. J.-L. Courthaudon, E. Dickinson, Y. Matsumura, and A. Williams, *Food Struct.*, 1991, **10**, 109.
7. E. Dickinson and J.-L. Gelin, *Colloids Surf.*, 1992, **63**, 329.
8. E. Dickinson, A. Mauffret, S. E. Rolfe, and C. M. Woskett, *J. Soc. Dairy Technol.*, 1989, **42**, 18.
9. J.-L. Courthaudon, E. Dickinson, Y. Matsumura, and D. C. Clark, *Colloids Surf.*, 1991, **56**, 293.
10. J.-L. Courthaudon, E. Dickinson, and D. G. Dalgleish, *J. Colloid Interface Sci.*, 1991, **145**, 390.
11. J.-L. Courthaudon, E. Dickinson, and W. W. Christie, *J. Agric. Food Chem.*, 1991, **39**, 1365.
12. P. J. Wilde and D. C. Clark, *J. Colloid Interface Sci.*, 1993, **155**, 48.
13. A. R. Mackie, P. J. Wilde, D. R. Wilson, and D. C. Clark, *J. Chem. Soc. Faraday Trans.*, 1993, **89**, 2755.
14. A. R. Mackie, S. Nativel, D. R. Wilson, S. Ladha, and D. C. Clark, *J. Sci. Food Agric.*, 1996, **70**, 413.
15. L. R. Fisher, E. E. Mitchell, and N. S. Parker, *J. Food Sci.*, 1985, **50**, 1201.
16. E. Dickinson and G. Iveson, *Food Hydrocolloids*, 1993, **6**, 533.
17. E. Dickinson, R. K. Owusu, S. Tan, and A. Williams, *J. Food Sci.*, 1993, **58**, 245.
18. P. Walstra and L. De Roos, *Food Rev. Intern.*, 1993, **9**, 503.
19. P. Walstra, *Dechema-Monograph*, 1974, **77**, 87.
20. W. D. Pandolfe, *J. Dispersion Sci. Technol.*, 1981, **2**, 459.
21. A. G. Gaonkar, *J. Am. Oil Chem. Soc.*, 1989, **66**, 1090.
22. A. Williams, J. J. M. Janssen, and A. Prins, *Colloids Surf. A*, 1997, **125**, 189.

Foam Formation by Food Proteins in Relation to their Dynamic Surface Behaviour

By H. K. A. I. van Kalsbeek and A. Prins

DEPARTMENT OF FOOD TECHNOLOGY AND NUTRITIONAL SCIENCES, FOOD PHYSICS GROUP, WAGENINGEN AGRICULTURAL UNIVERSITY, P.O. BOX 8129, 6700 EV WAGENINGEN, THE NETHERLANDS

1 Introduction

Proteins are usually very suitable to make and to stabilize foams and so they are often used in aerated food systems. The foaming properties are, however, dependent on the nature of the protein which is used.[1] Since foaming is a dynamic process in which surfaces of bubbles are strongly deformed, the dynamic surface properties of the protein solutions are considered to play an important role in the ability of the foaming agent to create foam. In this study the dynamic surface properties and the foam formation properties of proteins with different molecular structure are studied and mutually compared.

2 Theory

General Aspects

During the whipping of a surface-active solution, air is incorporated in the solution in the form of large bubbles. These large bubbles have to be broken down into smaller ones by mechanical agitation. Once the small bubbles are formed, rupture of the thin film between the bubbles resulting in the formation again of large bubbles (recoalescence) has to be prevented. In order to stabilize foam bubbles against recoalescence, the bubble surface must possess the appropriate mechanical properties. One of these required properties is the presence of a surface tension gradient at the bubble surface. This surface tension gradient slows down motion of the bubble surface when pairs of bubbles approach each other. Hereby, the flow of liquid between the bubbles is hampered, and so the bubbles can only approach each other slowly.[2] To create a surface tension gradient, the proteins must first adsorb at the interface of the

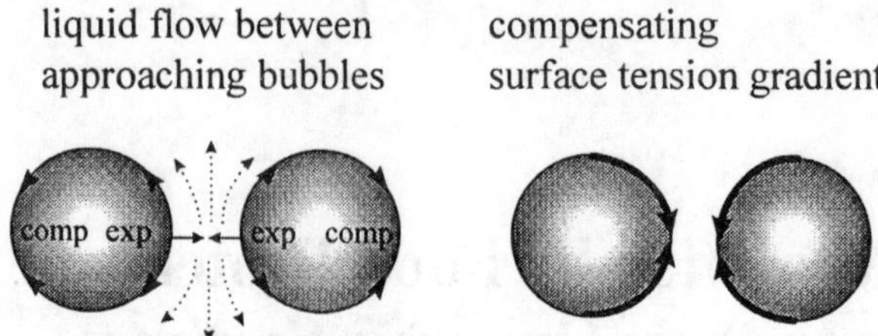

Figure 1 *Surface tension gradient present at a bubble surface and its stabilizing effect against flow of liquid out of the film between two bubbles*

bubbles and lower the surface tension. When this surface is deformed by motion of liquid along the surface, a surface tension gradient can be generated. The surface tension gradient which is generated by the motion of liquid is shown schematically in Figure 1.

Dynamic Surface Properties

The capability of a system to create surface tension gradients can be studied using the overflowing cylinder technique[3–5] shown schematically in Figure 2. With this technique an expanded surface far from equilibrium can be obtained in a steady-state situation. The relative expansion rate $\mathrm{d}\ln A/\mathrm{d}t$ (s^{-1}) of the top surface is related to the distance to the centre of the surface r (m) and to the radial surface velocity v_r ($\mathrm{m\,s}^{-1}$):

$$\frac{\mathrm{d}\ln A}{\mathrm{d}t} = \frac{\partial v_r}{\partial r} + \frac{v_r}{r}. \tag{1}$$

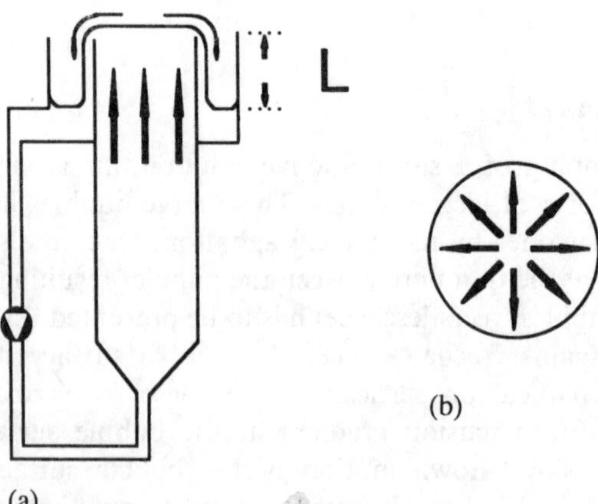

Figure 2 *Schematic representation of (a) cross-section of the overflowing cylinder and (b) top view of the expanding surface*

It is found that, in the central part of the surface, the velocity increases from the centre to the rim in a way proportional to r. This simplifies equation (1) to

$$\frac{\mathrm{d}\ln A}{\mathrm{d}t} = 2\frac{\partial v_r}{\partial r}. \tag{2}$$

The relative expansion rate is related to the average time τ (s) a molecule spends in the surface:[6]

$$\tau \approx \frac{1}{\mathrm{d}\ln A/\mathrm{d}t}. \tag{3}$$

The expansion rate of the surface can be varied by changing the length L of the falling film at the outside of the cylinder. The weight of this film pulls at the surface of the overflowing liquid, and by increasing the value of L the expansion rate can be increased. The expansion rate is an autonomous property of each system, and it depends on the type and concentration of the surfactant used. Some typical data for protein solution and pure water are given in Figure 3.[7] Compared to pure water, the protein system differs in two ways: at low height of the falling film, the surface expansion rate is slowed down by the formation of a stagnant surface layer, and at high falling film lengths the surface expansion rate is increased in comparison to pure water. Both phenomena are related to the formation of surface tension gradients at the surface (see Figure 4).

Outside the cylinder, at the end of the falling film, the surface of the falling film is compressed to the wall and the surface tension here (γ_{comp}) is lower than the surface tension of the expanded surface at the top (γ_{centre}). At low L values

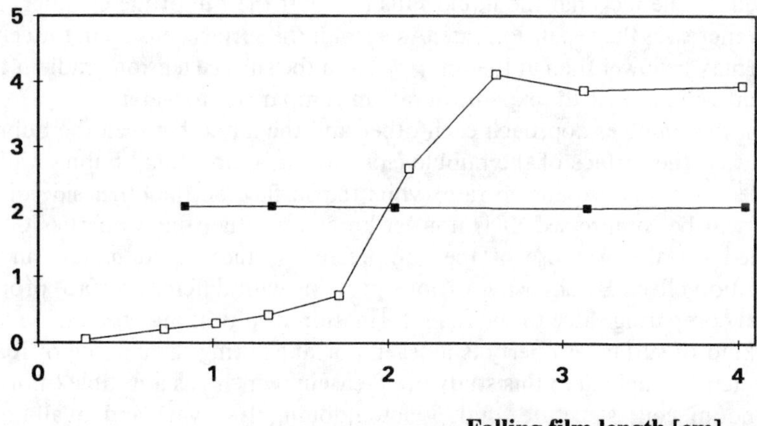

Figure 3 *Relative expansion rate as a function of the length of the falling film in the overflowing cylinder of pure water (■) and a β-casein solution (□) at 25 °C*

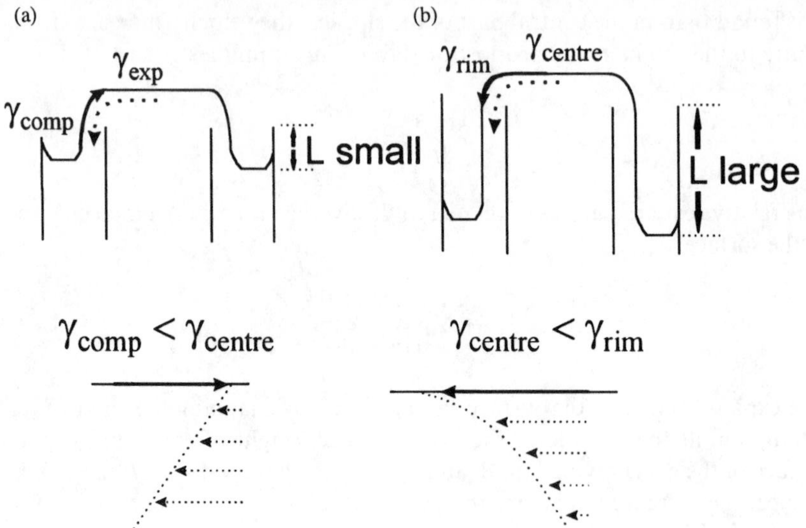

Figure 4 *Surface tension gradient present at the top of the overflowing cylinder for (a) a short length of the falling film and (b) a long length of the falling film*

this results in a surface tension gradient which slows down the expansion rate in comparison to water (Figure 4a). The compressed part of the surface is hardly accessible for measurements, but surface tensions of compressed surfaces can be examined in a Langmuir trough equipped with a caterpillar belt.[8]

On increasing *L*, the gradient between the compressed and expanded parts of the surface will not be high enough any more to slow down the surface expansion, and so the system will 'turn over' to the situation of Figure 4b. In this situation the surface may be accelerated in comparison to pure water. This is caused by the fact that the liquid falls down at the rim of the cylinder, and thereby increases the expansion rate. As a result the surface tension in the centre (γ_{centre}) may be lower than at the rim (γ_{rim}) and the surface tension gradient that is present will increase the expansion rate in comparison to water.

When two bubbles approach each other and the liquid between the bubbles flows away, the surface of the bubble close to the approaching bubble will be expanded at a high expansion rate, while the surface on the other side of the bubble will be compressed. It is not yet known whether the properties of the expanded surface or those of the compressed surface are dominant in the stabilization effect. By choosing various proteins with different surface properties, and comparing these to the foam formation results, it may become known which kind of surface property is most important during the making of foam. The proteins examined in this study are β-casein, which has a flexible (more or less random coil) structure, and β-lactoglobulin, lysozyme and ovalbumin, which all have a globular structure. Of these globular proteins lysozyme has a very stable structure[9] and ovalbumin is known to be sensitive to surface denaturation.[10] The pH of the solutions is standardized at 6.7, although this means that the proteins are not all away from their iso-electric points to the same extent.

3 Materials and Methods

Materials

The β-casein was purchased from Eurial (Rennes, France). The β-lactoglobulin was purified by a method described by Caessens.[11] Lysozyme was purchased by Nederlandse Industrie voor Eiprodukten (Nunspeet, The Netherlands, batch number 10396052) and ovalbumin from Sigma-Aldrich Chemie (Zwijndrecht, The Netherlands, A-5503 Lot nr.106H7070).

All proteins were studied in aqueous buffers of imidazole and NaCl in distilled water. The pH of the buffers was adjusted to 6.7 by a solution of HCl and the final ionic strength was 0.075 M adjusted using NaCl. The protein concentration used in this study was 0.25 g l^{-1}.

Dynamic Surface Properties

Dynamic surface properties were studied at 25 °C using the overflowing cylinder technique[3–5] and the caterpillar belt method.[8]

In the overflowing cylinder apparatus the liquid is pumped upwards from the bottom to the top of a vertical cylinder (height 60 cm, diameter 6 cm). At the top the liquid flows downwards over the rim of the cylinder and falls down at the outside wall. Then the liquid is collected in a wider cylinder that is concentric to the inner one. From here the liquid is pumped in a close loop to the vertical cylinder again. A flow rate of 35 $cm^3\ s^{-1}$ was used. The height of the falling film was increased by removing solution from the system. The expansion rate of the surface at the top of the cylinder was measured using inert talcum powder particles and a laser Doppler anemometer technique at relative expansion rates higher than about 0.6 s^{-1}. At lower relative expansion rates the velocity was measured using polypropylene particles and a stopwatch. The time was measured during which the particles moved a certain distance over the surface, and from the velocity the $d \ln A/dt$ values were calculated using equation (2).

The caterpillar belt trough is a Langmuir trough equipped with one fixed barrier and moving barriers connected to a caterpillar belt. The barriers can be moved in two directions and thereby can expand or compress the surface. In this paper only results of compression experiments are discussed. The relative compression rate of the surface, $d \ln A/dt$, at a certain velocity of the barriers v (m s^{-1}) and distance between the fixed and moving barriers l (= 16.5 cm) is given by:

$$\frac{d \ln A}{dt} = \frac{v}{l}. \tag{4}$$

The compression of the surface was started at the lowest dilational deformation rate and finished at the highest rate. Before the compression was started, the surfaces were expanded at a rate of 0.02 s^{-1}. Then the expansion was stopped, and the surfaces were allowed to relax for about 15 hours, after which time the systems were considered to be in equilibrium and the compression was started.

The surface tension measurements of the expanding surfaces of both the overflowing cylinder and the caterpillar belt apparatus were carried out using a Wilhelmy plate.

Foam Formation

Foams were formed by whipping aqueous protein solutions at 20 °C in a cylindrical glass vessel of diameter 6.1 cm. Whipping was done with a fan-shaped, stainless steel whisk connected to a 20 cm vertical rod of diameter 8 mm. The whisk was of thickness 0.1 mm and diameter 4.6 cm and was provided with radial spokes. The spokes of the whisk were 1.55 cm long and 0.7 mm wide, and the tips of the spokes were placed 5 mm from each other.[12]

The whisk was situated in the middle of 100 ml of protein solution, and each sample was whipped for 70 seconds at a frequency of 2500 min^{-1}. Directly after foam making, the sample was taken from the stirred dispersion. After dilution with the original solution, photographs were taken using a CCD-video camera connected to a microscope and a computer. The bubbles present in the video images were measured using Image Analysis software (PC-Image, Foster Findlay Associates, United Kingdom). The bubble size is expressed as the volume–surface average diameter d_{32}. The amount of foam present was measured with a ruler and expressed as percentage of the original amount of liquid.

4 Results and Discussion

Dynamic Surface Behaviour

The relative expansion rate of the top surface of the overflowing cylinder contains information about the ability of a system to create surface tension gradients. The expansion rate as a function of the height of the falling film at the outside of the cylinder is given in Figure 5 for pure water and for solutions of the different proteins.

At long lengths of the falling film, the relative expansion rate of the β-casein surface is approximately twice the value of pure water, whereas for β-lactoglobulin the value is *ca.* 1.4 times the rate for water, and for lysozyme and ovalbumin there is no significant difference from the water. The differences in relative expansion rates are caused by surface tension gradients which are present at the surfaces of adsorbed β-casein and β-lactoglobulin, but appear to be absent for lysozyme and ovalbumin. The presence of surface tension gradients is related to the extent that the surface tension of the expanded surface can be kept low. The surface at the rim of the overflowing cylinder is

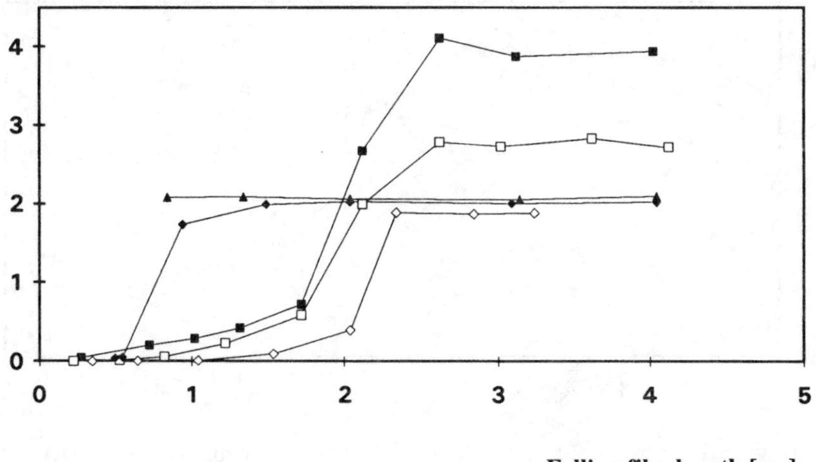

Figure 5 *Relative expansion rate of the surface of the overflowing cylinder for solutions of β-casein (■), β-lactoglobulin (□), lysozyme (◆), ovalbumin (◇) and pure water (▲) as a function of the length of the falling film*

expanded faster than in the centre, and it may be expected that the surface tension at the rim is close to the value at 25 °C for pure water, *i.e.*, 72 mN m^{-1}. As a consequence of the lowering of the surface tension in the centre of the top surface, a surface tension gradient is build up which accelerates the expansion rate of the surface (see Figure 4b). The surface tension in the centre of the overflowing cylinder ($\gamma_{centre} = \gamma_{dyn}$) of the different systems is given as a function of the relative expansion rate in Figure 6. The β-casein has by far the lowest surface tension in expansion, and at a d ln A/dt value of 4 s^{-1} the surface tension is still lower than 70 mN m^{-1}. For β–lactoglobulin a surface tension gradient is also present at the surface, but it is clearly lower than for β-casein. This seems to be in agreement with γ_{dyn} which is much higher for β-lactoglobulin than for β-casein at all expansion rates. At expansion rates higher than *ca.* 1 s^{-1}, the dynamic surface tension of lysozyme or ovalbumin is already close to 72 mN m^{-1}. Therefore no surface tension gradient is present at the surface, and so the relative expansion rate is about the same as the value for pure water. Comparable differences in dynamic surface tensions have been found before.[13]

The decrease in γ_{dyn} is caused by different processes.[14] First the proteins have to be transported to the surface by convection/diffusion, and this is followed by adsorption of the proteins at the surface. After adsorption, the protein is expected to unfold its cohesive structure, which makes it possible for different segments of the molecule to attach the interface and lower the surface tension. It is expected that transport of the proteins to the surface is of the same order of magnitude for the different proteins.[15] This means that differences in the rate of lowering of γ_{dyn} are mainly caused by different adsorption/unfolding rates. The

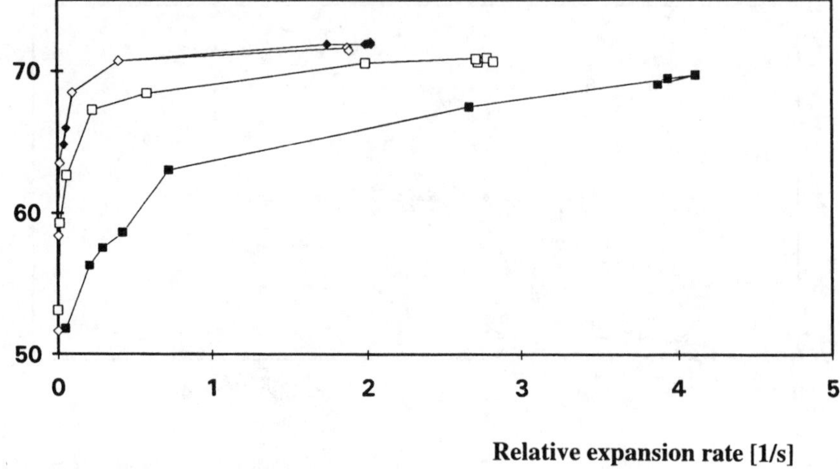

Figure 6 *Dynamic surface tension at 25 °C of β-casein (■), β-lactoglobulin (□), lysozyme (◆) and ovalbumin (◇) as a function of the relative expansion rate of the surface of the overflowing cylinder*

different sub-processes of the adsorption and unfolding have recently been studied in more detail by Boerboom.[6] In this study the extent of lowering of γ_{dyn} is considered to reflect the total effect of adsorption and unfolding, and the time needed to decrease γ_{dyn} is considered as a relaxation time. This relaxation time can be defined as the time during which a certain decrease of γ_{dyn} is obtained, and it can be considered to be equivalent to the parameter τ of equation (3). As an example, a decrease in γ_{dyn} of 3 mN m^{-1} (with respect to 72 mN m^{-1}) is present at d ln A/dt values of 2 s^{-1}, 0.2 s^{-1}, 0.07 s^{-1} and 0.07 s^{-1} for, respectively, β-casein, β-lactoglobulin, lysozyme and ovalbumin, according to Figure 6. The corresponding relaxation times are 0.5 s, 5 s, 15 s and 15 s, respectively. For most other values of γ_{dyn} the relaxation times of the different proteins are in the same order. The large differences in relaxation times, and hence the large differences in γ_{dyn}, are believed to cause the differences in surface tension gradients which can be created at the surface of the overflowing cylinder. The relaxation times are in the order that would be expected: the flexible β-casein has the shortest relaxation time, while the relaxation times of the globular proteins are more than an order of magnitude higher.

At low falling film heights the surface expansion rate is slowed down by the presence of a stagnant surface layer. The value of the film height at which this delay goes over into a faster expansion rate may be a measure of the degree to which a stagnant surface layer can be formed. In Figure 5 it can be seen that for lysozyme this transition is at a height of less then 1 cm, while for the other three proteins the jump in expansion rate is at *ca.* 2 cm height. In Figure 7 the expansion rate is expressed on a logarithmic scale. This figure indicates that the

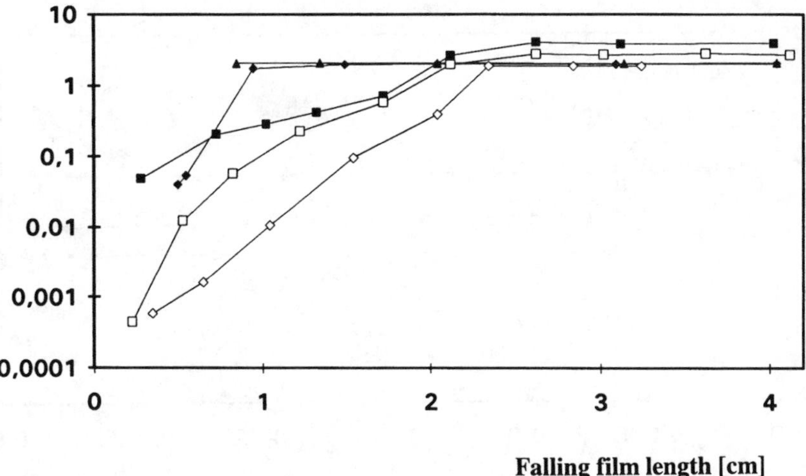

Figure 7 *Relative expansion rate of the surface of the overflowing cylinder at 25 °C for β-casein (■), β-lactoglobulin (□), lysozyme (◆) and ovalbumin (◇). The relative expansion rate is expressed on a logarithmic scale (same data as in Figure 5)*

delay in d ln $A/\mathrm{d}t$ is not the same for all proteins. Ovalbumin, and to a lesser extent β-lactoglobulin, slows down the expansion rate a few orders of magnitude more than does β-casein or lysozyme at the smallest values of L. The delay in expansion rate can be explained in part by the surface tension gradient which is present between the expanded top surface and the compressed surface at the end of the falling film. Because the compressed part of the film is hardly accessible, the surface compression has been studied using a caterpillar belt trough. In Figure 8 the surface tension in compression is given as a function of the relative compression rate. The differences between these surface tensions are rather small. Ovalbumin is found to have the lowest surface tension in compression. The differences between the surface tension in expansion (Figure 6) and in compression (Figure 8) are to be expected in the following order: ovalbumin > lysozyme > β-lactoglobulin > β-casein. This order coincides with the delay of the expansion rate at the top of the surface, except for lysozyme which slows down the expansion rate to the lowest degree. The cause of this may be that the compression in the caterpillar belt trough was started after relaxation of the surface for 15 hours, whereas each falling film height in the overflowing cylinder was changed after only 5 to 10 minutes. This implies that lysozyme needs much more time to build up a stagnant surface layer in the overflowing cylinder than do the other proteins. Furthermore, the compression rate in the compressed film of the overflowing cylinder may differ from the compression rate in the Langmuir trough, although the surface tension in compression is not very dependent on the compression rate (Figure 8).

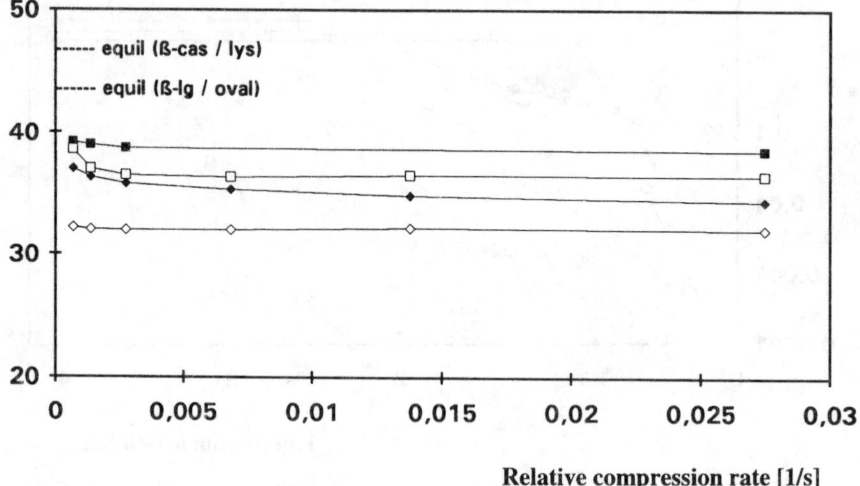

Figure 8 *Surface tension at 25 °C in compression of β-casein (■), β-lactoglobulin (□), lysozyme (◆) and ovalbumin (◇) as a function of the relative compression rate. The equilibrium surface tensions (equil) for the proteins are also given*

Besides the influence of a surface tension gradient on the delay in expansion rate, the formation of a protein network in the compressed surface may also contribute to the presence of a stagnant surface layer.[16] This can especially be the case for ovalbumin which was found to form a rigid, skin-like layer in the compressed surface. The other proteins also formed 'skins', but not to the same extent as ovalbumin.

Foam Formation

The foam produced by the whipping technique is characterized by the amount of foam generated and the average bubble diameter (d_{32}). From these two parameters can be calculated the total area of formed surface, which is taken as a measure of the foam forming capacity. From the results of Table 1 it follows that the foaming capacities of the proteins differ substantially. The β-casein and β-lactoglobulin both form much more foam with smaller bubbles than do lysozyme and ovalbumin; this has been found before for some of these proteins.[1]

Comparing dynamic surface properties of the protein solutions with the foam formation data, a relation can be found between the ability to form surface tension gradients at high expansion rates in the overflowing cylinder and the foam forming capacity. The β-casein and β-lactoglobulin are able to form these surface tension gradients and also have a good foam forming capacity. On the other hand, lysozyme and ovalbumin are not able to form surface tension

Table 1 *Foam formation properties of solutions of β-casein, β-lactoglobulin, lysozyme and ovalbumin expressed as the amount of foam produced, the average bubble diameter* d_{32}*, and the total bubble surface area formed*

Type of protein	*Fraction of incorporated air/%*	*Average bubble diameter* (d_{32})/*μm*	*Total surface area produced/m²*
β-casein	45	110	2.5
β-lactoglobulin	40	100	2.5
lysozyme	7.5	700	0.08
ovalbumin	7.5	850	0.05

gradients at high expansion rates and both have a relatively poor foam formation capacity. Ovalbumin is able to form a stagnant surface layer at low falling film lengths in the overflowing cylinder, but this appears not to be related to the extent of foam formation. It is therefore expected that the relevant surface tension gradient over a bubble surface, is mainly determined by the expanded part of the bubble and not by the compressed part.

Figure 9 illustrates a supposed mechanism for the way surface tension gradients are formed during whipping. When two bubbles approach each other, it is expected that the part of the surface where the bubbles are close to each other has the highest deformation rate, since the liquid motion is expected to be greatest here. The part of the surface which is furthest away from the other bubble will have the lowest deformation rate. Hence, a surface tension gradient will be present over the surface if the system is able to decrease γ_{dyn} more at the part of the surface where the deformation rate is lower than at the part where the deformation rate is higher. For surfaces produced from solutions of lysozyme and ovalbumin at low relative expansion rates ($<0.5\ s^{-1}$), γ_{dyn} has the value of pure water and these proteins are not able to form a surface tension gradient if different parts of the bubble surface are expanded faster than *ca.* $0.5\ s^{-1}$.

Furthermore, it is noticeable that differences in foam formation behaviour between β-casein and β-lactoglobulin are rather small, although β-casein creates

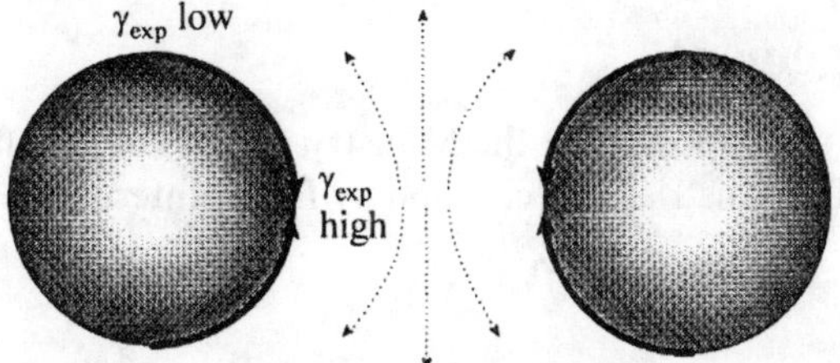

Figure 9 *Supposed mechanism for the effect on foam stability of the presence of a surface tension gradient between two parts of the surface which are expanded at different rates, where* γ_{exp} *is the dynamic surface tension of the expanded surface*

a higher surface tension gradient than β-lactoglobulin. This implies that, if a system is able to decrease the surface tension at relative expansion rates higher than *ca.* 0.5 s^{-1}, then the maximum d ln $A/\mathrm{d}t$ that can be obtained in the overflowing cylinder will exceed the certain threshold value which is necessary to stabilize the thin films between the bubbles during foam formation.

An order of magnitude calculation indicates that the maximum d ln $A/\mathrm{d}t$ values which are obtained in the overflowing cylinder for β-casein and β-lactoglobulin are able to stabilize the thin films between the bubbles to a thickness of about 1 μm. On the other hand, the very small maximum d ln $A/\mathrm{d}t$ values for lysozyme and ovalbumin means that these films may reach a thickness of 10 nm or less, which means that the films will become sensitive to rupture.

5 Conclusions

The ability to accelerate the expansion rate of the surface of the overflowing cylinder is strong for β-casein and β-lactoglobulin, and practically absent for lysozyme and ovalbumin. This ability is related to the time-scale over which a certain surface tension in expansion can be decreased. This time-scale increases in the order β-casein $<$ β-lactoglobulin $<$ lysozyme $\approx$ ovalbumin, for a wide range of dynamic surface tensions.

The delay in expansion rate at low falling film lengths of the overflowing cylinder is a measure of the extent of formation of stagnant surface layers. Ovalbumin and β-lactoglobulin slow down the expansion rate more than β-casein and lysozyme. This can partly be explained by the difference in surface tension between the expanded top surface and the compressed surface at the end of the falling film.

The β-casein and β-lactoglobulin have a high foam formation capacity as determined by the amount of created foam and the smallest average bubble diameter. Lysozyme and ovalbumin form much less foam and much larger bubbles than do β-casein and β-lactoglobulin.

Comparing the increase and decrease of the expansion rate of the surface to foam formation, it is concluded that the ability to increase the expansion rate is related to good foam formation capacity. It is concluded that the influence of the compressed part of the surface is of less importance during foam formation.

Acknowledgement

This research was supported by the Ministry of Economic Affairs through the programme IOP–Industrial Proteins and by DMV International.

References

1. D. E. Graham and M. C. Phillips, in 'Foams', ed. R. J. Akers, Academic Press, New York, 1976, p. 237.
2. P. Walstra and A. L. de Roos, *Food Rev. Int.*, 1993, **9**, 503.

3. D. J. M. Bergink-Martens, Ph.D. Thesis, Wageningen Agricultural University, 1993, ISBN 90-5485-128-7.
4. D. J. M. Bergink-Martens, H. J. Bos, A. Prins, and B. C. Schulte, *J. Colloid Interface Sci.*, 1990, **138**, 1.
5. D. J. M. Bergink-Martens, H. J. Bos, and A. Prins, *J. Colloid Interface Sci.*, 1994, **165**, 221.
6. F. J. G. Boerboom, Ph.D. Thesis, Wageningen Agricultural University, in preparation.
7. A. Prins, M. A. Bos, F. J. G. Boerboom, and H. K. A. I. van Kalsbeek, in 'Studies of Interface Science', ed. D. Möbius and R. Miller, Elsevier, Amsterdam, 1998, p. 221.
8. A. D. Ronteltap and A. Prins, *Colloids Surf.*, 1990, **47**, 285.
9. W. Norde and J. P. Favier, *Colloids Surf.*, 1992, **64**, 87.
10. E. Doi, N. Kitabatake, H. Hatta, and T. Koseki, 'Food Proteins: Structure and Functional Relationships', ed. J. E. Kinsella and W. Soucie, American Oil Chemists Society, Champaign, 1989, p. 252.
11. P. W. J. R. Caessens, S. Visser, and H. Gruppen, *Int. Dairy J.*, 1997, **7**, 229.
12. P. W. J. R. Caessens, H. Gruppen, S. Visser, G. A. van Aken, and A. G. J. Voragen, *J. Agric. Food Chem.*, 1997, **45**, 2935.
13. G. A. van Aken and M. T. E. Merks, *Colloids Surf. A*, 1996, **114**, 221.
14. D. E. Graham and M. C. Phillips, *J. Colloid Interface Sci.*, 1979, **70**, 403.
15. S. Damadoran, *Adv. Food Nutr. Res.*, 1990, **34**, 1.
16. A. Prins, F. J. G. Boerboom, and H. K. A. I. van Kalsbeek, *Colloids Surf. A*, 1998, accepted for publication.

Effect of High Pressure on Protein–Polysaccharide Complexes

By Vanda B. Galazka, Drummond Smith[1], Eric Dickinson, and Dave A. Ledward[2]

PROCTER DEPARTMENT OF FOOD SCIENCE, UNIVERSITY OF LEEDS, LEEDS LS2 9JT, UK
[1]BBSRC INSTITUTE OF FOOD RESEARCH, EARLEY GATE, READING RG6 2EF, UK
[2]DEPARTMENT OF FOOD SCIENCE AND TECHNOLOGY, UNIVERSITY OF READING, WHITEKNIGHTS, P.O. BOX 226, READING RG6 6AP, UK

1 Introduction

Food products often contain a mixture of two or more macromolecules which together improve their processing and/or eating qualities. The desired effects of enhancing structure, stability and textural characteristics may be achieved by mixing proteins with polysaccharides.[1] The functional properties of proteins are affected by the nature and strength of their interactions with other biopolymers. Depending on solvent conditions (pH, temperature, ionic strength), and the structure of the biopolymers, these interactions may be specific or non-specific, attractive or repulsive, weak or strong. For instance, the anionic dextran sulfate (DS) forms a strong electrostatic complex at low ionic strength and neutral pH with bovine serum albumin (BSA),[2] whereas there is no evidence to suggest that β-lactoglobulin (β-Lg) forms such a complex.[3]

Carrageenans are commonly used in milk based products as thickeners, gelling agents and stabilizers.[4] Like dextran sulfate, carrageenan is a sulfated polysaccharide. The main difference between the various forms of carrageenan is in the number and position of sulfate groups on the polygalactose backbone. κ-Carrageenan (κ-CAR) carries about 25% ester sulfate groups by weight, and ι-carrageenan (ι-CAR) about 32%.[4]

Many studies have been carried out on the interactions of carrageenans with proteins, especially with the caseins[5–7] but also with other proteins.[8,9] Recently, it was reported[10,11] that the carrageenans, κ-CAR and ι-CAR, form an electrostatic complex with adsorbed BSA in emulsion systems, the strength of the interaction being dependent upon ionic strength and pH. In both systems, the strength of interaction increased as the pH was reduced from pH 7 towards pH 6. The destabilization behaviour in oil-in-water emulsions at pH 6 was interpreted[10] in terms of bridging flocculation leading to a gel-like emulsion network over a certain carrageenan concentration range. The system containing BSA + κ-CAR was found to give a weaker emulsion gel network under similar pH and ionic strength conditions,[11] which indicates that the BSA + κ-CAR interactions are weaker than the BSA + ι-CAR interactions. In systems containing dextran sulfate, a large effect was seen at pH 7, and this induced strong bridging flocculation in a concentrated BSA stabilized emulsion.[12] This can be attributed to a net attractive electrostatic protein–polysaccharide interaction at the emulsion droplet surface.[3] The strength of the BSA–polysaccharide interaction and the associated bridging flocculation behaviour appears to be related to the density of charged sulfate groups on the polysaccharide (DS > ι-CAR > κ-CAR).[13]

It is now well established that high-pressure treatment has a disruptive effect on the tertiary and quaternary structure of most globular proteins, with relatively little effect on the secondary structure. BSA and β-Lg at pH 7 and low ionic strength are both denatured and aggregated during/after high pressure treatment,[14–16] which is mainly considered to be due to the formation of intermolecular disulfide bonding via —SH/—SS— linkages. Due to the pressure-induced changes in protein conformation and electrostatic interactions, it is suggested that non-covalent protein–polysaccharide interactions could also be altered as a result of pressure treatment.

Recently we have demonstrated[15] that the anionic polysaccharide dextran sulfate (DS) inhibits the degree of BSA unfolding during pressurization. Moreover, complexation of the protein with the polysaccharide has the effect of protecting the BSA against extensive aggregation after pressure treatment. Other studies[12] have shown that, for oil-in-water emulsions containing BSA + DS, pressure treatment of the protein before homogenization affects the flocculation and rheology of the BSA-stabilized emulsion containing the added polysaccharide.

In this study we are interested in the effects of high-pressure treatment on the properties of mixtures containing (i) β-Lg with DS and (ii) BSA with polysaccharides of differing charge densities (DS, ι-CAR and κ-CAR). The purpose of this paper is to link recent surface tension and emulsion studies on the BSA + carrageenan systems[10–11] with their solution properties, and to relate changes in surface tension with previous solution work on the (a) β-Lg + DS[17] and (b) BSA + DS systems.[14,15] Size exclusion chromatography is used to show the presence and nature of protein–protein and BSA–polysaccharide complexation, under various pH and ionic strength conditions.

2 Materials and Methods

Bovine serum albumin (product A-7638, lot 16H9314, globulin-free lyophilized powder, ⩾99% purity) and dextran sulfate (product D-6001, lot 112H0372,

500 kDa) were purchased from Sigma Chemical Co. (St Louis, MO, USA). The food-grade carrageenan samples (ι-CAR and κ-CAR) were gifts from Systems Bio Industries (Carentan, France). The ι-CAR was in almost pure sodium form with about 5% contamination by κ-CAR, with a weight-average molecular weight given by the suppliers as 560 kDa and a z-average hydrodynamic radius of 80 nm. The κ-CAR was 60% in the potassium form and 40% in the sodium form. The weight-average molecular weight was given by the suppliers as 720 kDa and the z-average hydrodynamic radius as 100 nm.

1-Anilinonaphthalene-8-sulfonate (ANS) ammonium salt (316.4 Da) was obtained from SERVA (Feinbiochemica, Heidelberg, Germany). Buffer solutions (20 mM imidazole, pH 6.5–7; 20 mM bis-Tris, pH 6.5; and 20 mM Tris-HCl, pH 7–8) for the size exclusion chromatography and surface tension experiments were prepared from analytical grade reagents and HPLC grade water.

The carrageenan solutions (2 g l^{-1}) for structural analysis were prepared by dispersing the powder in HPLC grade water and continuously stirring for 20 min at 70 °C. The resulting solution was cooled to 25 °C and 5 g l^{-1} BSA was dissolved in it, and adjusted to pH 7 by addition of 0.05 M HCl or 0.05 M NaOH. BSA or β-Lg alone (5 g l^{-1}) or a mixture of BSA or β-Lg (5 g l^{-1}) + DS (2 g l^{-1}) was dissolved in HPLC grade water, and the pH was adjusted to 7 by addition of 0.05 M NaOH or 0.05 M HCl. These solutions, as well as those for surface tension (β-Lg 5×10^{-4} wt%; BSA 1×10^{-3} wt%) experiments, were sealed in Cryovac bags (Cryovac-W.R. Grace Ltd., London), and subjected to high-pressure treatment as previously described.[14,18] Imidazole buffer was used for the surface tension experiments and all size exclusion chromatography experiments were performed at pH 6.5.

Conformational changes of both the proteins in the absence and presence of polysaccharide were compared using the reaction with 1-anilinonaphthalene-8-sulfonate (ANS)[14,15] ammonium salt to assess surface hydrophobicity, thermal stability by differential scanning calorimetry (DSC),[15] and size exclusion chromatography.[13,15] The time-dependent change in the air–water surface tension was measured by the Wilhelmy plate technique (Krüss K_{12} Processor Tensiometer) at 25 °C.

3 Results and Discussion

Protein Surface Hydrophobicity

Probe spectrofluorimetry studies at pH 7 for pressure-treated β-Lg indicate a substantial increase in protein surface hydrophobicity (Table 1) in the absence and presence of DS. At higher treatment regimes (350 and 550 MPa for 20 min) the surface hydrophobicity determined for the mixtures does not increase to the same extent as for β-Lg alone. In contrast, under similar experimental conditions, BSA shows a consistent trend of reduction in fluorescent intensity I with increasing pressure. Addition of DS to BSA greatly reduces the surface hydrophobicity for both the native and the pressure-treated samples.

Table 1 *Effect of high pressure on the surface hydrophobicity of β-Lg, BSA and mixtures of β-Lg + DS (1:1 by weight) and BSA + DS (2:1 by weight) at pH 7 (1 g l^{-1} protein, 4 × 10^{-5} mol dm^{-3} ANS). Fluorescence intensity (I) (arbitrary units) is monitored at 470 nm from excitation at 350 nm. Duration of treatment is 20 min. Values are the averages of triplicate measurements*

Treatment pressure (MPa)	Fluorescence intensity (I)			
	β-Lg	β-Lg + DS	BSA	BSA + DS
0	22	22	153	85
200	39	34	150	78
350	66	45	127	64
550	85	59	88	48

The protein hydrophobicity (S_o) values determined for BSA in the absence and presence of sulfated polysaccharides (DS > ι-CAR > κ-CAR) are compared in Table 2. The reduction of surface hydrophobicity for pure BSA is in agreement with the previous work of Galazka *et al.*[14,15] and Hayakawa *et al.*[19,20] who have suggested that this could be due to the lower number of hydrophobic groups binding to the ANS because of increased intermolecular interactions,[19,20] or due to the burying of some binding sites in the partially denatured and refolded protein.

Addition of DS (BSA:DS weight ratio = 2:1) greatly reduces the measured protein hydrophobicity (S_o) [untreated BSA, 275 AU/mol ANS; untreated

Table 2 *Influence of high-pressure processing on the protein hydrophobicity (S_o) of native and pressure-treated BSA (0.17–16 μM) and mixtures of BSA (0.17–16 μM) + polysaccharide (2.5:1 by weight) in aqueous solution (2.31 μM ANS, pH 7). The duration of pressure treatment was 20 min. All experiments were performed in triplicate with an estimated error of ±10%*

Sample	Pressure (MPa)	S_o (absorbance units/mol of ANS)
BSA	0	275
BSA	600	126
BSA:dextran sulfate	0	27
BSA:dextran sulfate	300	20
BSA:ι-carrageenan	0	67
BSA:ι-carrageenan	600	35
BSA:κ-carrageenan	0	148
BSA:κ-carrageenan	600	85
Polysaccharide alone	600	0

BSA + DS, 27 AU/mol ANS]. Replacement of DS with ι-CAR (BSA:ι-CAR weight ratio = 2.5:1) leads to an increase in S_o [67 AU/mol ANS], and the presence of κ-CAR (BSA:κ-CAR weight ratio = 2.5:1) instead of ι-CAR leads to a further increase in S_o [148 AU/mol ANS] for the untreated mixtures, but this is still lower than for the BSA alone. In all cases, S_o for the pressure-treated mixture is lower than for the untreated mixture at the same weight ratio. The ANS probe has been shown not to bind to the polysaccharide alone ($I \sim 0.8$), and it is assumed that the reduction in surface hydrophobicity is due to electrostatic repulsion between the two negatively charged molecules and/or to possible blocking of ANS from the β-Lg or BSA binding sites as a result of protein–polysaccharide complexation. In the pressurized mixtures it is thought that some pressure-induced modification of β-Lg or BSA also occurs in the complexed state.

Denaturation

From the data presented in Table 3 we note that the values of the endothermal peak temperature T_m (70.5 °C) for the untreated and treated BSA (600 MPa for 20 min) are the same at neutral pH (within experimental error). However, there is a 45% reduction in the total calorimetric enthalpy ΔH after pressure processing. Our DSC data are consistent with previous work from Galazka *et al.*[15] and Hayakawa *et al.*[20] which indicates that high-pressure treatment causes a major loss of tertiary structure, which is not recovered after pressure release.

Addition of κ-CAR to BSA leads to a small reduction in T_m with a reduction in ΔH (20%), and pressurization of the mixture reduces the value of ΔH by 60%. In systems containing ι-CAR, the T_m shifts to a slightly lower value with a corresponding decrease in ΔH (~5%) for the native mixture; pressure proces-

Table 3 *Influence of high-pressure treatment on the endothermic peak temperatures (T_m) and total calorimetric enthalpies (ΔH) of BSA ($5\,g\,l^{-1}$) and mixtures of BSA (5 mg/mL) + polysaccharide ($2\,g\,l^{-1}$) in aqueous solution at pH 7. Duration of treatment was 20 min. Quoted values are the averages of duplicate measurements with estimated experimental errors of ±0.5 °C in T_m and ±10% in ΔH*

Sample	*Pressure (MPa)*	*T_m (°C)*	*ΔH (kcal mol^{-1})*
BSA	0	70.5	1.75×10^5
BSA	600	70.5	0.97×10^5
BSA + κ-carrageenan	0	69.0	1.41×10^5
BSA + κ-carrageenan	600	69.9	0.73×10^5
BSA + ι-carrageenan	0	68.5	1.33×10^5
BSA + ι-carrageenan	600	69.5	0.31×10^5
BSA + dextran sulfate	0	52.6	0.22×10^5
BSA + dextran sulfate	600	46.9	0.25×10^5
Polysaccharide alone	600	0	0

sing induces a large reduction in ΔH ($\sim$80%) at the same weight ratio. Replacement of ι-CAR with DS, induces such a large drop in T_m and ΔH that pressure treatment has little additional effect. The polysaccharides alone do not deviate from the baseline. Comparison of the three systems suggests that weak complexes are formed in the BSA + carrageenan systems, and that pressure treatment causes further denaturation of the protein.

Protein–Protein and Protein–Polysaccharide Complexes

Size exclusion chromatography has been used to assess the molecular sizes and nature of protein–protein and protein–polysaccharide complexes in the absence and presence of NaCl. Figure 1 shows the effect of high-pressure treatment on the elution profile of BSA ($5\,g\,l^{-1}$) and mixtures of BSA + DS ($2.5\,g\,l^{-1}$) at pH 7 in the absence of NaCl. The major peak for the native BSA (Figure 1a) elutes at approximately 6.6×10^4 Da, and pressurization at 600 MPa (Figure 1b) induces the formation of dimers with an approximate molecular weight of 1.5×10^5 Da. This corresponds to earlier studies[13–15] which have suggested that during high-pressure treatment these units are stabilized by —S—S— bridges. In the mixed systems, we note, that the addition of DS (Figure 1c) introduces a strong narrow peak corresponding to distinct protein–polysaccharide complex(es) of molecular weight 5×10^5 Da, together with a small peak corresponding to residual monomeric BSA. Pressurization at 600 MPa for 20 min (Figure 1d) shows no significant change in the complex molecular weight.

To investigate the ionic nature of the BSA–DS interactions, 100 mM NaCl was added to the system. In Figure 2a we note that the presence of NaCl in the untreated mixture gives a very strong peak at 6.6×10^4 Da corresponding to monomeric BSA together with a minor high-molecular-weight peak at 5×10^5 Da. Chromatogram (b) for the pressure-treated mixture in the presence of 100 mM NaCl shows a larger peak at 5×10^5 Da and a smaller peak at 6.6×10^4 Da. These results suggest that NaCl induces reversible dissociation of some of the complexed BSA into the monomeric form. Moreover, in the situation at high ionic strengths, where the electrostatic BSA–DS interaction is screened, high pressure induces protein aggregation both in the presence and absence of DS.

Figure 3 shows chromatograms for BSA ($5\,g\,l^{-1}$) + ι-CAR ($2\,g\,l^{-1}$) at pH 6.5 and 7.0 in the absence of NaCl. Chromatogram (a) for the untreated mixture at neutral pH shows a strong peak at short elution times corresponding to the distinct protein–polysaccharide complex(es) of molecular weight of approximately 6×10^5 Da with a marginally smaller peak at 6.6×10^4 Da. Reducing the pH to 6.5 (Figure 3c) leads to a larger peak at the shorter elution times (6×10^5 Da) with a corresponding smaller peak at 6.6×10^4 Da. Pressurization of the mixtures at pH 7 (Figure 3b) and pH 6.5 (Figure 3d) induces even broader and stronger second peaks corresponding to the stronger protein–polysaccharide complex(es). Addition of 20 mM NaCl (not shown) leads to the presence of one peak at 6.6×10^4 Da which is consistent with the

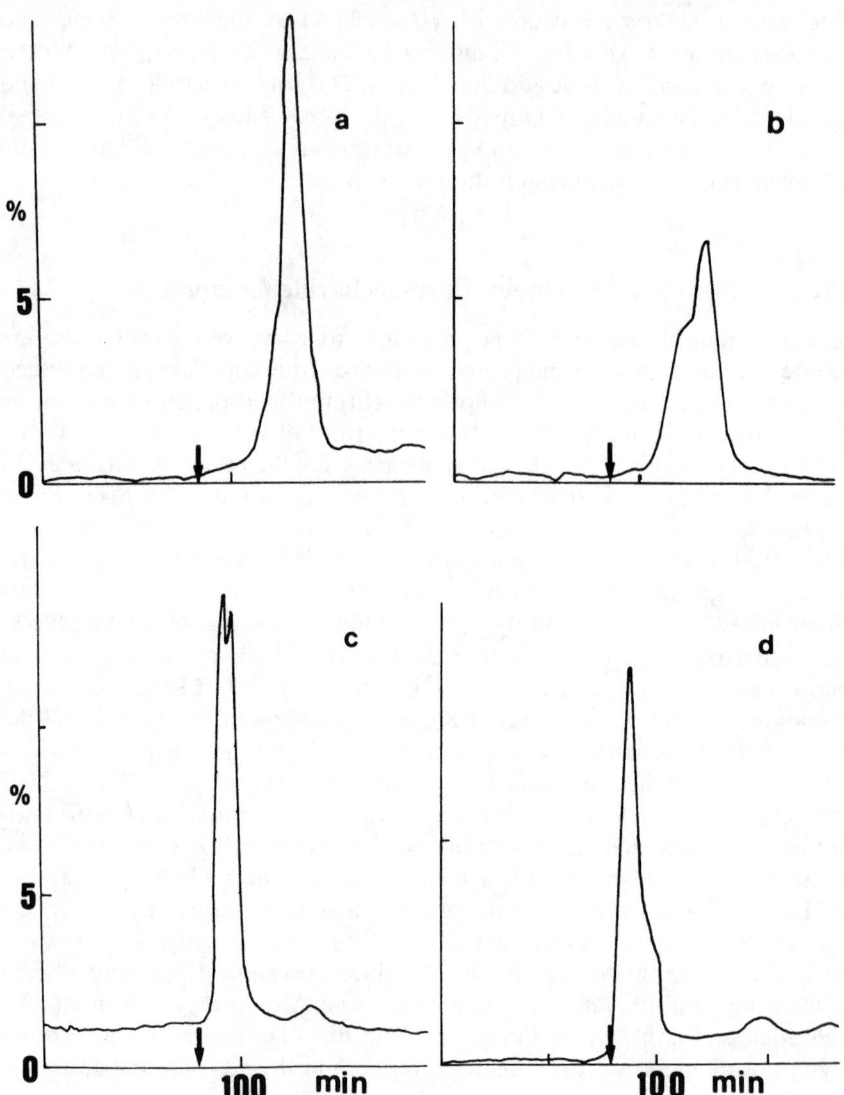

Figure 1 *Size exclusion chromatography profiles of native and pressure-treated (600 MPa for 20 min) BSA (5 g l^{-1}) and mixtures of BSA (5 g l^{-1}) + DS (2.5 g l^{-1}) at pH 7, in absence of NaCl. The absorbance at 280 nm is plotted against elution time: (a) native BSA; (b) pressure-treated BSA; (c) untreated BSA + DS; (d) pressure-processed BSA + DS. Arrow indicates position of the void volume*

concept that the protein–polysaccharide complex is held together by electrostatic interactions.

Chromatograms for the mixed system BSA + κ-CAR at pH 7 and 6.5 in the absence of NaCl are presented in Figure 4. Chromatogram (a) for the mixed system at pH 7 shows a large narrow peak corresponding to monomeric BSA

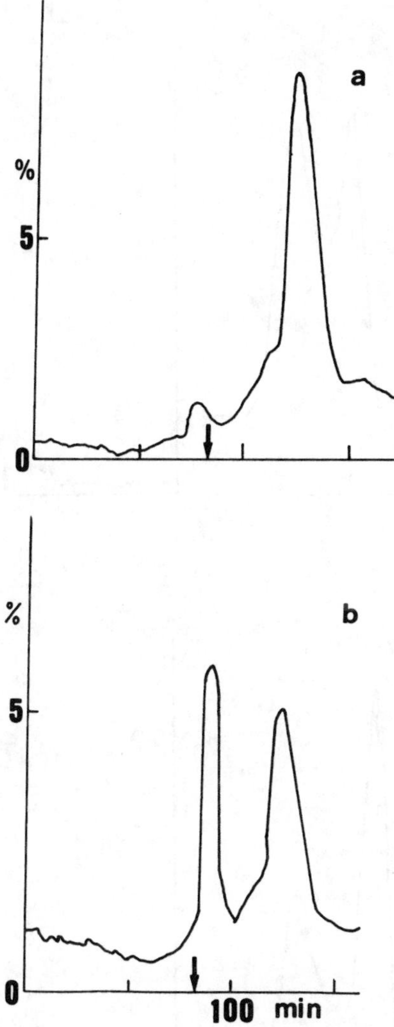

Figure 2 *Size exclusion chromatography profiles of native and pressure-treated (600 MPa for 20 min) mixtures of BSA (5 g l^{-1}) + DS (2.5 g l^{-1}) at pH 7, in the presence of 100 mM NaCl. The absorbance was measured at 280 nm as a function of the elution time: (a) untreated; (b) treated. Arrow indicates position of the void volume*

with a very small second peak, which elutes at approximately 7×10^5 Da, indicating a very weak interaction between protein and polysaccharide. Decreasing the pH to 6.5 (Figure 4c) gives a strong narrow main peak and a second small peak which indicates the formation of a weak protein–polysaccharide complex. After pressure treatment at pH 7 and 6.5 (Figures 4b and 4d) we note that the major peak (6.6×10^4) becomes smaller with a corresponding increase in size of the second peak. Experiments repeated in the presence of a

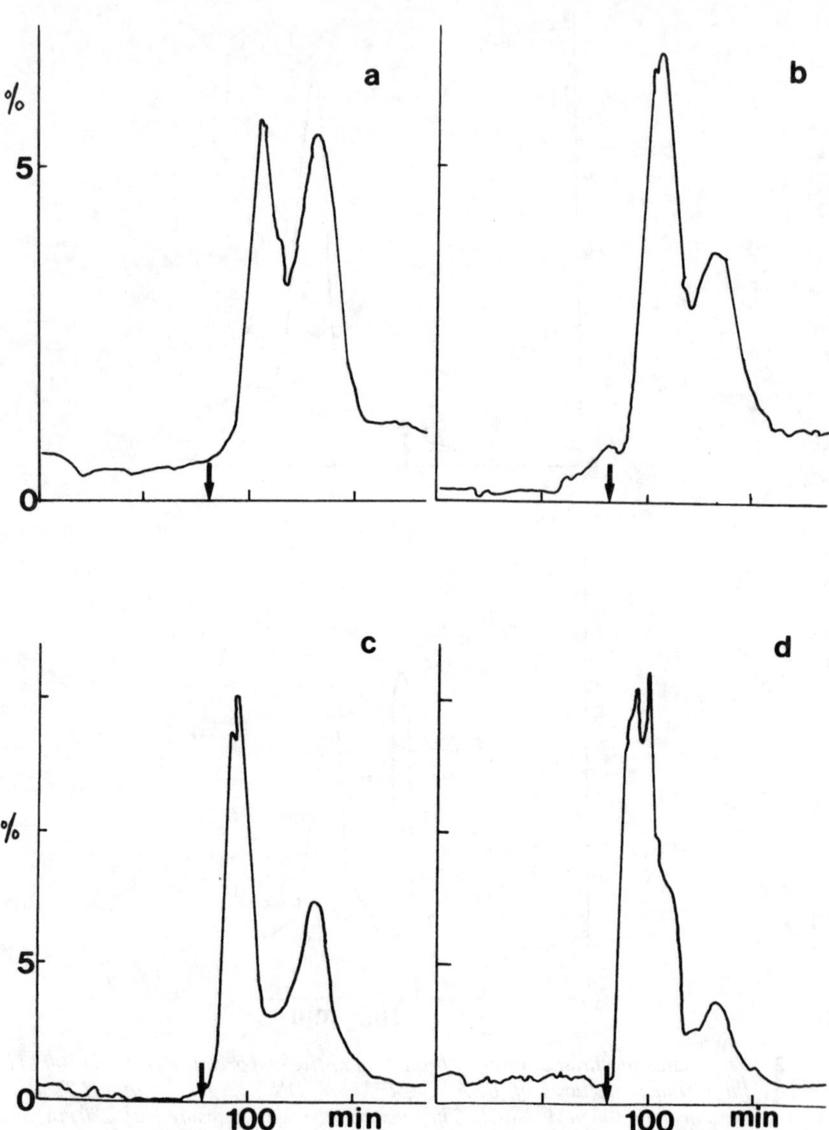

Figure 3 *Influence of high-pressure processing (600 MPa for 20 min) and pH (6.5 and 7.0) on size exclusion chromatograms of mixtures of BSA (5 g l^{-1}) + ι-CAR (2.0 g l^{-1}) in the absence of NaCl. The absorbance was measured at 280 nm as a function of the elution time: (a) untreated at pH 7; (b) treated at pH 7; (c) untreated in the presence of imidazole buffer, pH 6.5, 20 mM; (d) pressure treated at pH 6.5 in imidazole buffer. Arrow indicates position of the void volume*

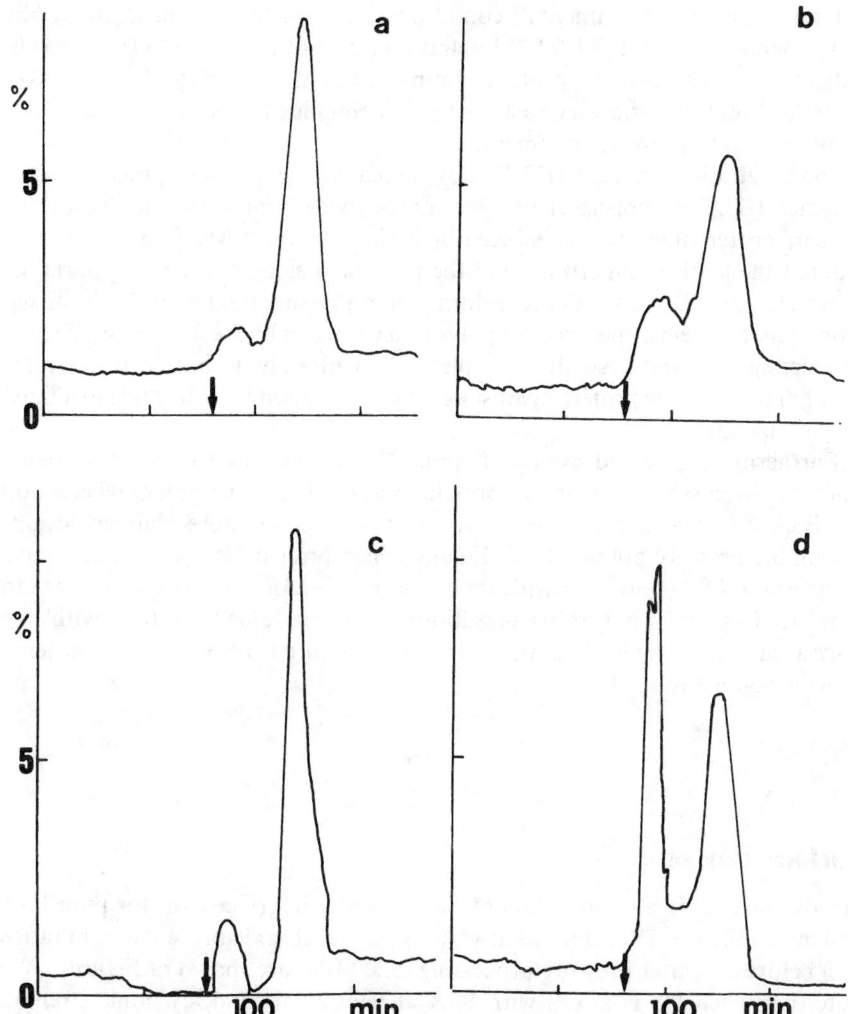

Figure 4 *Influence of high-pressure processing (600 MPa for 20 min) and pH (6.5 and 7.0) on size exclusion chromatograms of mixtures of BSA (5 g l^{-1}) + κ-CAR (2.0 g l^{-1}) in the absence of NaCl. The absorbance at 280 nm is plotted as a function of the elution time: (a) untreated at pH 7; (b) treated at pH 7; (c) untreated in the presence of imidazole buffer, 20 mM, pH 6.5; (d) pressure treated at pH 6.5 in imidazole buffer. Arrow indicates position of the void volume*

pressure-resistant buffer (bis-Tris HCl, pH 7, 20 mM) gave identical data to those performed in water, which indicates that the changes seen during the high-pressure processing were not simply due to the induced change in pH during pressurization. In the BSA + ι- or κ-carrageenan systems, however, it cannot be excluded that additional 'free' denatured protein becomes stuck to the complex during treatment.

Under similar experimental conditions, size exclusion chromatography experiments at neutral pH for β-Lg and mixtures of β-Lg + DS (not shown) indicate that the protein–protein complexation is influenced by —S—S— bridging[14] and that the very weak β-Lg–DS complex is electrostatic in nature before and after pressure treatment.

Some DL-dithiothreitol (DTT) was added to the pressure-processed biopolymer (BSA + polysaccharide) samples to determine the nature of the protein–protein interactions induced by high pressure. It was found that DTT reduced the level of dimerization in the pure BSA sample, which supports the view that —S—S— linkages occur during high-pressure treatment. In the mixed biopolymer systems, the protein–polysaccharide complex(es) were unaffected, supporting the hypothesis that complexation with polysaccharide at low ionic strength protects the protein against aggregation caused by a disulfide mediated polymerization.

Furthermore, it would seem that application of pressure treatment probably leads to the dissociation of the protein–polysaccharide complex, whereupon the BSA becomes partially denatured and so exposes more charged groups during the pressure holding time. Finally, when pressure is released, attractive electrostatic BSA–polysaccharide interactions are reformed more strongly than ever, leading to rapid recomplexation of the unfolded protein with the polysaccharide, which then protects the denatured BSA against protein–protein aggregation.[13,15]

Surface Tension

Steady-state surface tension data at the air–water interface (6 h) for pure BSA and mixed BSA + DS solutions at various polysaccharide:protein weight ratios (W) before and after pressure processing (520 MPa) are shown in Figure 5. We note that when DS is mixed with BSA at neutral pH and low ionic strength (20 mM), the mixture BSA + DS (1:3 by weight) gives a higher surface tension value (γ = 57.4 $\pm$ 0.2 mN m^{-1}) than does BSA alone (γ = 52.6 $\pm$ 0.2 mN m^{-1}). The data indicate the formation of complexes which prevent the reduction in surface activity after high pressure treatment (for pressure-treated BSA + DS (1:3 by weight) we have γ = 57.4 mN m^{-1}).

In the set of experiments for β-Lg alone and mixtures of β-Lg + DS (1:3 by weight) the surface tension after 6 h for the pressure-processed (400 MPa for 20 min) β-Lg is lower (native β-Lg, γ = 51.7 $\pm$ 0.03 mN m^{-1}; treated β-Lg, γ = 50.7 $\pm$ 0.02 mN m^{-1}). The presence of DS does not lead to any significant changes in surface activity (untreated mixture β-Lg + DS (1:3 by weight), γ = 51.7 $\pm$ 0.2 mN m^{-1}; pressure treated β-Lg + DS (1:3 by weight), γ = 50.7 mN m^{-1}). However, at shorter measurement times (up to 2 h), the slightly lower tension of the pure β-Lg solution suggests that weak electrostatic complexation occurs between β-Lg and DS.

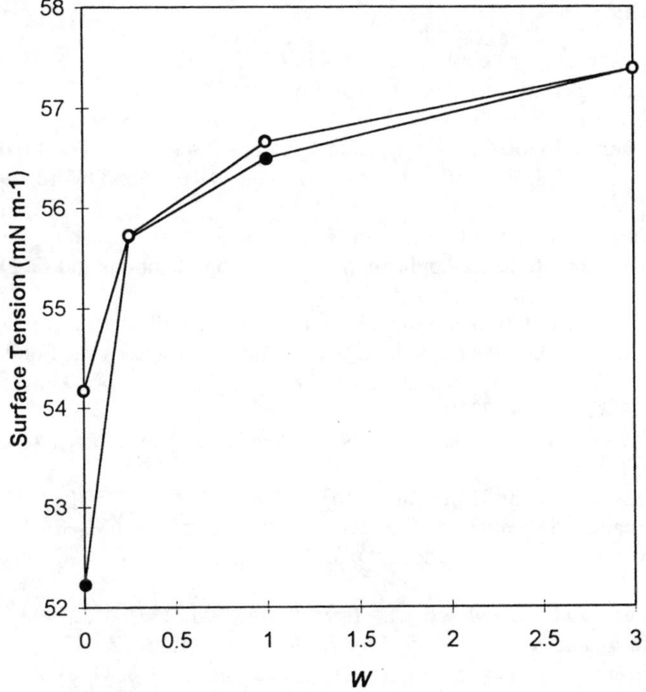

Figure 5 *Surface tension after 6 hours of solutions of BSA + DS as a function of polysaccharide:protein weight ratio (W): ●, untreated samples; ○, pressure processed samples (520 MPa for 20 min). Total BSA content 0.01 g l^{-1}, pH 7, 20 mM imidazole buffer, 25 °C*

4 Conclusions

We have shown that sulfated polysaccharides (DS, ι-CAR and κ-CAR) can form electrostatic complexes with BSA at pH ⩽7.0 at low ionic strength. The strength of the interaction increases considerably on lowering the pH from 7 to 6.5. The strength of BSA–polysaccharide complexation seems to be correlated with the density of sulfate group on the polysaccharide chain (DS > ι-CAR > κ-CAR).

Pressure treatment affects the surface activity of both β-Lg and BSA. The increase in surface activity for β-Lg can be attributed to an increase in protein surface hydrophobicity. Although DS appears to form weak electrostatic complex(es) with β-Lg, the surface activity after several hours is not affected to any significant extent. In the case of BSA, it appears that complexation with DS has the effect of protecting the protein against loss of functionality due to —S—S— linkage formation during or after high pressure processing.

Acknowledgement

The authors are grateful to the BBSRC for funding this research.

References

1. D. A. Ledward, in 'Protein Functionality in Food Systems', ed. N. S. Hettiarachchy and G. R. Ziegler, Marcel Dekker, New York, 1996, p. 225.
2. E. Dickinson and V. B. Galazka, *Food Hydrocolloids*, 1991, **5**, 281.
3. E. Dickinson and V. B. Galazka, in 'Gums and Stabilisers for the Food Industry', ed. G. O. Phillips, D. J. Wedlock, and P. A. Williams, IRL Press, Oxford, 1992, vol. 6, p. 351.
4. L. G. Enriquez and G. J. Flick, in 'Food Emulsifiers: Chemistry, Technology, Functional Properties and Applications', ed. G. Charalambous and G. Doxastakis, Elsevier, Amsterdam, 1989, p. 235.
5. T. H. M. Snoeren, P. Both, and D. G. Schmidt, *Neth. Milk Dairy J.*, 1976, **30**, 132.
6. M. G. Lynch and D. M. Mulvihill, in 'Gums and Stabilisers for the Food Industry', ed. G. O. Phillips, P. A. Williams, and D. J. Wedlock, Oxford University Press, Oxford, 1994, vol. 7, p. 323.
7. D. D. Drohan, A. Tziboula, D. McNulty, and D. S. Horne, *Food Hydrocolloids*, 1997, **11**, 101.
8. B. K. Chakraborty and H. E. Randolph, *J. Food Sci.*, 1972, **37**, 719.
9. P. B. Fernandes, in 'Gums and Stabilisers for the Food Industry', ed. G. O. Phillips, P. A. Williams, and D. J. Wedlock, Oxford University Press, Oxford, 1996, vol. 8, p. 171.
10. E. Dickinson and K. Pawlowsky, *J. Agric. Food Chem.*, 1997, **45**, 3799.
11. E. Dickinson and K. Pawlowsky, *Food Hydrocolloids*, 1998, **12**, 417.
12. E. Dickinson and K. Pawlowsky, *J. Agric. Food Chem.*, 1996, **44**, 2992.
13. V. B. Galazka, D. Smith, D. A. Ledward, and E. Dickinson, *Food Chem.*, 1998, in press.
14. V. B. Galazka, I. G. Sumner, and D. A. Ledward, *Food Chem.*, 1996, **57**, 393.
15. V. B. Galazka, D. A. Ledward, I. G. Sumner, and E. Dickinson, *J. Agric. Food Chem.*, 1997, **45**, 3465.
16. E. M. Dumay, M. T. Kalichevsky, and J.-C. Cheftel, *J. Agric. Food Chem.*, 1994, **42**, 1861.
17. V. B. Galazka, J. Varley, D. Smith, D. A. Ledward, and E. Dickinson, in 'High Pressure Food Science, Bioscience and Chemistry', ed. N. S. Isaacs, Special Publication No. 222, Royal Society of Chemistry, Cambridge, 1998, p. 175.
18. V. B. Galazka, D. A. Ledward, and J. Varley, in 'Food Colloids: Proteins, Lipids and Polysaccharides', ed. E. Dickinson and B. Bergenståhl, Royal Society of Chemistry, Cambridge, 1997, p. 127.
19. I. Hayakawa, J. Kajihara, K. Morikawa, M. Oda, and Y. Fujio, *J. Food Sci.*, 1992, **57**, 288.
20. I. Hayakawa, Y.-Y. Linko, and P. Linko, *Lebensm.-Wiss. Technol.*, 1996, **29**, 756.

On the Stability of Oil-in-Water Emulsions Formed using Highly Hydrolyzed Whey Protein

By C. Ramkumar, S. O. Agboola, H. Singh, P. A. Munro, and A. M. Singh[1]

INSTITUTE OF FOOD, NUTRITION AND HUMAN HEALTH, MASSEY UNIVERSITY, PALMERSTON NORTH, NEW ZEALAND
[1]NEW ZEALAND DAIRY RESEARCH INSTITUTE, PALMERSTON NORTH, NEW ZEALAND

1 Introduction

Proteins provide stability when used in food emulsions by adsorbing at the oil–water interface and forming a strong interfacial layer around the emulsion droplets. It is known[1–3] that the emulsifying properties of proteins improve upon limited hydrolysis through more efficient adsorption, but extensive hydrolysis, which produces many short peptides, results in a drastic loss of emulsifying properties. It has been reported[4,5] that there is an optimum chain length for peptides to provide good emulsifying properties.

Extensively hydrolyzed proteins are used in the formulation of many nutritional and health applications such as hypoallergenic infant formulas and enteral formulations, which are essentially oil-in-water emulsions.[6,7] Because of the poor emulsifying properties of extensively hydrolyzed proteins, these formulations usually require the addition of emulsifiers and stabilizers to facilitate the formation of a stable emulsion. In addition, these formulations are often fortified with various minerals, such as sodium, potassium, calcium, magnesium, phosphorus and chloride, some of which are known to influence emulsion stability. The effect of heat on emulsion stability is particularly important, as these formulations are usually sterilized by retorting; therefore the ingredient mixture must be able to withstand high temperatures. Because of the range and complexity of ingredients used, poor emulsification properties and poor heat stability are some of the main difficulties encountered in the manufacture of these products. There have been very few studies of the stability of emulsion systems in which these kind of hydrolyzates have been used,[5,8] and the effects of external factors, (*e.g.*, calcium ions, emulsifier addition, high temperature) on emulsion stability have not been properly established.

The objectives of this study were to extend the findings of Agboola *et al.*,[8] and to examine the formation and properties of emulsions formed with a highly hydrolyzed commercial whey protein product under a range of conditions, including the addition of calcium ions and different types of lecithin combined with heat treatment under retort conditions.

2 Materials and Methods

A whey protein product hydrolyzed to 27% degree of hydrolysis (WPH 931) was supplied by the New Zealand Dairy Board, Wellington, New Zealand. The dried product contained 90.5% protein, 4.5% moisture, 2.8% ash, 0.1% fat and 0.2% lactose. Soya oil was purchased from Davis Trading Co., Palmerston North, New Zealand. All other reagents were of analytical grade and were supplied by BDH Chemicals Ltd., Poole, England.

A typical oil-in-water emulsion (4 wt% soya oil) was formed from a solution of WPH in deionized water at room temperature (20 $\pm$ 2 °C). In some cases, a known quantity of calcium chloride solution was added to the WPH solution prior to mixing with soya oil and emulsification. In each case, the pH of the solution was adjusted to 6.8. Soya oil, heated to 60 °C, containing dispersed unmodified or hydroxylated lecithin (2.5 g l^{-1} of emulsion), in some experiments, was added to the WPH solution. This mixture was then passed through a two-stage homogenizer at no input pressure and homogenized with first and second stage pressures of 20.6 and 3.4 MPa. This process was repeated for a more efficient degree of homogenization.

The droplet-size distribution and the volume–surface average particle diameter (d_{32}) were determined by light scattering using a Malvern Mastersizer E. The presentation factor was 2NAD (*i.e.* refractive index and absorption of emulsion particles of 1.456 and 0 respectively) and a polydisperse model was chosen for size distribution. Emulsion droplets were sized using distilled water as the dispersant.

Nile blue (a fluorescent dye for staining the fat phase) at a level of 0.1 wt% was added to the samples, which were then mounted on a Leica TCS 4D confocal laser scanning microscope. The laser source was Ar/Kr, and the excitation wavelength was 488 nm. Samples were viewed under oil immersion using $\times$ 100 objective (numerical aperture = 1.4).

3 Results and Discussion

In previous work[8] the effects of WPH concentration on the particle-size distributions of emulsions prepared at different homogenization pressure were determined. At a given WPH concentration, the average droplet diameter (d_{32}) was found to decrease with increasing homogenization pressure (Figure 1). In emulsions formed using homogenization pressures of 10.3 and 20.6 MPa, the value of d_{32} did not vary much with increase in WPH concentration from 0.5 to 5%. At 34.3 MPa, the d_{32} of emulsions formed with 0.5% and 1% WPH were

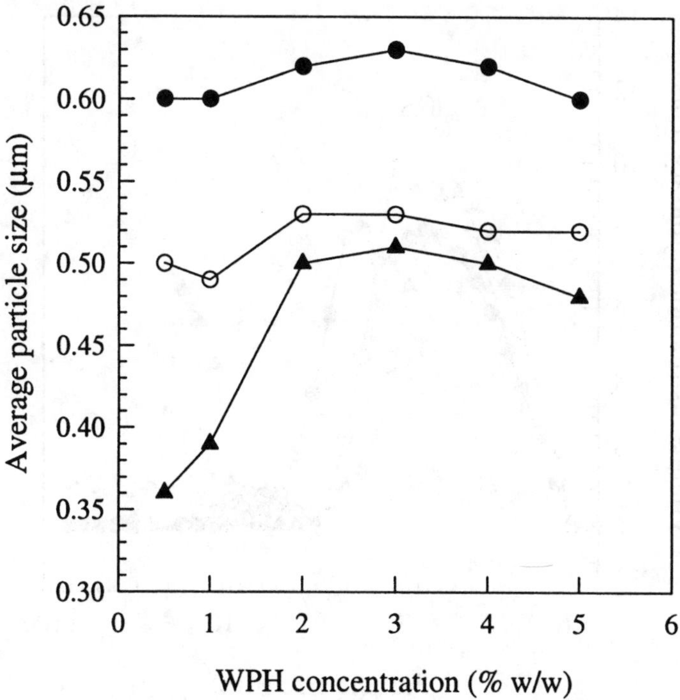

Figure 1 *Average particle sizes d_{32} as a function of WPH concentration in emulsions (4% soya oil) formed by homogenization with first-stage pressures of 10.3 (●), 20.6 (○) or 34.3 MPa (▲)*

lower than those formed at higher concentrations. Particle-size distributions of emulsions prepared with 2% WPH (Figure 2) show that increasing the homogenization pressure from 10.3 to 20.6 MPa leads to the formation of a higher proportion of small particles. However, there was found to be an increase in the proportion of particles in the range 6–20 μm. When the pressure was increased to 34.3 MPa, there was a further increase in the proportions of droplets above 10 μm in size. The proportion of these large droplets was apparently not enough to increase the d_{32} beyond those formed at lower pressures (Figure 1). Examination of these emulsions by confocal microscopy showed some very large oil droplets and free non-emulsified oil which apparently were not detected by the light scattering.

The size of emulsion droplets in general depends to a very large extent on the power density of the emulsifying equipment, and the concentration and surfactant properties of the constituents of the emulsion.[9–11] In this work the droplet size was found to depend upon WPH concentration and homogenization pressure; large droplets and 'free oil' were present in emulsions formed at low WPH concentrations and high homogenization pressures. This situation probably arises from the relatively small number of large peptides present in the WPH solution.[8] At low WPH concentrations and high homogenization pres-

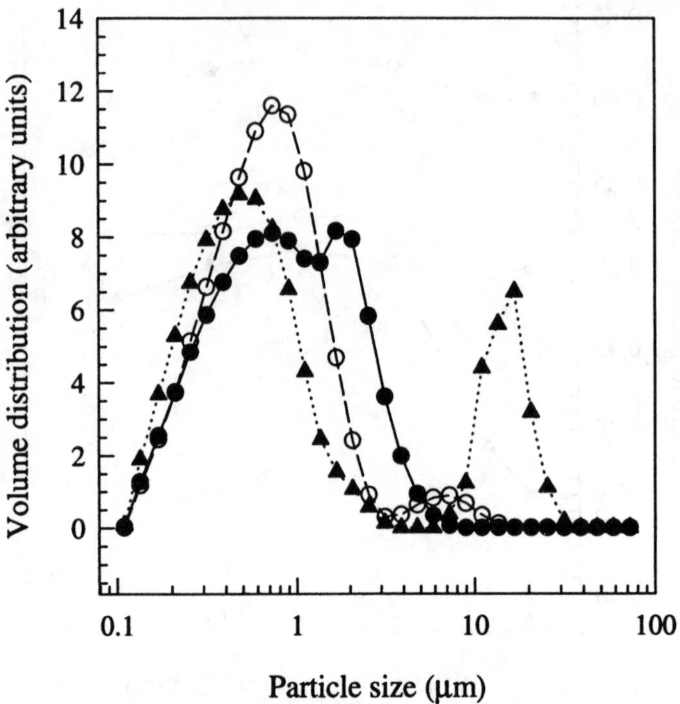

Figure 2 *Droplet-size distributions for emulsions (4% soya oil, 2% WPH) formed using first-stage homogenization pressures of 10.3 (●), 20.6 (○) or 34.3 MPa (▲)*

sures, instability of the emulsions is due to the inability of the predominantly short peptides in the WPH product to adsorb and adequately stabilize the increased surface area. Because of the lack of secondary and tertiary structures, the short peptides will have poor ability to provide steric stabilization and to form strong interfacial films, even if they are adsorbed. The end result is an immediate recoalescence of oil droplets, leading to the formation of some very large droplets and some 'free' oil. Increasing the concentration of the WPH provides sufficient quantities of large 'suitable' peptides to cover the interface completely, and so causes an improvement in the integrity of the interfacial layer, which is able to resist coalescence. A lower homogenization pressure does not create so much surface area for the WPH peptides to have to stabilize, and so it can produce more stable emulsions.

Addition of Lecithin and Calcium Chloride

The effect of addition of different lecithins at a concentration of 0.25% prior to homogenization at 20.6 MPa was examined in emulsions formed with 4% WPH. Addition of unmodified lecithin was found to result in the formation of

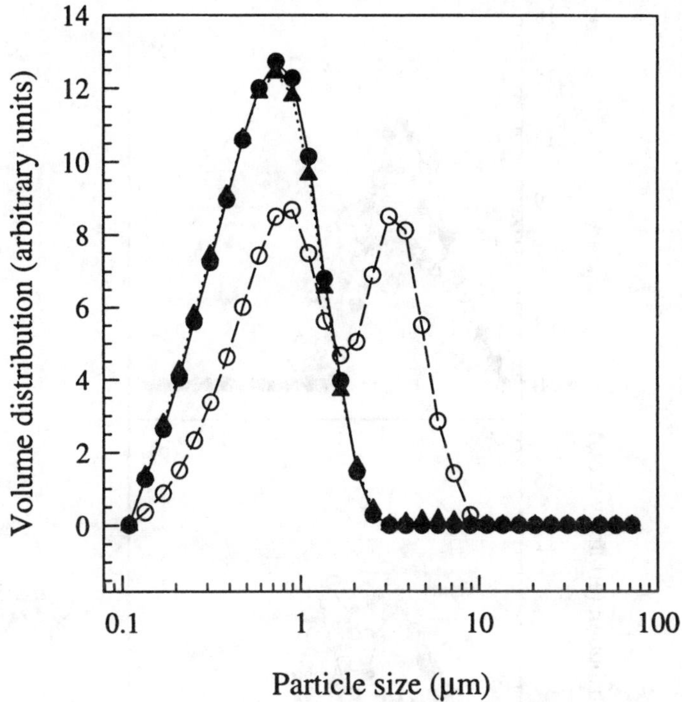

Figure 3 *Droplet-size distributions for emulsions (4% soya oil, 4% WPH) containing no lecithin (●), 0.25% unmodified lecithin (○) or 0.25% hydroxylated lecithin (▲)*

a significant population of large droplets (1–10 μm), indicating destabilization of the emulsion (see Figure 3). In contrast, when hydroxylated lecithin was added, the size distributions of emulsion droplets were found to remain unchanged. The destabilization caused by unmodified lecithin may be attributed to the adsorption of lecithin in preference to WPH peptides at the oil–water interface, because of competition between the two species during homogenization. Previous studies have shown that lecithin, in comparison with proteins, can promote greater coalescence via a reduction in both electrostatic and steric repulsion potentials.[12–14] It is also possible that unmodified lecithin may complex in the aqueous phase with the larger, more surface-active peptides, and consequently reduce their concentration at the interface. It is not immediately clear why hydroxylated lecithin behaves differently. It is probable that a reduction in the hydrophobic properties of lecithin caused by the substitution of the hydrophilic hydroxyl groups influences its ability to adsorb during emulsification and/or to interact with the larger peptides in the aqueous phase.

Figure 4 shows the effect of addition of calcium chloride (to give calcium ion concentrations in the range 2.5–37.5 mM) on the droplet-size distribution for

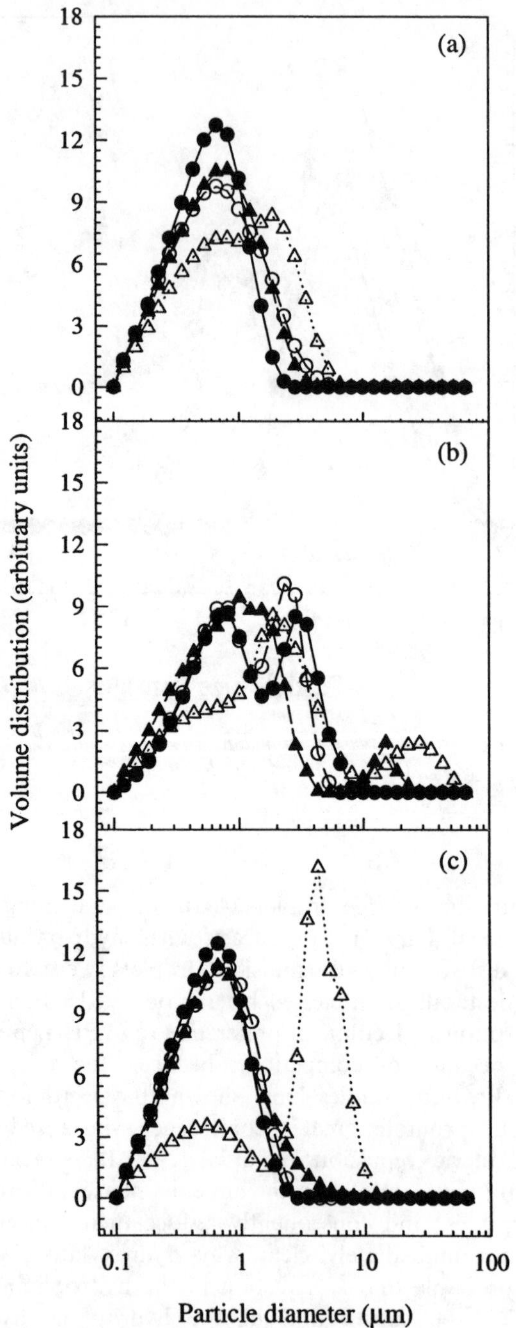

Figure 4 *Droplet-size distributions for emulsions (4% soya oil, 4% WPH) containing (a) no lecithin, (b) 0.25% unmodified lecithin, and (c) 0.25% hydroxylated lecithin. The emulsions contained varying levels of calcium chloride: 0 (●), 2.5 (○), 12.5 (▲) or 37.5 mM (△)*

emulsions made with or without lecithin. All the lecithin-free emulsions containing up to 12.5 mM ionic calcium showed monomodal distributions of droplet sizes, but emulsions containing higher concentrations of ionic calcium showed a broad population of particles, with a large proportion of particles between 1 and 10 μm (Figure 4a). All emulsions containing unmodified lecithin showed bimodal distributions and at ionic calcium concentrations above 12.5 mM showed a peak containing particles $>10\,\mu$m (Figure 4b). The size distribution of emulsions prepared with hydroxylated lecithin showed only a slight shift towards higher particle sizes with calcium ion additions up to 12.5 mM, but at 37.5 mM there was a very large increase in the proportion of particles between 1 and 10 μm (Figure 4c).

The destabilization caused by addition of ionic calcium to the control emulsions (no added lecithin) may be due to the binding of calcium ions to negatively charged peptides in the WPH solution, which may cause aggregation of the larger, surface-active peptides. This situation would tend to reduce the effective concentration of these 'suitable' peptides, resulting in insufficient quantities being available to cover the interface completely and provide stability. In addition, calcium ions may bind to the peptides present at the droplet surface which would reduce the interdroplet electrostatic repulsion and so enhance the likelihood of flocculation.

Confocal microscopic examination of emulsions prepared without added calcium chloride (Figure 5a) was found to show uniformly sized droplets in the range 1–3 μm, and the sizes of the droplets tended to be larger in the emulsions containing added calcium chloride (in the range 1 and 10 μm). Some flocculated particles were also visible in emulsions containing 30.0 and 37.5 mM calcium (Figure 5b). This indicates that instability of the emulsion in the presence of calcium arises from both coalescence and flocculation. Flocculation normally occurs before droplets can coalesce. After droplet flocculation, coalescence depends on the probability of the interfacial film rupturing in finite time which depends on the thickness of the adsorbed layer.[1,9,11]

Confocal micrographs of emulsions prepared with hydroxylated lecithin, containing 37.5 mM ionic calcium, show even larger droplets and more pronounced flocculation (Figure 5c) as compared to the control emulsions, essentially confirming the particle size distribution results. The greater extent of destabilization of lecithin-containing emulsions in the presence of calcium ions may be attributed to possible binding of calcium ions to negatively-charged phosphate groups of certain phospholipid components, such as phosphatidyl inositol and phosphatidic acid, which may consequently influence its ability to adsorb and compete with WPH peptides at the interface.

Heat Treatment

Heat treatment of the control emulsion (no lecithin, no calcium chloride), in a retort at 121 °C for 16 min, resulted in a large increase in d_{32}, with the size distribution of the emulsion becoming bimodal (Figures 4a, 6a), and the development of a population of droplets in the size range from 6 to 30 μm.

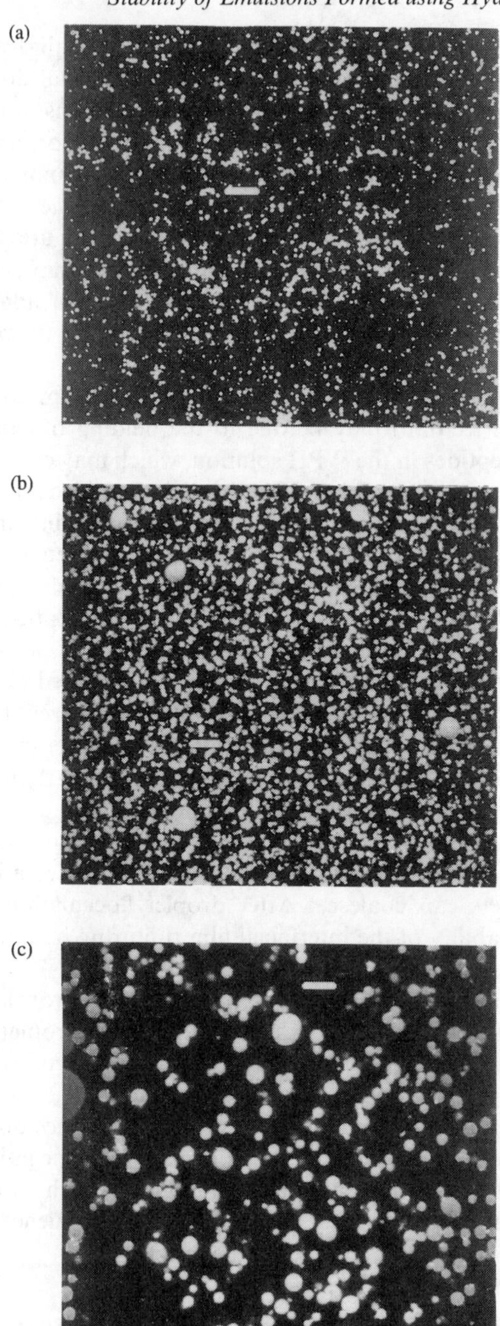

Figure 5 *Typical confocal laser micrographs of emulsions (4% soya oil, 4% WPH) containing (a) no added calcium ions or lethicin, (b) 37.5 mM ionic calcium, and (c) 37.5 mM ionic calcium and 0.25% hydroxylated lecithin. Scale bar = 10 μm*

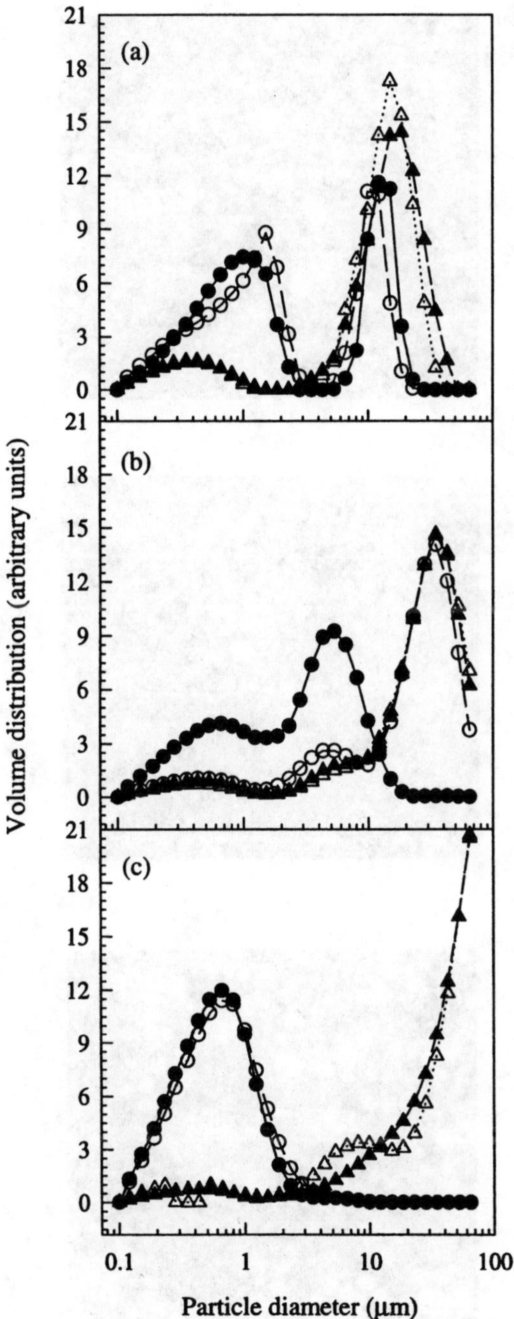

Figure 6 *Droplet-size distributions for emulsions (4% soya oil, 4% WPH), heated at 121 °C for 16 min, containing (a) no lecithin, (b) 0.25% unmodified lecithin, and (c) 0.25% hydroxylated lecithin. The emulsions contained varying levels of calcium: 0 (●), 2.5 (○), 12.5 (▲) or 37.5 mM (△)*

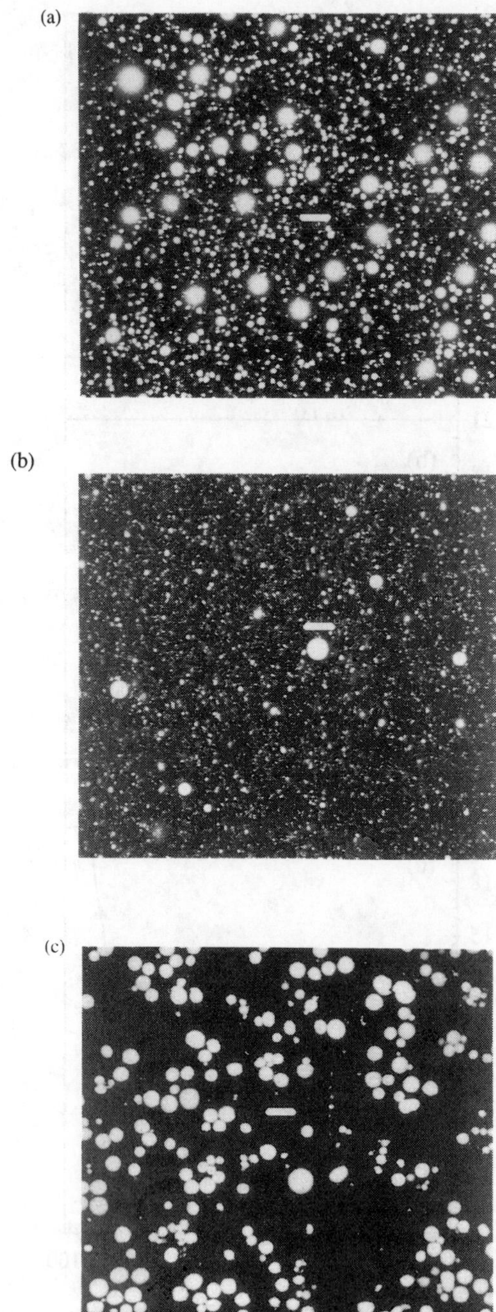

Figure 7 *Typical confocal laser micrographs of emulsions (4% soya oil, 4% WPH), heated at 121 °C for 16 min, containing (a) no added calcium ions or lecithin, (b) 0.25% hydroxylated lecithin, (c) 37.5 mM ionic calcium, and (d) 37.5 mM ionic calcium and 0.25% hydroxylated lecithin. Scale bar = 10 μm*

(d)

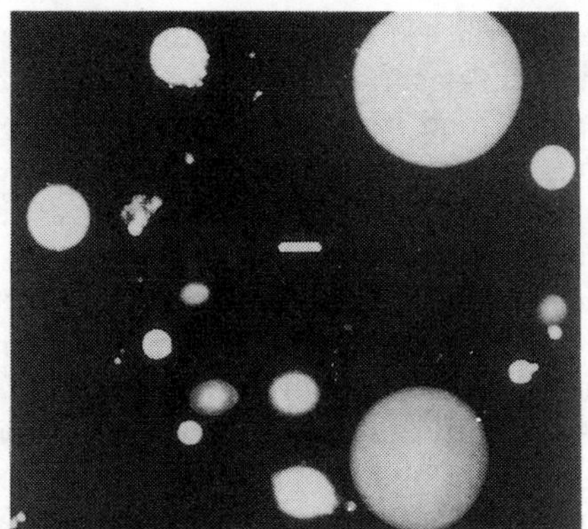

Figure 7 *(continued)*

When emulsions containing unmodified lecithin were heated, the d_{32} value also increased, and the size distribution shifted towards higher particle size ranges, but the effect was smaller than that observed in the control (no lecithin). In contrast, the size distribution of emulsions containing hydroxylated lecithin remained unchanged after heating (Figures 4c, 6c). Confocal laser microscopy on a heated control emulsion shows the presence of a distinct population of larger droplets with some evidence of flocculation (Figure 7a). However, it was observed that emulsions containing hydroxylated lecithin remain unaffected by heat treatment (Figure 7b); unheated emulsions with added hydroxylated lecithin were similar to those shown in Figure 4a; (*i.e.* to the unheated emulsion), confirming the particle size distribution results which suggest that hydroxylated lecithin provides better stability towards heating.

When lecithin-free emulsions containing $>2.5\,\text{mM}$ ionic calcium were heated, the size of the second peak, representing droplets in the size range 10–50 μm, was found to increase, with a parallel decrease in the size of the first peak (Figure 6a). Heat treatment of emulsions containing both calcium chloride ($>12.5\,\text{mM}$) and lecithin, especially those containing hydroxylated lecithin, led to much greater destabilization of the emulsions, as evidenced by the presence of a large population of droplets with diameter $>10\,\mu\text{m}$ (Figures 6b, c). Some of these particles were outside the range of the 45 mm (focal length) lens used in the Mastersizer. Confocal micrographs of these emulsions showed extensive flocculation of droplets and some very large particles (Figures 7c, d).

These results suggest that heat treatment promotes aggregation/flocculation of droplets which then leads to coalescence of droplets. This mainly arises from the inability of the adsorbed peptide layers to provide effective steric or charge stabilization. It is also possible that desorption of some loosely adsorbed peptides may also occur during heating, which is also likely to enhance droplet aggregation and coalescence. The reasons for continued stability after retorting of the emulsions containing hydroxylated lecithin (but not unmodified lecithin) are not very clear. Adsorption of hydroxylated lecithin at the droplet surface may increase the overall charge and degree of hydration at the oil droplet surface, thus preventing close approach of the droplets during heating and reducing the incidence of aggregation and coalescence. Binding of ionic calcium by hydroxylated lecithin, however, would reduce the interdroplet repulsions and promote heat-induced aggregation and coalescence of droplets.

In conclusion, the stability of emulsions formed by extensively hydrolyzed WPH is influenced by a number of processing and compositional factors. Emulsions formed using relatively high concentrations of WPH at low homogenization pressures are fairly stable, but they are destabilized by retorting at 121 °C for 16 min. This instability may be reduced, to a large extent, by incorporating hydroxylated lecithin into the emulsion. However, in the presence of calcium ions, hydroxylated lecithin is incapable of preventing emulsion destabilization during heat treatment.

Acknowledgements

This work was funded by the New Zealand Dairy Board. We are grateful to Professor Douglas Dalgleish for useful discussions and to Neill Cropper & Co. Ltd., Auckland, New Zealand, for supplying the Central Soya lecithin samples. We thank the New Zealand Dairy Research Institute for providing facilities for particle-size measurements.

References

1. J. E. Kinsella, *CRC Crit. Rev. Food Sci. Nutr.*, 1984, **21**, 197.
2. J.-M. Chobert, C. Bertrand-Harb, and M.-G. Nicolas, *J. Agric. Food Chem.*, 1988, **36**, 883.
3. S. O. Agboola and D. G. Dalgleish, *J. Agric. Food Chem.*, 1996, **44**, 3631.
4. S. W. Lee, M. Shimizu, S. Kaminogawa, and K. Yamaguchi, *Agric. Biol. Chem.*, 1987, **51**, 161.
5. A. M. Singh and D. G. Dalgleish, *J. Dairy Sci.*, 1998, **81**, 918.
6. M. K. Schmidl, S. L. Taylor, and J. A. Nordlee, *Food Technol.*, 1994, **48**(10), 77.
7. I. M. Mahmoud, *Food Technol.*, 1994, **48**(10), 89.
8. S. O. Agboola, H. Singh, P. A. Munro, D. G. Dalgleish, and A. M. Singh, *J. Agric. Food Chem.*, 1998, **46**, 84.
9. D. F. Darling, in 'Food Structure and Behaviour', ed. J. Blanshard and P. J. Lillford, Academic Press, London, 1987, p. 107.
10. P. Walstra, in 'Food Structure and Behaviour', ed. J. Blanshard and P. J. Lillford, Academic Press, London, 1987, p. 87.
11. P. Walstra, *Chem. Eng. Sci.*, 1993, **48**, 333.
12. L. Rhydag and I. Wilton, *J. Amer. Oil Chem. Soc.*, 1981, **58**, 830.
13. P. van der Meeren, J. van der Deelen, and L. Baert, *Prog. Colloid Polym. Sci.*, 1995, **98**, 136.
14. H. Cruijsen, Ph.D. Thesis, Wageningen Agricultural University, 1996.

Effect of Cholesterol Reduction from Hen's Egg Yolk Low-Density Lipoprotein on its Emulsifying Properties

By Yoshinori Mine and Marie Bergougnoux

DEPARTMENT OF FOOD SCIENCE, UNIVERSITY OF GUELPH, GUELPH, ONTARIO, CANADA N1G 2W1

1 Introduction

Hen's egg yolk provides excellent emulsifying properties to a variety of food products such as mayonnaise, ice cream, bakery items and salad dressings. Egg yolk contains various emulsifying agents such as hydrophobic and hydrophilic proteins, phospholipids and cholesterol. Egg yolk consists of soluble plasma (78% of the total liquid yolk) which is composed of livetins and low-density lipoprotein (LDL).[1] The LDL contains 12.5% protein and 80% lipids. The lipid in LDL consists of 70% neutral lipid, 26% phospholipids (71–76%, phosphatidylcholine (PC), 16–20% phosphatidylethanolamine (PE), and 8–9% sphingomyelin and lysophospholipids), and 4% free cholesterol.[2] The LDL has been considered as the major factor governing the emulsifying properties of egg yolk. Protein–phospholipid complexes (lipoproteins) are the components of egg yolk responsible for stabilizing an emulsion.[3,4] In an emulsion prepared with egg yolk, the contribution of proteins to emulsifying activity is higher than that of phospholipids.[5] Egg yolk proteins exhibit a higher adsorbing capacity than globular proteins, because they have a more flexible structure and a greater surface hydrophobicity.[6] The emulsifying properties and heat stability of the protein-stabilized emulsion are improved substantially through the formation of a complex between lysolecithin and free fatty acids.[7,8] Emulsifying capacity and heat stability of egg yolk are also improved by fermentation with pancreatic phospholipase.[9] These results indicate that the emulsifying properties of egg yolk lipoproteins might be closely related to the structure of phospholipid–protein complexes and their interactions at an oil–water interface.

Concerns regarding the relationship between cholesterol (or oxidized cholesterol products) and coronary heart disease have resulted in attempts to develop various technologies for reducing the cholesterol content of egg yolk.[10] These techniques include solvent extraction, supercritical fluid extraction, enzymatic degradation, and complexation with β-cyclodextrin (CD).[11] The removal of cholesterol by adsorption to CD is an alternative approach. A few studies have been reported[5,12,13] on functional properties of low-cholesterol egg yolk, but no information is available on adsorption behaviour or on the effect of cholesterol reduction from egg yolk on the phospholipid–apoprotein interactions at an oil–water interface. The objective of this article is to present emulsifying properties of cholesterol-reduced LDL (CR-LDL) and phospholipid–protein interactions at the interface.

2 Experimental

Preparation of Cholesterol-Reduced Egg Yolk Low-Density Lipoprotein (CR-LDL)

The LDL was prepared from fresh egg according to the modified method of Raju and Mahadevan.[14] The LDL concentration was determined from the protein concentration measured by a modified Lowry procedure.[15] Egg yolk LDL contains about 4% of cholesterol. Extraction of cholesterol from LDL was carried out using CD. The LDL solution (6.0 wt%) was heated to 50 °C in a water bath and cyclodextrin (Wacker Chemicals, CT) was added at a CD:cholesterol molar ratio of 2 or 4. The sample was mixed for 45 min at 50 °C and cooled at 4 °C for 1 h. The slurry was centrifuged for 30 min at 8000 *g* at 10 °C. The supernatant containing the CR-LDL was decanted and used for the preparation of emulsions. The determination of egg yolk composition and the reduction ratio of cholesterol in CR-LDL were measured using flame ionization (TLC-FID) on an Iatroscan (Iatroscan MK-5, Iatron Laboratories, Tokyo, Japan).

Determination of Emulsifying Properties

The LDL and CR-LDL preparations were diluted with various buffers (50 mM acetate and imidazole buffers containing 0.1 M and 1.5 M NaCl, pH 3.5 and 7.0, respectively) to give a final LDL concentration 0.4–4.0 wt% in the aqueous phase. Emulsions were prepared by homogenizing 2.0 mL for each LDL or CR-LDL solution with 0.5 mL of pure triolein (>99%) for 1 min at a speed of 22,000 rpm using a Polytron PT 2000 homogenizer (Kinematica AG, Switzerland). The droplet-size distribution of emulsions was determined on a Mastersizer X (Malvern Instruments, Malvern, UK) with optical parameters defined by the manufacturer's presentation code 0303. The emulsions were centrifuged at 20 °C and 5000 *g* for 30 min and the cream washed with 5 mL of appropriate buffer with each washing following by centrifugation. The subnatants were pooled together and filtered through a 0.22 μm filter. The protein contents were determined according to the modified Lowry method.[15] The surface concentration was estimated as a difference between the protein concentration of the subnatant solution and the total protein used to make the emulsion. The washed cream was treated with a solution of 10 wt% SDS in 0.1 M Tris buffer, pH 8.0, containing 5 vol% 2-mercaptoethanol (ME). The protein composition of the supernatant was determined by SDS-polyacrylamide gel electrophoresis (SDS-PAGE) on 4–15% gradient gels using the Bio-Rad Mini Protean II electrophoresis cell at a constant voltage of 20 mA/gel.

Analysis of Phospholipids and Cholesterol Present at the Interface

The creams were extracted with approximately 5 volumes of chloroform:methanol (2:1) and the solvent layer was evaporated using a rotary evaporator, the residue being dissolved in 5 ml of 5 vol% ethyl acetate in hexane and loaded onto a prepacked Sep-Pack silica cartridge (Waters Co., Milford, MA). The column had been previously dehydrated in succession with 5 mL ethyl acetate, 10 ml 50 vol% ethyl acetate in acetone, 5 ml ethyl acetate, and 20 mL hexane. The unadsorbed lipid (triolein) was washed off the column with 20 mL of 5 vol% ethyl acetate in hexane and a portion of the eluate collected and analyzed as described below for phospholipids and cholesterol. The phospholipids and cholesterol were then eluted from the column with 20 ml of methanol:water (98:2) and the eluate collected and transferred into evaporation flasks. The contents of the flasks were evaporated to dryness on a rotary evaporator and the residue dissolved in 0.1 mL of chloroform:methanol (2:1) solvent. The phospholipid contents were then analysed by TLC-FID using the Iatroscan system. The samples were developed for FID as follows: one μL of sample was applied onto the Chromarods (Chromarod-S3), dried and developed first in chloroform-methanol-water (70:30:3) for 10 min. The Chromarods were dried and developed in a second solvent system containing petroleum ether-diethyl ether-acetic acid (80:30:0.2) for 30 min. After drying, the Chromarods were loaded onto the Iatroscan system and the area under each peak determined and used to calculate the concentration of PC, PE and cholesterol. Egg PC (QP Corporation, Tokyo, Japan), cholesterol (Sigma Chemicals) and bovine liver PE (Avanti Polar Lipids, CA) were used as standards.

3 Results

Cholesterol Reduction from Egg Yolk LDL

Table 1 shows the effect of CD:cholesterol molar ratio on the extraction of cholesterol from egg yolk LDL. The reduction of cholesterol in LDL was 48.5 and 92.7% for values of 2 and 4 of the CD:cholesterol molar ratio, respectively. The removal of cholesterol from liquid egg yolk by adsorption to CD has been described by several researchers.[16] The most important factors influencing cholesterol reduction were dilution of egg yolk to a defined water:solid ratio (2.9), and CD concentration at a CD:cholesterol molar ratio of 4.0 and pH (10.5).[16] We have done all experiments at pH 6.5, because we could not find any effects of pH on cholesterol reduction efficiency with CD. The CD remaining in the sample after centrifugation was determined using a freshly prepared

Table 1 *Effect of β-cyclodextrin (CD):cholesterol molar ratio on the reduction of cholesterol from egg yolk low-density lipoprotein (LDL)*

Samples	*Ratio of CD:cholesterol (molar ratio)*	*Cholesterol reduction (%)*	*Residual CD (%)*
LDL (Control)	0	0	0
CR-LDL (1)	2	48.5	0.025
CR-LDL (2)	4	92.7	0.074

phenolphthalein solution.[16] The value for each sample was found to be negligibly small.

Emulsifying Properties of Cholesterol-reduced LDL

The changes in particle size of the emulsions as a function of LDL concentration and pH are shown in Figure 1. The mean particle size of emulsions decreases with increasing LDL concentration used to make the emulsion. Generally, there was a decrease in particle size with increasing amount of emulsifier. The latter observation is supported by the fact that an increase in the concentration of surface-active agents generally leads to a reduction in interfacial surface tension of the droplets, facilitating their break-up into smaller droplets.[17] At the low concentrations of surfactant, the system produces larger emulsion droplets because insufficient surfactant is present to cover all of the freshly created oil surface. At high surfactant concentration, the system produces the maximum interfacial area under the proceeding thermodynamic conditions. At concentrations above 2.4% of LDL, the average droplet sizes of emulsions reached a plateau at 0.8–0.9 μm, and there was no noticeable difference in particle size

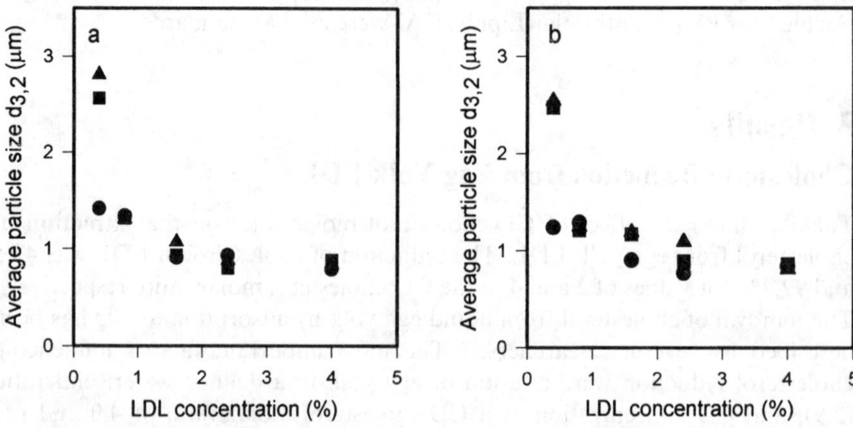

Figure 1 *Mean droplet size of emulsions (20 vol% oil) stabilized by egg yolk LDL (●), 48.5% CR-LDL (■) and 92.7% CR-LDL (▲) as a function of LDL concentration at (a) pH 7.0 and (b) pH 3.5*

between pH 7.0 and 3.5. The CR-LDL formed larger emulsion droplets than LDL at 0.4% at pH 7.0 and 3.5. The results would indicate that LDL is a better emulsifier at low concentrations than CR-LDL at neutral and acid pH values. Figure 2 shows the amount of protein present on the surface of the emulsion as a function of LDL concentration. At pH 7.0, the surface protein concentration of LDL emulsion increases from 0.24 to 1.25 mg m^{-2} with increasing LDL concentration. The concentration of protein at the interface was greater for emulsions stabilized with CR-LDL at pH 7.0. However, there was no noticeable difference between 48.5% CR-LDL and 92.7% CR-LDL. On the other hand, the surface protein concentration was markedly different at pH 3.5. The LDL surface protein ranged from 0.32 to 1.08 mg m^{-2}. However, the 48.5% CR-LDL and 92.7% CR-LDL formed much thicker films, ranging from 0.67–1.56 and 0.74–3.01 mg m^{-2}, respectively. These data suggest that the adsorption behavior of CR-LDL at an oil–water interface is different from LDL at pH 3.5. The higher surface protein concentration of 92.7% CR-LDL at pH 3.5 when compared to the control LDL can be explained on the basis of increased coagulation of lipoproteins at low pH values. More recently, we reported that egg yolk LDL micelles breakdown when the micelles come into contact with the interface, and rearrangement of lipoproteins, cholesterol and phospholipids takes place following adsorption at the oil–water interface.[18] Recent ^{31}P NMR and enzymatic hydrolysis studies have shown[19] that the membrane fluidity of egg yolk LDL is high and that the interactions of protein with phospholipids may not be so strong[19] as proposed by Burley.[20] There is no information regarding the role of cholesterol on the structural change of egg yolk LDL. In general, it is believed that cholesterol is an important component in stabilizing biological cell membranes. Removing the cholesterol from egg yolk LDL may cause a structural change of the phospholipid–protein interaction in the LDL micelle.

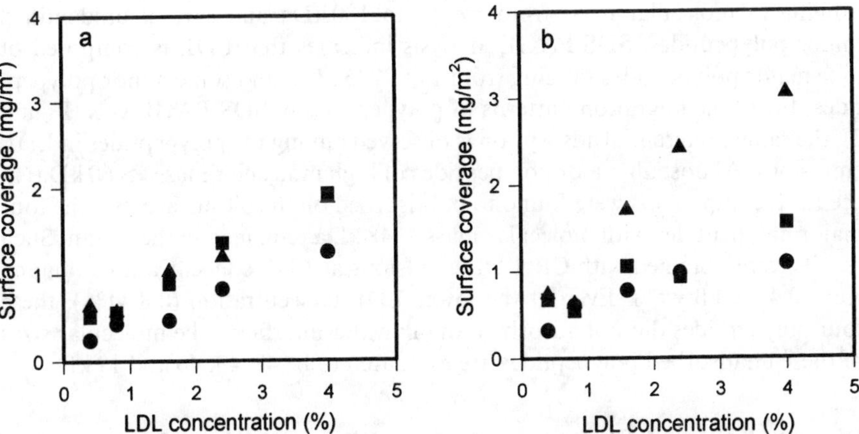

Figure 2 *Surface protein coverage of emulsions (20 vol% oil) stabilized by egg yolk LDL (●), 48.5% CR-LDL (■) and 92.7% CR-LDL (▲) as a function of LDL concentration at (a) pH 7.0 and (b) pH 3.5*

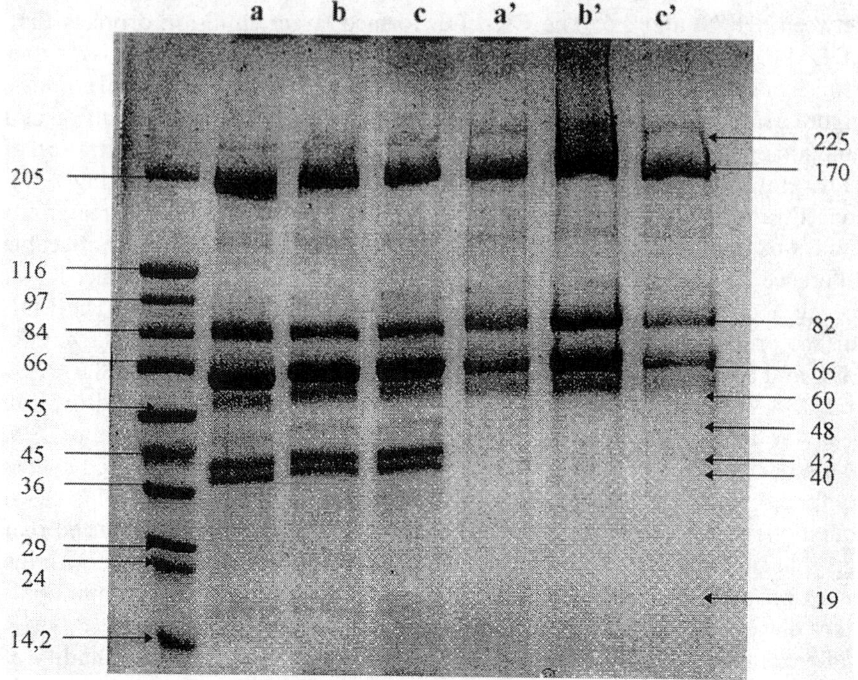

Figure 3 *SDS-PAGE profiles of egg yolk lipoproteins. Lane a,a': LDL; Lane, b,b': 48.5% CR-LDL; Lane c,c': 92.7% CR-LDL; Lane a–c: control; Lane a'–c': emulsions stabilized by 2.4 wt% LDL or CR-LDL*

Protein Composition at the Oil–Water Interface

It has been reported[14,21] that LDL consists of about six major polypeptides ranging in molecular mass from *ca.* 10 to 180 kDa and several unidentified minor polypeptides. SDS-PAGE analysis indicates that LDL is composed of nine major polypeptides ranging from 19 to 225 kDa, and some minor polypeptides. From the migration patterns of polypeptides in SDS-PAGE gels (Figure 3), the same preferential adsorption is observed among the polypeptides in LDL emulsions. Almost all major polypeptides of high molecular mass (>60 kDa) in the LDL components were found to be adsorbed on the oil surface, but the four major polypeptides with molecular mass <48 kDa remained in the serum. Such results were obtained with CR-LDL emulsions at LDL concentrations ranging from 0.4 to 4.0 wt%. Even at the lower LDL concentration (0.8 wt%), these four polypeptides did not adsorb at an oil–water interface. The molecular sizes of these unadsorbed polypeptides are estimated to be 48, 43, 40 and 19 kDa.

Phospholipid and Cholesterol Composition at the Interface

The compositions of phospholipids and cholesterol at the interface in oil–water emulsions are shown in Figure 4. The emulsifying properties of egg yolk LDL

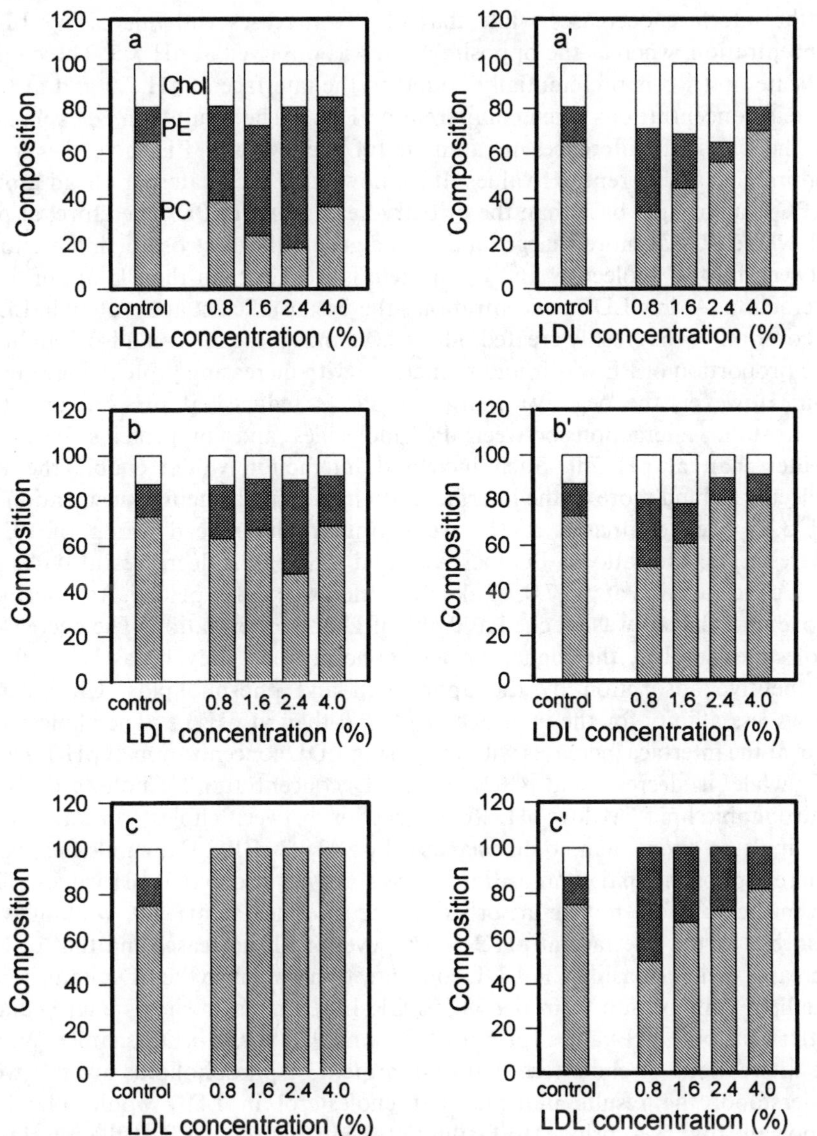

Figure 4 *Composition of cholesterol and phospholipids at oil–water interface stabilized by egg yolk LDL (a,a'), 48.5% CR-LDL (b,b') and 92.7% CR-LDL (c,c') as a function of LDL concentration at pH 7.0 (a–c) and pH 3.5 (a'–c')*

are attributed in part to the phospholipid–protein complex which can interact with the oil phase through hydrophobic groups and also with the aqueous phase through the charged phospholipid groups. The proportions of phospholipids and cholesterol in the LDL are PC, 66.6%, PE, 19.0% and cholesterol, 16.4%. These results are similar to previously reported values.[2] At pH 7.0, the PC level at the interface decreases while that of PE increases with increasing LDL concentration, whereas the opposite trend was observed at pH 3.5. These data indicate that PC is preferentially bound to the interface at pH 7.0 and at low protein concentrations when compared to pH 3.5. The results can be explained on the basis of differences in affinities of the PC and PE molecules with apoproteins at different pH values. It is known that the quaternary head group of PC is a stronger base than the primary head group of PE. Therefore, at pH 7.0, where PC has more charge on the head group, the electrostatic interactions between the PC molecules and apoprotein is weaker than the PE–apoprotein interaction. At low LDL concentrations, there is insufficient apoprotein in LDL to cover all of the freshly created oil surface, and so the PC can adsorb tightly. The proportion of PE was found to increase with increasing protein concentration. However, the negative charge on PC is reduced at pH 3.5, and the electrostatic interaction between PC molecules and apoproteins becomes greater than at pH 7.0. Such increased interaction would enable the PC molecule to bind more to the interface with increasing protein concentration at pH 3.5, when compared to the weak primary base head group of PE.[22] Interestingly, the ratio of PC increases and that of PE decreases at 4.0% of LDL concentration at pH 7.0, unlike the trend observed at pH 3.5. As described above, the plateau was reached at 2.4% of LDL concentration. The decreased cholesterol level at the higher protein concentration may be as a result of competitive adsorption by the apoprotein and phospholipids. Cholesterol shows less affinity for the interface at pH 7.0 than at pH 3.5. The cholesterol level at the interface increases with increasing LDL concentration at pH 7.0 and 3.5, while it decreases at 4.0% of LDL concentration. Cholesterol is a hydrophobic lipid. At low pH, the interaction between cholesterol and phospholipids or apoprotein could increase. For 92.7% CR-LDL emulsions, very different phospholipid compositions were observed at different pH values. The PE molecule could not be adsorbed to the interface at pH 7.0, while it was adsorbed at the interface at pH 3.5. The level of PE decreases and that of PC increases with increasing CR-LDL concentration similar to the LDL emulsions. The lipid composition from the 48.5% CR-LDL emulsion shows intermediate patterns at both pH values for both LDL and 92.7% CR-LDL sample. While the interactions of cholesterol with apoprotein or phospholipids are not well understood, the results indicate that cholesterol in LDL would play an important role as a 'bridge' to facilitate the interaction of PE at the interface. However, PE can penetrate towards the interface at low pH values because of reduced electrostatic interaction with the PC and apoproteins. The differences in lipids composition at the interface can be related to stability properties of LDL and CR-LDL emulsions.

Stability of LDL and CR-LDL Emulsions

The stability properties of emulsions containing 0.8 and 4.0% LDL or CR-LDL under different conditions are shown in Figure 5. Whatever the pH and NaCl concentration, the emulsions stabilized with high LDL concentration were

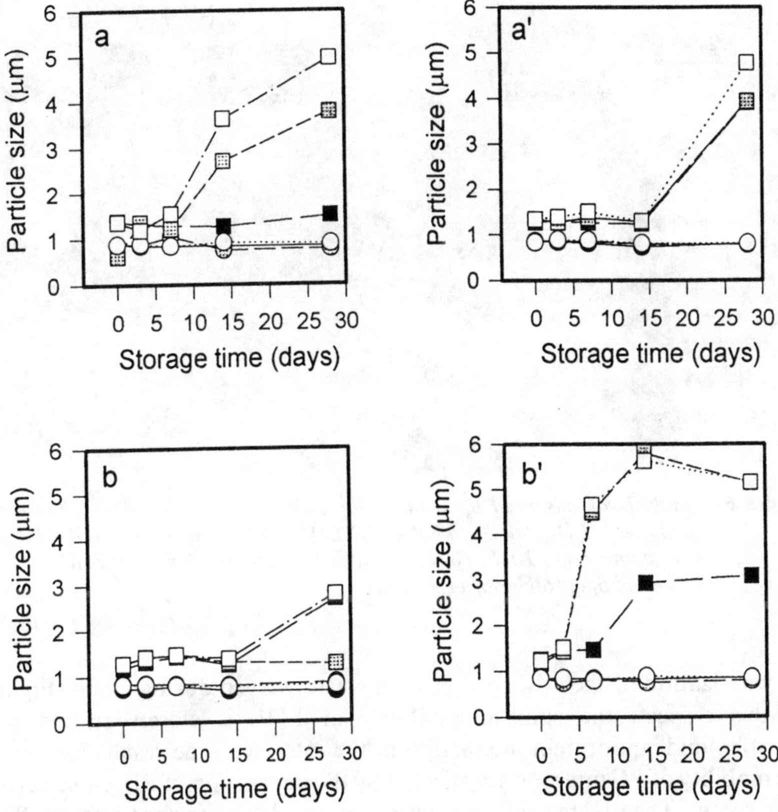

Figure 5 *Changes of mean particle size of emulsions (20 vol%) stabilized by egg yolk LDL (●, ■), 48.5% CR-LDL (●, ■) and 92.7% CR-LDL (○, □) as a function of storage time at pH 7.0 (a,a') and pH 3.5 (b,b') containing 0.1 M NaCl (a,b) and 1.5 M NaCl (a',b'). (●, ●, ○): 4.0% LDL or CR-LDL concentration (■, ■, □): 0.8% LDL or CR-LDL concentration*

stable for up to 1 month. At pH 7.0 and 0.1 M NaCl, the CR-LDL systems were found to be unstable after 2 weeks and the mean particle size of the emulsion was observed to increase with increasing aging time. However, the CR-LDL emulsions are stable at high NaCl concentration at pH 7.0 for up to 2 weeks. Interestingly, CR-LDL emulsions are stable at pH 3.5 and 0.1 M NaCl with low surface concentrations, while coalescence/flocculation of the emulsions increased at 1.5 M NaCl, resulting in an increase in emulsion particle size. It has been suggested that emulsions made with proteins should be more unstable at low pH. However, the present result is not consistent with this fact. The lipid composition was also analyzed in emulsions of 3 weeks old. The PC was found to be more dissociated from the interface of CR-LDL emulsions when compared to the control LDL emulsions. The PC molecules were retained on the emulsions at pH 7.0 and 1.5 M NaCl, whereas they were dissociated from the interface at pH 3.5 and 1.5 M NaCl. From the SDS-PAGE results, the

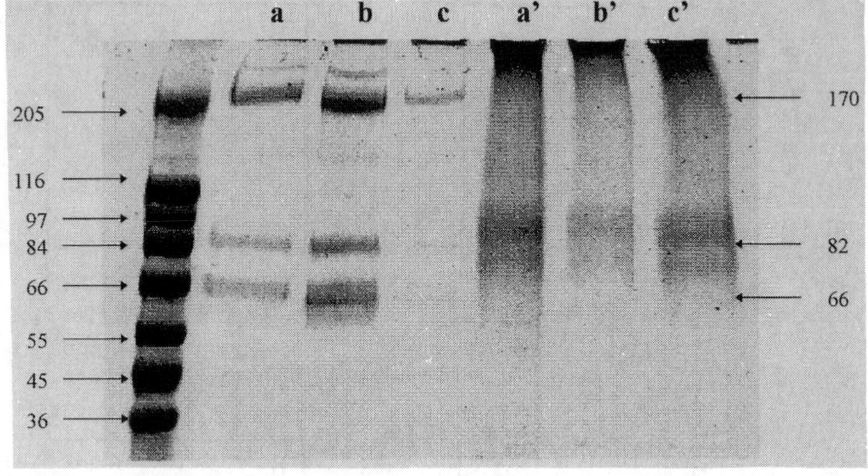

Figure 6 *Analysis of adsorbed lipoproteins by SDS-PAGE. Emulsions (20 vol% oil) made with LDL (a,a'), 48.5% CR-LDL (b,b') and 92.7% CR-LDL (c,c') containing 2.4% LDL. (a, b, c): adsorbed lipoproteins after 24 h. (a', b', c'): adsorbed lipoproteins after 3 weeks*

polymerization of the adsorbed protein was observed during aging (Figure 6). In this respect, the emulsion stability of LDL is closely related to the phospholipid–apoprotein interaction in the interface. The results indicate that the mobility of PC molecules is affected by the interaction of PE, cholesterol and apoprotein. The pH and salt concentration are also important factors affecting the phospholipid–protein interaction at the interface. At pH 7.0 and low NaCl concentration, the mobility of PC molecules in CR-LDL emulsions is relatively high and the PC may be easily dissociated from the interface during aging. On the other hand, the effective charge on PC is reduced at high NaCl concentration or low pH values, resulting in less dissociation of PC from the interface. The PE molecules at the interface may be an important bridge between the PC and the oil phase or apoproteins. At pH 3.5 and 0.1 M NaCl, the results are similar to those at pH 7.0 and 1.5 M NaCl, which shows that lowering the pH reduces the charge of PC and apoproteins, whereas the presence of high salts at low pH may decrease the binding affinity of the phospholipids at the interface, resulting in the breakdown of emulsion droplets.

4 Conclusions

The effect of cholesterol reduction from egg yolk LDL on its emulsifying properties was investigated. Our results demonstrate that cholesterol is an important component in the stabilization of LDL emulsions. Removing cholesterol from LDL causes the formation of larger particle sizes at the low protein concentration and also leads to the formation of much thicker films at the interface as a result of apoprotein aggregates. We have also found that removing cholesterol from LDL changes the phospholipid–apoprotein interactions at the interface. These changes may be responsible for the instability of CR-LDL emulsions. Further studies on the relationship of phospholipid–protein interactions to structure–function properties would be useful for better understanding of egg yolk lipoprotein functionality.

Acknowledgements

This research was supported by the Ontario Egg Producer's Marketing Board (Ontario, Canada) and the Natural Sciences and Engineering Research Council of Canada (NSERC).

References

1. K. A. McCully, C. C. Mok, and R. H. Common, *Can. J. Biochem. Physiol.*, 1962, **40**, 937.
2. W. G. Martin, W. G. Tattrie, and W. H. Cook, *Can. J. Biochem. Physiol.*, 1963, **41**, 657.
3. R. Vincent, W. D. Powrie, and O. Fennema, *J. Food Sci.*, 1966, **31**, 643.
4. R. Mizutani and R. Nakamura, *Lebensm.-Wiss. Technol.*, 1985, **18**, 60.
5. N. A. Bringe, D. B. Howard, and D. R. Clark, *J. Food Sci.*, 1996, **61**, 19.
6. V. D. Kiosseoglou and P. Sherman, *Colloid Polym. Sci.*, 1983, **26**, 502.
7. Y. Mine, H. Kobayashi, K. Chiba, and M. Tada, *J. Agric. Food Chem.*, 1992, **40**, 1111.
8. Y. Mine, K. Chiba, and M. Tada, *J. Agric. Food Chem.*, 1993, **41**, 157.
9. C. E. Dutilh and W. Groger, *J. Sci. Food Agric.*, 1981, **32**, 451.
10. G. W. Froning, 'Egg Science and Technology', 4th edition, Food Products Press, Binghamton, N.Y., 1994, p. 483.
11. Z. M. Merchant, G. G. Anilkumar, and R. G. Krishnamurthy, 1991, U.S. patent 5037661.
12. A. G. Awad, M. R. Bennink, and D. M. Smith, *Poultry Sci.*, 1997, **76**, 649.
13. A. Paraskevopoulou and V. Kisseoglou, *Food Hydrocolloids*, 1995, **9**, 205.
14. K. S. Raju and S. Mahadevan, *Anal. Biochem.*, 1974, **61**, 538.
15. M. A. K. Markwell, S. M. Haas, L. L. Bieber, and N. E. Tolbert, *Anal. Biochem.*, 1978, **87**, 206.
16. D. M. Smith, A. C. Awad, M. R. Bennink, and J. L. Gill, *J. Food Sci.*, 1995, **60**, 691.
17. N. S. Parker, *CRC Crit. Rev. Food Sci. Nutr.*, 1987, **25**, 285.
18. Y. Mine, *J. Agric. Food Chem.*, 1998, **46**, 36.
19. Y. Mine, *J. Agric. Food Chem.*, 1997, **45**, 4564.
20. R. W. Burley, *CSIRO Food Res. Q.*, 1975, **35**, 1.
21. R. W. Burley and R. W. Sleigh, *Aust. J. Biol. Sci.*, 1980, **33**, 255.
22. C. R. Scholfield, 'Lecithins—Sources, Manufacture & Uses', Am. Oil Chem. Soc., Champaign, IL, 1989, p. 7.

Foaming of Glycoprotein Alcoholic Solutions

By J. Senée, B. Robillard,[1] and M. Vignes-Adler

LABORATOIRE DES PHÉNOMÈNES DE TRANSPORT DANS LES MÉLANGES DU CNRS, 4 TER ROUTE DES GARDES, F-92190 MEUDON, FRANCE
[1]LABORATOIRE DE RECHERCHE MOËT & CHANDON, 20 AVENUE DE CHAMPAGNE, F-51205 EPERNAY CEDEX, FRANCE

1 Introduction

Beers and sparkling wines are amongst the most celebrated examples of foaming biological liquids. Beer foams are usually stable and thick, with very small spherical bubbles which may be disproportionated and become polyhedral during their lifetime. On the contrary, sparkling wine foams are transient and light, with spherical bubbles of about 1 mm diameter which collapse very rapidly. This essential transient and even evanescent character of sparkling wine foam has an important consequence: tiny changes in composition, concentration, or quality of the various compounds can impair, inhibit or enhance the foamability. Commercial consequences can be severe, and so reliable information on the factors and compounds controlling foamability and foam stability of champagne are of considerable interest from both the practical and theoretical points of view.

Foams are random dispersions of gas bubbles separated by liquid films. It has long been stated that a direct link should exist between the lifetime of the foam film and the overall foam stability. This has initiated a large number of studies on isolated liquid films since the pioneering work of Mysels[1] and Sheludko.[2] Likewise, foam film stability is dependent on the surface properties of the solution and on the disjoining pressure effect arising from the balance of intermolecular forces. As far as we know, the effects of the change of scale between the various levels of investigation of foamability properties (*i.e.*, from the surface and film scale towards the scales of a few bubbles and a foam column) are still unclear; and so foam properties are to some extent unpredictable from the surface and film properties, although qualitative correlations can be derived. The aim of our research is to tackle this problem for the case of sparkling alcoholic beverages.

Wines are multicomponent systems containing, in addition to alcohol, many organic compounds which may show surface activity by themselves (proteins) or by association with other compounds (polysaccharides associated with proteins, for example). Alcoholic solutions of proteins can be used as models of sparkling alcoholic beverages with respect to their foaming properties. The positive role of proteins on foam behaviour has long been attributed to their adsorbability at liquid interfaces. Since in the native structure most proteins have hydrophobic and hydrophilic domains, the presence of an air–water interface can induce conformational change that maximizes exposure of the hydrophobic domains to the air and the hydrophilic domains to the aqueous medium, thereby favouring the protein adsorption at the surface.

In general, the surface properties of a solution depend on the state of the surface-active agents within the underlying bulk solution. The protein adsorbability is severely modified when ethanol is added to the solvent.[3] Codissolved protein and ethanol interact specifically in bulk solution, and the effects of ethanol on protein stability in aqueous solution have been reviewed by Franks.[4] Ethanol has two main actions on proteins at and above room temperatures.[5] (i) It acts as a poor solvent, the more so at the protein isoelectric point when the ionic forces are weak. (ii) It is also known as a protein denaturing agent, acting primarily by weakening the hydrophobic bonds, and exposing the hydrophobic side chains to ethanol in the denatured state.

On the other hand, some recent studies have shown that glycoproteins rather than proteins are the dominant macromolecules in the foam of sparkling wines.[6] Carbohydrates (*e.g.*, sugar) tend to protect proteins against denaturation by ethanol and to increase their solubility; the degree of protection is, to a first approximation, a function of the number of hydroxyl groups. Hence, one can expect that interactions between glycoproteins and the alcoholic solvent will be intricately dependent on the sugar content of the glycoproteins. Likewise, the structure of the glycoproteins, which depends on the process of production, will also influence their interactions and consequently their adsorbability.

Against this background, we have investigated the surface, film and foam properties of alcoholic solutions of two glycoproteins: the first one, maltosyl bovine serum albumin (MBSA), is a commercial sample, while the second one is a mixture of yeast glycoproteins (YGP) extracted from wine. The two samples differ dramatically in their sugar content and in their history with respect to previous coexistence with ethanol. The glycoprotein molecular conformational changes that are induced by salts and ethanol result from both the modification of solvent physical properties and the specific interactions between ethanol and the protein and sugar moieties. The analysis at the molecular level is a very intricate matter beyond the scope of this study. We have only characterized them by their global influence on the solution surface and film properties.

2 Materials and Methods

Materials

A model solvent (MS) was prepared using Milli-Q water with a resistivity of 18 MΩ cm^{-1} and a concentration of organic contaminant lower than

Table 1 *Model solvent composition*

Constituent	*Concentration/g dm^{-3}*
Ethanol	94.7
Tartaric acid	3.7
Lactic acid	4.8
Glycerol	4.7
K^+	0.45
Ca^{2+}	0.083
Mg^{2+}	0.0782
Na^+	0.0214

30 mg m^{-3}. The composition of the buffer solution is presented in Table 1. The ethanol concentration and the ionic strength of the solutions were 12 vol% and 0.02 M, respectively. This particular composition of the model solvent partially mimics the composition, pH, and ionic strength of Champagne wine.[7] It does not contain any lipid, protein, and volatile substances (other than ethanol) which are usually present in wines.

The MBSA (from Sigma Chemical lot 98F8120) was prepared from coupling of bovine albumin and maltose via reductive amination.[8] One mole of albumin contained 14 moles of disaccharide linked to lysine residues. The sugar content represented 5% of the molecule. The MBSA molecular weight and p*I* were 71 kDa and ~4, respectively. It was used without further purification.

The YGP was provided by R. Marchal from the Oenology Laboratory of the University of Reims who obtained the sample from alcoholic fermentation in a synthetic medium. After fermentation, the medium was ultrafiltered on a 10 kDa molecular weight cut-off membrane and the retentate was freeze-dried. The lyophilizate was essentially a mixture of mannoproteins and glucans. The proportion of sugar in this mixture was higher than 90% and the molecular weight of each structure ranged between 40 kDa and >100 kDa. The p*I* of the YGP ranged between 3 and 4.3. (Since the sugars are hydrophilic, the more sugar that is present, the more hydrophilic are the domains in the glycoprotein.)

The MBSA and YGP differ mainly in their sugar content; both are positively charged at the solvent pH.

Methods

The surface tensions were monitored by an apparatus of the du Noüy type in a temperature regulated cell at 20.0 °C. The precision of the measurements was estimated to be ± 0.1 mN m^{-1}.

Single thin liquid films were formed above a bubble attached to an air–water surface in a tightly closed cell (Figure 1), placed in the field of a metallographic Nikon microscope ($\times 200$), and illuminated by reflected heat-filtered monochromatic light. Interferometric pictures of the films were visualized by means

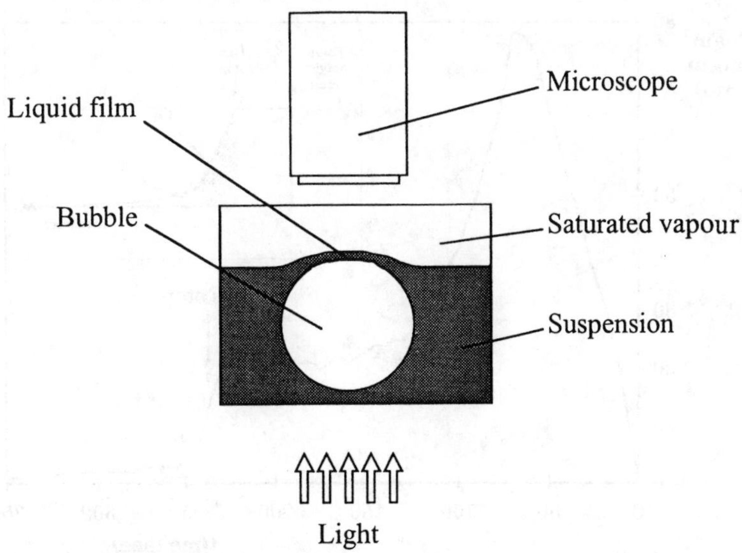

Figure 1 *Schematic representation of a film at an attached bubble (not to scale)*

of a CCD Lhesa camera connected to a tape recorder. Film thicknesses were determined by the classical microinterferometric technique, developed by Sheludko.[2] Details can be found elsewhere.[9]

The first bubble was created 2 hours after pouring the liquid in order to ensure saturation of the cell. About 10 bubbles were created for the liquids which make films of infinite lifetime, and more than 50 for the other ones. The bubble and film mean diameters are $\Phi = 1.21 \pm 0.05$ mm and $d = 0.42$ mm, respectively.

Foam parameters were measured by a sparging method.[10] The liquid was poured into a 1 litre test tube whose bottom was a sintered glass plate (porosity 40–60 μm), and then nitrogen gas sparging was carried out at constant flow-rate ($Q = 5$ l h^{-1}) and pressure (300 kPa). Usually, the foam height increased to a maximum and decreased to a steady height. Then the gas flow was stopped (see Figure 2). Each sample was analysed in triplicate.

The foam expansion is defined as the ratio of the maximum foam volume V_{max} to the initial liquid volume V_{liq}, *i.e.*,

$$E_{max} = \frac{V_{max}}{V_{liq}} . \tag{1}$$

The lifetime is defined as

$$L_f = \frac{1}{H_0} \int_{t_{max}}^{\infty} H \, dt , \tag{2}$$

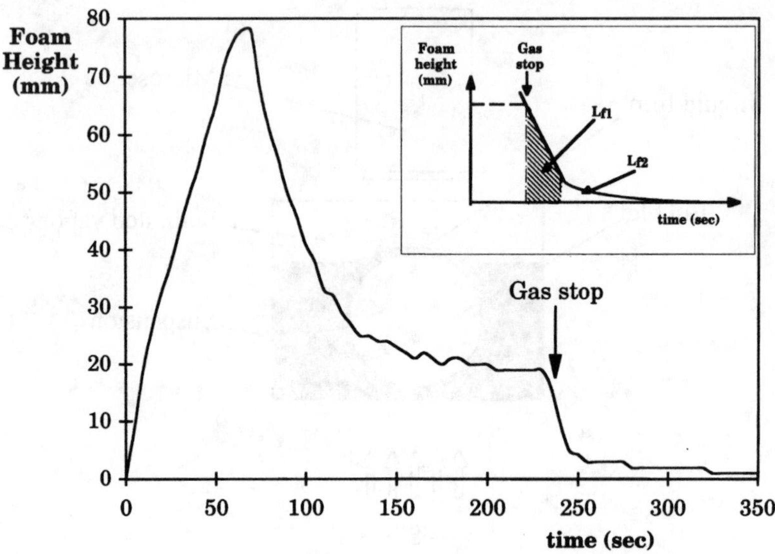

Figure 2 *Typical evolution of foam height with time. Insert shows definition of* L_{f1} *and* L_{f2} *lifetimes*

where t_{max} is the time at which the gas flow was stopped, H is the foam height at time t, and H_0 is the initial foam height.

3 Results

Surface Tension

We have checked that the MS surface tension was equal to 48.2 mN m^{-1}, *i.e.*, the surface tension of the equivalent aqueous alcoholic solution at same ethanol content. This indicates that, except for ethanol, the other organic compounds are not surface active at pH 3 and ionic strength 0.02 M. The surface tensions of the solutions were measured as a function of time. A true plateau was reached almost instantaneously with YGP solutions and after 60 min with MBSA solutions. A quasi-equilibrium surface tension $\sigma_{q.eq.}$ could be measured and the surface pressure Π of the surface active materials could be calculated from

$$\Pi = \sigma_{water} - \sigma_{q.eq.} \cdot \tag{3}$$

The results are reported in Table 2. For both glycoproteins we find $\Pi \approx 26 \pm 1$ mN m^{-1} whatever the concentration. Except for the more dilute YGP solution, the surface pressure is slightly higher than obtained for MS. Moreover, it was also found[9] that, in the absence of ethanol, we have $\Pi^{MBSA}_{eth\ free} = 13.1$ mN m^{-1} and $\Pi^{YGP}_{eth\ free} = 9$ mN m^{-1} when the glycoprotein concentration is 10 mg dm^{-3}.

Films

The drainage rates and lifetimes t_f of the films depend sensitively on the nature and concentration of the glycoproteins. Thick films (TF) formed from very dilute MBSA solution (C = 0.2 mg dm^{-3}) are very mobile and unstable with formation of a shallow dimple (Figure 3a). At intermediate MBSA concentrations (0.5 mg dm^{-3} $\leqslant C \leqslant$ 10 mg dm^{-3}) there is again formation of an axisymmetrical dimple, with a much larger curvature, which is sucked into the film Plateau border leaving behind a constant thickness film (transition TF → GF) near the bottom edge (Figure 3b). When the grey film (GF) is 120 nm thick,

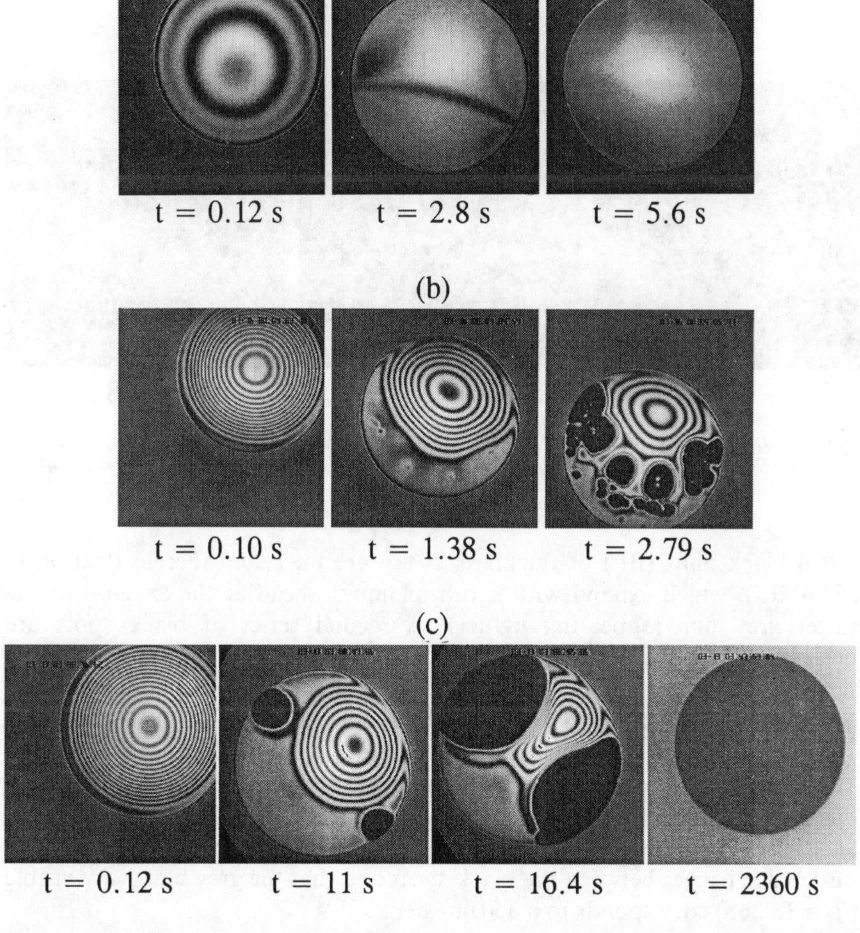

Figure 3 *Interference pictures of draining films formed from MBSA solutions at various bulk glycoprotein concentrations. A change in grey colours corresponds to a thinner film. (a) C = 0.2 mg dm^{-3}; (b) C = 10 mg dm^{-3}; (c) C = 40 mg dm^{-3}*

(a)

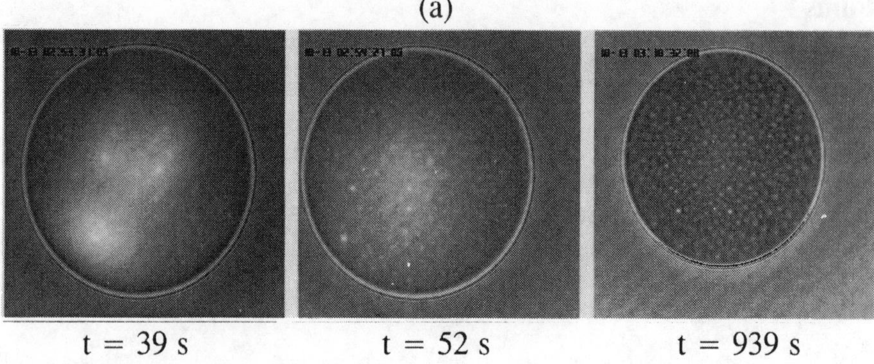

t = 39 s t = 52 s t = 939 s

(b)

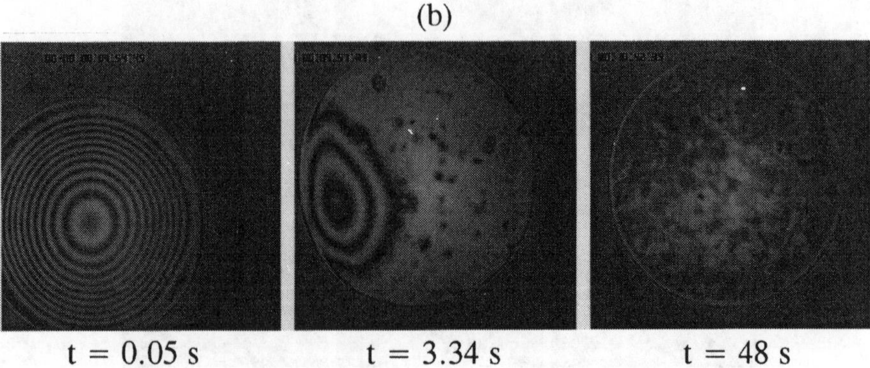

t = 0.05 s t = 3.34 s t = 48 s

Figure 4 *Interference pictures of draining films formed from YGP solutions at various bulk glycoprotein concentrations: (a) $C = 3\,mg\,dm^{-3}$; (b) $C = 10\,mg\,dm^{-3}$*

several black films (BF) of thickness 20 nm are nucleated therein (transition GF → BF), which expand with a rim of liquid ahead at the expense of the thicker grey film (appearing lighter). A second series of black spots are nucleated behind the first series. The rim thickens by steps, and its maximum thickness is 500 nm, which means that the black spots are like craters 480 nm deep. These films are very unstable with lifetime of *ca.* 3 s. At a large MBSA concentration ($C = 40\,\text{mg dm}^{-3}$) the phenomenon is essentially the same as at $C = 10\,\text{mg dm}^{-3}$, but now everything happens much slower. Only two black films with thick rims are formed and these finally coalesce into a single stable black film (Figure 3c). It is remarkable that the rim can be as thick as three interference fringes between the black spots and that the grey film (very visible at $t = 16.36$ s) corresponds to a 350 nm step.*

*Actually, for the same MS concentration, the film aspects for the native BSA solution are very similar.[9]

The drainage of the YGP films is very different from the MBSA ones (Figure 4). At $C = 3\,\mathrm{mg\,dm^{-3}}$, organized small white spots appear in the grey film (t = 55 s) and the whole film gets a regularly organized granulous aspect, the typical grain diameter being 6 μm. The film lifetime is infinite with a 15 nm thickness, and the bubble remains in equilibrium. At the higher concentration $C = 10\,\mathrm{mg\,dm^{-3}}$, one can observe the formation of darker and more irregular clusters which obviously are not black films. Microscopic inspection with a ($\times$400) enlargement shows that they are polydisperse (mostly bidimensional) aggregates ($\sim$15 μm). The film ruptures at a finite time, equal to 50 s when the film thickness is *ca.* 25 nm at its periphery.

Foams

For most of the liquids investigated here, two characteristic foam timescales could be measured after gas sparging. There is a fast decay over a timescale L_{f1}; then a layer of 1 to 5 bubbles thick remains at the surface which collapses after a much longer time denoted L_{f2}. This second step does not exist with MS or with less concentrated YGP solutions. Note that it does exist with foams formed from wines.[11]

Foam parameters are reported in Table 2. The highest values were obtained with the MBSA solutions. Foaming of the concentrated solution was so extensive that it overflowed the test tube. Then, it did not totally decay, and a foam ring remained indefinitely stuck on the tube wall looking rather like a gel foam.

4 Discussion

Solution Surface Activity

There is no obvious correlation between the surface, film and foam properties of the present solutions. The quasi-equilibrium surface pressures of the two surface-active agents in the presence of 12% alcohol content are almost equal to each other and to the value for MS. The interaction forces between the macromolecules and the ethanol molecules at the interface can be understood from the following expression deduced from thermodynamic considerations:

$$\Delta\Pi = \Pi - (\Pi^{\mathrm{macromol}}_{\mathrm{eth\ free}} + \Pi^{\mathrm{macromol\ free}}_{\mathrm{eth}}).$$

With the values of Table 2 and section 3, we have $\Delta\Pi^{\mathrm{MBSA}} = -11\,\mathrm{mN\,m^{-1}}$ and $\Delta\Pi^{\mathrm{YGP}} = -8.2\,\mathrm{mN\,m^{-1}}$. The quantity $\Delta\Pi$ is non-vanishing and negative for both macromolecules. This means either that the ethanol molecules and the glycoproteins interact attractively at the surface, or that there is a competition between them in the adsorption process, the alcohol hindering, albeit not preventing, the glycoprotein adsorption, like in MS solutions of native bovine serum albumin.[12] Other measurements would be necessary to conclude which model is valid in the present case. Actually, it was previously shown that, in the

Table 2 *Surface, film and foam properties of the glycoprotein solutions (n.m. = non-measurable)*

Solution	$C/mg\ dm^{-3}$	$\Pi/mN\ m^{-1}$	t_f/s	E_{max}	L_{f1}/s	L_{f2}/s
MS 0	24.8	5	0.14	8	0	
MBSA	0.2		6	0.2	7.5	11
	10	26.9	2	2.3	38	1980
	40	27	∞	≫5	n.m.	∞
YGP	3	24.5	∞	0.15	10	0
	10	25.6	48	0.5	16.4	54

mixed layer adsorbed from a 12% ethanol solution, a small amount of BSA molecules coexists with a significant amount of ethanol. Surface rheology measurements showed that the adsorbed protein molecules are completely unfolded, extended flat in the surface, and irreversibly adsorbed like in an insoluble monolayer.[13–14]

Although we have not yet made equivalent measurements for the MBSA and YGP solutions, we expect quite similar results for MBSA because its primary structure, with the exception of the sugar moiety (5%) that is not surface-active, is very close to the BSA one, and also because their film properties are very similar. We have not yet a clear explanation for the behaviour of the YGP solutions.

Films

Since all the bubbles have the same diameter, the film drainage occurs at the same constant capillary pressure P_c generated by the Plateau border. Drainage stops if the disjoining pressure arising from some colloidal forces is able to counterbalance the applied capillary pressure P_c. Obviously, the MS does not contain any stabilizing agent since its films are unstable. The two major differences between the macromolecular systems are the TF → GF → BF transitions occurring in the MBSA draining films, and the formation of aggregates in the YGP films which can be small or large depending on the concentration.

Actually, the behaviour of the MBSA film can also be explained by a micellization process. Similar transitions in film thinning have been obtained with very concentrated surfactant solutions[15–16] above the cmc and with colloidal solutions containing monodisperse latex particles of submicron size.[17] Recently, Bergeron and Radke[16] modelled the dynamics of spot formation and sheeting in films formed from micellar solutions by accounting for an equilibrium oscillatory structural component in the disjoining pressure isotherm. We could not measure the disjoining pressure isotherms, but it is probable that they do exhibit at least two oscillations, which are necessary to explain that grey and black films can coexist. As mentioned in the introduction, ethanol is a poor solvent of proteins, and micellization is expected. When micelles are locked into an ordered structure, intermicellar repulsive forces give rise to a strongly repulsive disjoining pressure. During film drainage whole micellar structures are expelled layer by layer (a depletion effect). Simultaneously, excess liquid from the growing black hole accumulates in a thicker rim because the viscous dissipation in the grey film (proportional to h^{-3}) prevents easy escape. As soon as the crater shape exists, the film curvature effect generates a local capillary suction in addition to P_c which sustains the black film expansion. For the dilute solution, the coexistence of many black holes destabilizes the film. For the concentrated MBSA solution, only two far-distant black holes were formed and the film lifetime was found to be infinite. There are then probably many more micelles, and it is possible that some very viscous stabilizing gel can be formed as was observed in the foam column experiment.

In the YGP solutions, the aggregates are apparently much larger than the MBSA micelles and they remain attached to the film. They appear when the film thickness h has decreased to 100 nm, which is about comparable to, or smaller than, the hydrodynamic diameter (120 nm) of the YGP in MS.[9] Now, the film is a confined system; this hinders Brownian motion due to its reduced dimensionality, and it squeezes the polymeric sugar chains during film thinning, which results in solvent depletion and increase of the local concentration. It appears that the macromolecular concentration (C_s) increases to a limiting value (C_l) corresponding to a supersaturation level where spontaneous aggregation and even precipitation occurs. The size of the aggregates increases with the macromolecule concentration. In the dilute solution, aggregates (6 μm) are formed which are trapped in the film in an almost crystalline state. In this ordered structure, intermicellar repulsive forces give rise again to a strongly repulsive and stabilizing disjoining pressure, explaining the infinite film lifetime. In the concentrated solution, the aggregates are so very large that they can be viewed as precipitates which form a percolating network in the plane of the film. Transversely to the plane of the film, a bridging effect that occurs between segments of the yeast glycoproteins in the same aggregate and the adsorbed molecules can reasonably exist since the aggregates remain attached to the film. This bridging effect is known to generate attractive destabilizing forces which reduce the film thickness. The origin of the shorter lifetime of the concentrated YGP films is not yet completely clear, although it is likely to be due to a dewetting effect following bridging.

Foams

Except for the concentrated MBSA solution, comparison between the single film lifetime and the foam lifetime is not rewarding. In unstable or metastable foams formed from liquids with low dynamic viscosity, gravity drainage is balanced neither by the bulk shear stresses nor the interfacial stresses. As long as the foam is young, with films of micrometre thickness, flows in the Plateau border network can be so important that they drag along the adsorbed molecules, which can completely dominate the foam lifetimes. The question is with which of the foam relaxation lifetimes, L_{f1} or L_{f2}, the film lifetime t_f should be compared. *A priori* the quantity L_{f2} should be used when the foam is old and the films are drained. The comparison is only reasonable with the very dilute MBSA solution and the more concentrated YGP solution. It seems amazing that the 10 mg dm^{-3} MBSA solution gives very unstable MBSA films and stable foams, while it is exactly the opposite for the dilute YGP solution. At this stage of the research, we can only conclude that the foaming process itself can be so violent, with consequences for glycoprotein folding/unfolding, that it modifies their further adsorption and micellization/aggregation in the films of the foam.

5 Conclusion

Films and foam properties of aqueous alcoholic solutions of glycoproteins are very dependent of the nature of the macromolecule. For both the investigated glycoproteins here, the interaction with ethanol generates a micellization/aggregation process. The physical properties and size of the micelles/aggregates are very different in the two cases, and the behaviour is dependent on their solution concentration. This can be attributed to a dramatically different history of the glycoproteins and to their very different sugar contents.

References

1. K. J. Mysels, K. Shinoda, and S. Frankel, 'Soap Films', Pergamon, New York, 1957.
2. A. Sheludko, *Adv. Colloid Interface Sci.*, 1967, **1**, 391.
3. M. Ahmed and E. Dickinson, *Colloids Surf.*, 1990, **47**, 353.
4. F. Franks, in 'Characterization of Proteins', ed. F. Franks, Humana Press, New Jersey, 1988.
5. J. F. Brandts and L. Hurt, *J. Am. Chem. Soc.*, 1967, **89**, 4826.
6. R. Marchal, S. Bouquelet, and A. Maujean, *J. Agric. Food Chem.*, 1996, **44**, 1716.
7. B. Duteurtre, *La Recherche*, 1986, **17**, 1478.
8. B. A. Schwartz and G. R. Gray, *Arch. Biochem. Biophys.*, 1977, **181**, 542.
9. J. Senée, Ph.D. Dissertation, Institut National Polytechnique de Lorraine, Nancy, 1996.
10. A. Rudin, *J. Inst. Brew.*, 1957, **63**, 506.
11. J. Senée, B. Robillard, and M. Vignes-Adler, *Food Hydrocolloids*, 1998, **12**, in press.
12. A. Dussaud, G. B. Han, L. Ter-Minassian-Saraga, and M. Vignes-Adler, *J. Colloid Interface Sci.*, 1994, **167**, 247.
13. A. Dussaud and M. Vignes-Adler, *J. Colloid Interface Sci.*, 1994, **167**, 256.
14. A. Dussaud and M. Vignes-Adler, *J. Colloid Interface Sci.*, 1994, **167**, 266.
15. J. Perrin, *Ann. Phys.*, 1913, **35**, 329.
16. V. Bergeron, A. I. Jiménez-Laguna, and C. J. Radke, *Langmuir*, 1992, **8**, 3027.
17. A. D. Nikolov and D. T. Wasan, *J. Colloid Interface Sci.*, 1989, **133**, 1.

On the Stability of the Gas Phase in Ice-Cream

By Susie Turan, Mark Kirkland, and Rod Bee

UNILEVER RESEARCH, COLWORTH LABORATORY, SHARNBROOK, BEDFORD MK44 1LQ, UK

1 Introduction

Ice-cream in its simplest form is a frozen and aerated mixture of water, cream and sugar, but the physics of the resulting product is complex. The presence of a fine ice-cream microstructure is critical to produce the desired texture and quality of ice-cream. Organoleptic evaluation of ice-cream has shown that small air cells and ice crystals are associated with increased creaminess and reduced iciness, our criteria for good quality ice-cream.[1] Ice-cream consists of about 50% gas by volume, and, like all dispersions of gas cells, the ice-cream foam structure is thermodynamically unstable and may be destabilized by a number of physical processes such as disproportionation, drainage and coalescence.[2] Whether these processes actually take place, and at what rate, depends on the physical properties of both the continuous liquid phase and the gas phase. Coalescence of bubbles is an important destabilizing mechanism in ice-cream. It occurs when the film between two bubbles ruptures and the bubbles merge.

It is possible to increase foam stability by reducing drainage via adsorption of partially coalesced fat droplets to gas cells[3–5] or through increasing the product viscosity by 'hardening' the product.[1,5] Ice-cream is hardened by reducing the temperature, typically from the manufacturing temperature of -5 to -25 °C. This results in an increased ice content, typically from 30 to 50 wt%. Hardening reduces the rate of structural change with respect to both air and ice phases.[6] However, during hardening and storage the gas phase coarsens, resulting in distorted or interlinked gas cells. This can lead to the situation where gas is lost from the product resulting in an irreversible loss of product volume, known as shrinkage. There is much disagreement in the literature as to the mechanism of shrinkage.[7–12] Our current hypothesis is that shrinkage is a two-stage process of channel formation followed by product collapse. It is believed[11] that coalescence of discrete gas cells to form interconnected gas channels is a pre-requisite for shrinkage to occur. If channelling does occur, the structure is prone to collapse, especially at elevated storage temperatures.[11–12] The driving force for collapse is the resulting minimization of surface area. The ability to quantify the degree of channelling before collapse is thus an important step in understanding the shrinkage mechanism. In this study, gas channelling is measured based on the ability of an aerated ice-cream to change in volume as a result of external pressure change.[11–15]

This paper will discuss the factors determining the stability of the gas phase in ice-cream, for example, the hardening conditions employed, and the interdependency of the fat, ice and gas phases in determining ice-cream microstructure.

2 Materials and Methods

Ice-cream samples were produced from premixes with the following formulation: 12% anhydrous butteroil (Meadow Foods), 12% spray dried skim milk powder (Express Foods), 15% sucrose (Tate & Lyle), and 0 to 1% MGP (monoglycerol palmitate containing 50 wt% saturated monoglyceride and 40 wt% diglyceride) (Quest International). The premix emulsions were aged overnight at 5 °C. Ice-cream was produced in a continuous freezer (Crepaco K118) at a mix flow-rate of 120 $dm^3 h^{-1}$. Ice-cream was aerated at 4 bar barrel pressure to a range of gas phase volumes (0.1–0.7). All ice-cream was extruded at −5.7 °C, collected in 500 ml waxed paper cartons, and hardened at −35 °C for two hours. Samples with gas volume fraction Φ_{gas} = 0.47 were also hardened at temperatures of −25 and −10 °C. The structure of ice-cream at extrusion was determined by plunging 5 ml aliquots of ice-cream into liquid nitrogen.

The microstructure of ice-cream was visualized by low temperature scanning electron microscopy (LT-SEM). Samples were stored at −80 °C prior to analysis using a JSM 6310F scanning electron microscope fitted with an Oxford Instruments ITC4 controlled cold stage. The samples were prepared using the Hexland CP2000 preparation equipment. At −80 °C, a 5 × 5 × 10 mm sample was mounted onto an aluminium stub using OCT mountant on the point of freezing and plunged into nitrogen slush. The sample was warmed to −98 °C, fractured and etched for 2 min, and then cooled to −115 °C. The surface was coated with Au/Pd at −115 °C, 6 mA and 2 × 10^{-1} mbar argon. The sample was transferred in vacuum to the LT-SEM and examined under microscope conditions of −160 °C and 1 × 10^{-8} Pa.

The gas phase structure in ice-cream was quantified by measuring the gas cell-size distribution from SEM micrographs using the AnalySIS 2.11—package AUTO (SIS Munster, Germany) with ‘B’ version software. The AnalySIS program may be run using SEM data direct from the JEOL microscope or as images scanned from Polaroids. The optimum magnification was such that there were <300 gas cells per image. The program was used semi-automatically in that particle edges were calculated automatically (by difference in grey-scale) and manually refined (by deleting and redrawing around particle boundaries not selected correctly). Since ice crystals may also have been selected, the gas cells were manually picked and the distribution analysed using the maximum

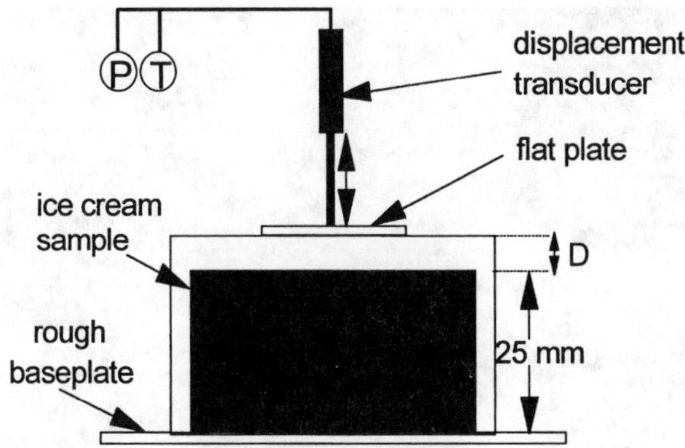

Figure 1 *Apparatus to measure gas channelling via expansion under reduced pressure. Pressure (P), displacement (D) and temperature (T) were monitored*

diameter parameter. All gas cells present on an SEM micrograph were counted, and generally >1000 gas cells were counted. The average size was determined as the number average, $d(1,0)$, of the individual cell sizes. Ice crystals were also sized from SEM micrographs in a similar manner, although all particle boundaries were drawn manually.

The degree of partial coalescence of fat (de-emulsified fat) after freezing was assessed by a solvent extraction technique.[16] The assay involved agitation of 50 ml of petroleum spirit (40 : 60) with 10 g of melted ice-cream for one minute. The solvent (containing fat) was decanted and, after evaporation, the weight of fat extracted was used to calculate the % de-emulsified fat.

The apparatus to measure gas channelling via pressure response is shown in Figure 1. A 25 mm slice of ice-cream was equilibrated at the test temperature (−15 °C) in a 40 cm vacuum desiccator. Any height change of the ice-cream upon 200 mbar pressure reduction was measured by a displacement transducer. Sample gas phase volume was measured using the Archimedes principle.

3 Results

Stability of the Gas Phase of Ice-Cream during Hardening

Stability of fresh ice-cream. The ice-cream structure immediately upon exit of the freezer at −5.7 °C may be visualized by plunging 5 ml aliquots into liquid nitrogen followed by SEM analysis.[17,18] As reported by other workers,[17,18] gas cells in fresh ice-cream were spherical in shape and *ca.* 20 μm in diameter (Figure 2). However, this foam structure is very unstable and many changes to both gas and ice phases are observed during hardening. Gas coarsening occurs due to a combination of the time taken to increase viscosity and the growth of ice crystals pushing bubbles together resulting in distorted and interlinked gas cells (Figure 3).[6,11,17,18] In order to separate the effect of the inherent instability from changes which occur due to ice growth, ice-cream (Φ_{gas} = 0.47) was held

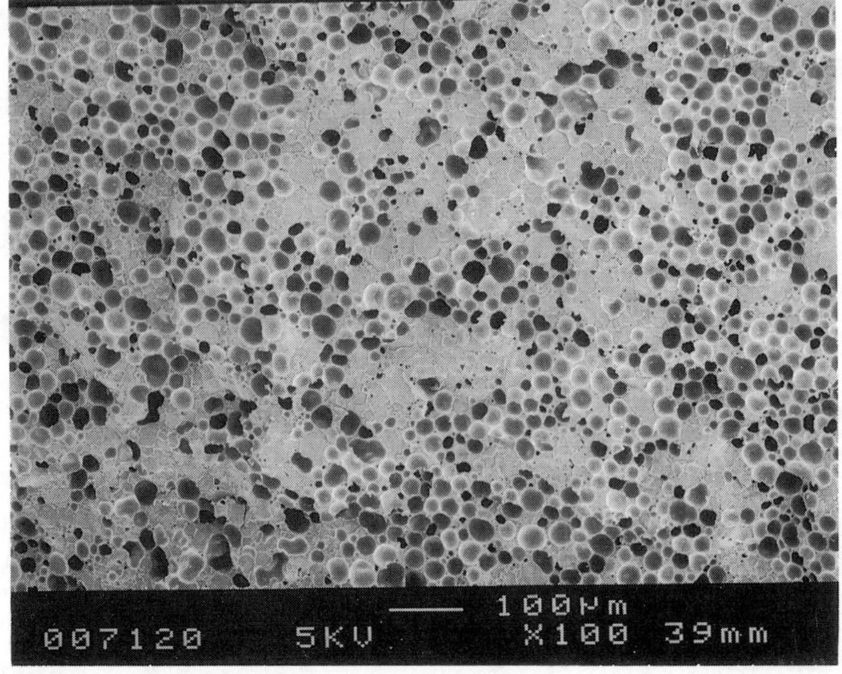

Figure 2 *Scanning electron micrograph of ice-cream quenched at extrusion in liquid nitrogen*

for up to thirty minutes at −5.7 °C. At regular time intervals, a small sample was taken and quenched in liquid nitrogen to trap the structure and halt any further changes. The structure was analysed by SEM and the gas phase quantified. Figure 4 clearly shows the increase in mean gas cell size. A corresponding decrease in number of gas cells per unit volume indicated that coalescence had occurred. The presence of the emulsifier MGP gave similar structures within the freezer but greatly increased the stability upon extrusion. MGP promotes partial coalescence of fat under shear. The fat aggregates adsorbed at the air–water interface stabilize gas cells against coalescence.[3,4,19] The de-emulsified fat level is a critical factor controlling the coalescence rate. However, there is insufficient fat in ice-cream to stabilize the foam at extrusion via a complete fat network, such as that in a whipped cream;[2,5] hence hardening is employed to reduce drainage.

Gas structure profile during hardening. The changes to the gas phase during hardening were monitored by quenching ice-cream from the centre of a block undergoing hardening at −35 °C at given time intervals, followed by SEM analysis. A large increase in mean gas cell size was observed upon hardening. The curve shape in Figure 5 indicates that most change occurred in the first few minutes when the sample was at a relatively high temperature and the foam structure was very fine, and the rate of coarsening decreased as the temperature

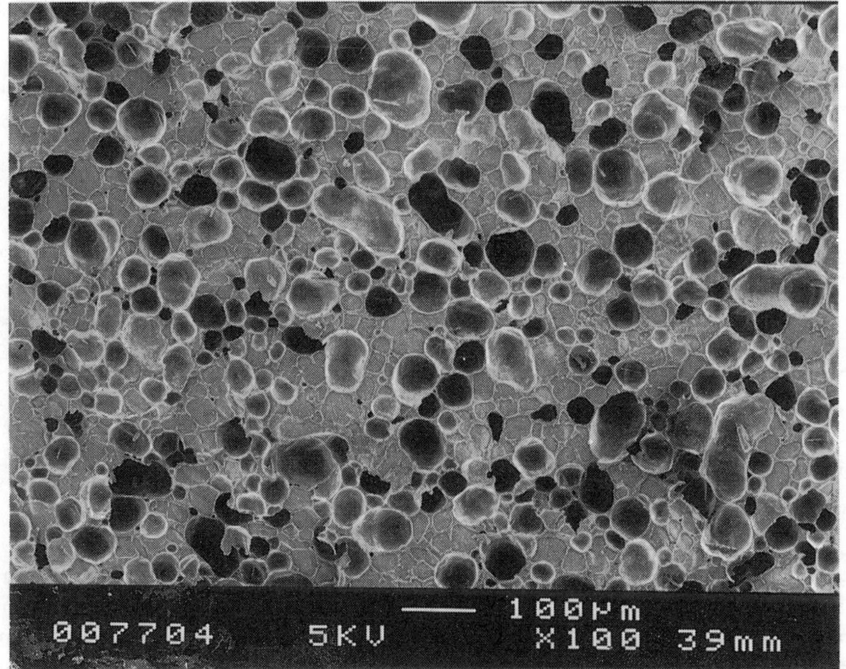

Figure 3 *Scanning electron micrograph of ice-cream hardened in a blast freezer at −35 °C*

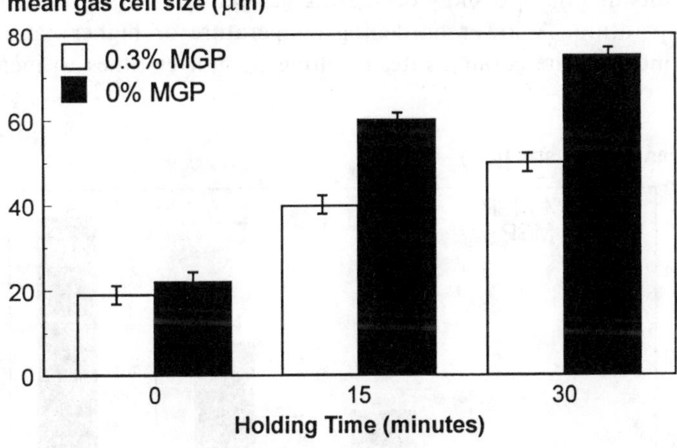

Figure 4 *Stability of ice-cream with 0 and 0.3% MGP at extrusion*

decreased. The extent of gas coarsening was similar to that found on holding the ice-cream at its extrusion temperature for the same time. Although ice growth will contribute to gas cell distortion, the dominant factor controlling gas cell coalescence is the low product viscosity.

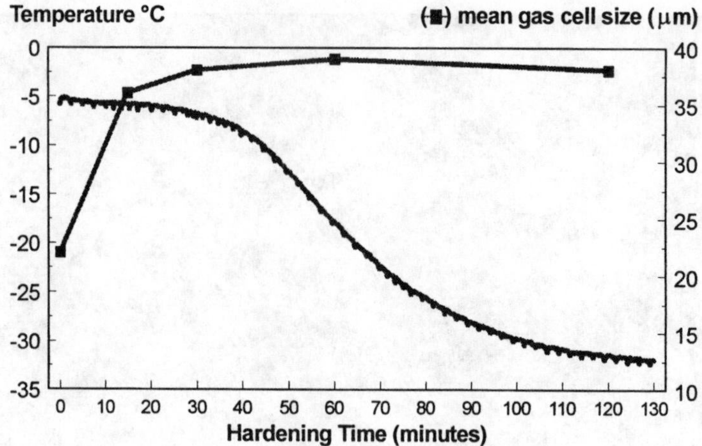

Figure 5 *Profile of mean gas cell size (■) and corresponding product temperature (—) during hardening for ice-cream with 0.3% MGP*

Effect of hardening rate. A key factor determining the extent of structural change during hardening is the time taken to increase ice content to such an extent that the coalescence rate is minimized. The effect of hardening time was studied by hardening samples at different rates (blast freezer at $-35\,°C$ and 3 m s^{-1} gas flow rate, and cold stores at $-25\,°C$ and $-10\,°C$ with no gas flow). The changes to the gas structure were monitored and compared to nitrogen quenched samples, equivalent to the fastest hardening rate.

The results in Figure 6 show decreasing gas cell size with decreasing hardening temperature. A lower hardening temperature or higher gas flow was found to increase the cooling rate, resulting in shorter times to increase the

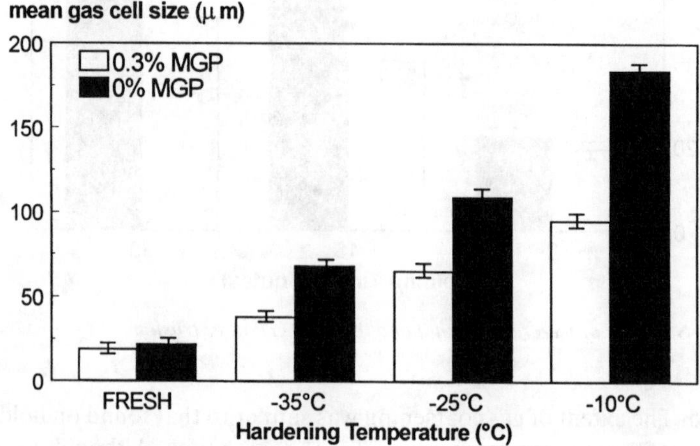

Figure 6 *Influence of the hardening temperature on the gas cell size of ice-cream with 0 and 0.3% MGP*

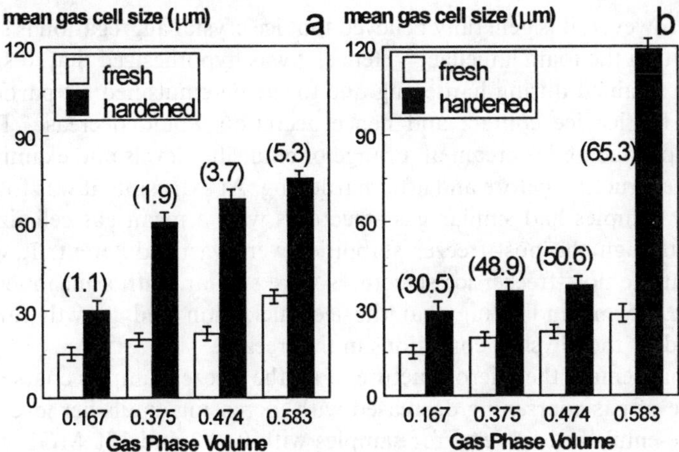

Figure 7 *Effect of gas phase volume on gas cell size of ice-cream with (a) 0% and (b) 0.3% MGP before and after hardening. Figures in parentheses indicate de-emulsified fat level (% w/w of total fat)*

viscosity to such an extent that coalescence was arrested. This trend was more easily observed for less stable 0% MGP samples with low de-emulsified fat levels. This implies that fat aggregates play a key role in reducing drainage rates until the viscosity is increased by hardening.

Effect of gas phase volume. In order for coalescence of gas cells to occur, they must come into contact. Hence it is logical that this phenomenon will be dependent on gas phase volume. The coalescence behaviour as a function of gas phase volume and de-emulsified fat level was investigated by comparing the gas structures at extrusion and after hardening (Figure 7). In quenched ice-cream, a small degree of coalescence was observed at high gas phase volumes giving rise to some large gas cells. Gas cells, however, remained spherical. The increased levels of de-emulsified fat in the 0.3% MGP system stabilized the foam at extrusion and reduced the coalescence. Upon blast freezing, all samples underwent coalescence resulting in an increased gas cell size. The gas structure within a hardened product was strongly dependent on the gas phase volume. As the gas phase volume was increased, the degree of coalescence and gas cell size increased. However, the presence of de-emulsified fat stabilized gas cells against coalescence and thus reduced the gas cell size at a given gas phase volume. Moreover, a combination of high de-emulsified fat and a high gas phase volume leads to a high level of coalescence, increased gas cell size and channel formation.

Influence of the Gas Phase Stability on Ice Structure

Whilst emulsifier addition can clearly influence the gas phase structure via its ability to promote fat de-emulsification, it should have no direct effect on ice growth. However, it is generally believed that ice crystal aggregation is sterically limited within the foam lamellae.[20] Hence, it was hypothesized that, if small gas cells were retained during hardening due to the de-emulsified fat particles, the potential for ice–ice contact and hence accretion would decrease. This was tested by producing ice-cream at a range of emulsifier levels and examining the gas and ice structure before and after hardening. At extrusion, it was found that all freezer samples had similar gas structures with a mean gas cell size of *ca.* 20 μm, although the post-freezer stabilities were vastly different. It was also found that the post-freezer ice structures were similar, with a mean ice crystal size of *ca.* 25 μm, indicating that the ice nucleation and growth rates were dominated by the physical conditions in the freezer.

Upon hardening, the microstructure of all the freezer samples coarsened and the degree of gas coarsening decreased with increasing emulsifier level (Figure 8). The de-emulsified fat level for samples with 0, 0.3 and 1% MGP were 3.7, 50.6 and 18.8%, respectively. Hence the extent of gas coalescence was not directly related to de-emulsified fat level since the 1% MGP sample underwent the least gas coalescence, yet had a relatively low de-emulsified fat value. It can be concluded that the presence of MGP can influence gas cell stability independently of de-emulsified fat level.[19] This increased stability is believed to be due to direct adsorption of MGP at the air–water interface. The ice crystal size was also monitored and it was found that although the samples underwent the same temperature/time profile, there was a small reduction (*ca.* 9%) in ice crystal size as the air cell size decreased at increasing MGP levels (Figure 8). This indicated that a highly dispersed gas phase could limit ice accretion upon hardening.

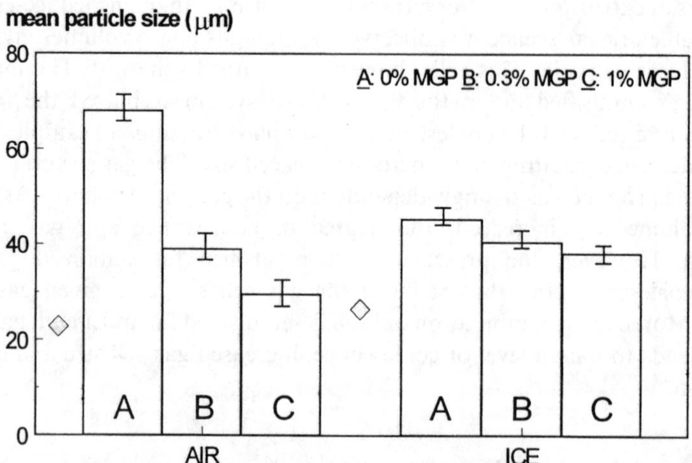

Figure 8 *Influence of emulsifier level (0–1% MGP) on the mean gas and ice particle sizes in ice-cream upon hardening. Symbol (◇) indicates mean particle size at extrusion*

Measurement of Gas Channelling in Ice-Cream

The previous sections have utilized a two-dimensional SEM technique to visualize changes occurring to the foam structure. However, any changes to the three-dimensional interconnectivity of bubbles cannot be easily distinguished in such a representation. Such interconnectivity or channelling would be expected to have important implications on product stability, for example, the shrinkage behaviour upon storage. The degree of channelling may be measured via a pressure response technique using the apparatus in Figure 1. Only discrete gas cells expand under reduced pressure, and any gas in continuous channels will cause no volume response in the ice-cream. If the volume (or height) change measured is compared to the predicted value, the proportion of gas present as discrete gas cells can be calculated.

The results shown in Figure 9 compare the predicted height increase with that measured experimentally for a 200 mbar pressure reduction for ice-cream at a range of gas phase volumes. The 0.3% MGP samples expand almost as much as predicted, indicating that they contain predominantly discrete gas cells. Samples without emulsifier were generally found to be more channelled. The increased gas stability with MGP was attributed to the higher levels of de-emulsified fat which can reduce coalescence and channel formation. The combination of high de-emulsified fat level and gas phase volume shown to produce large gas cell sizes also gave a small expansion, indicating that the sample contained a large proportion of channels. Whilst this technique can give no indication of gas cell size, there was a high correlation between gas cell size and the degree of channelling, indicating that coalescence is an important process determining gas cell size.

The hardening temperature, and hence the hardening rate, also plays an

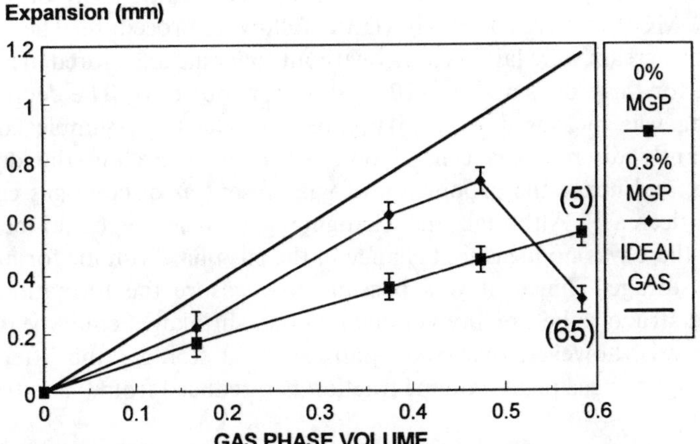

Figure 9 *Influence of gas phase volume on the pressure expansion of ice-cream at −15 °C with (■) 0% MGP and (◆) 0.3% MGP compared to (—) an ideal gas obeying Boyle's Law. Figures in parentheses indicate de-emulsified fat level (% w/w of total fat)*

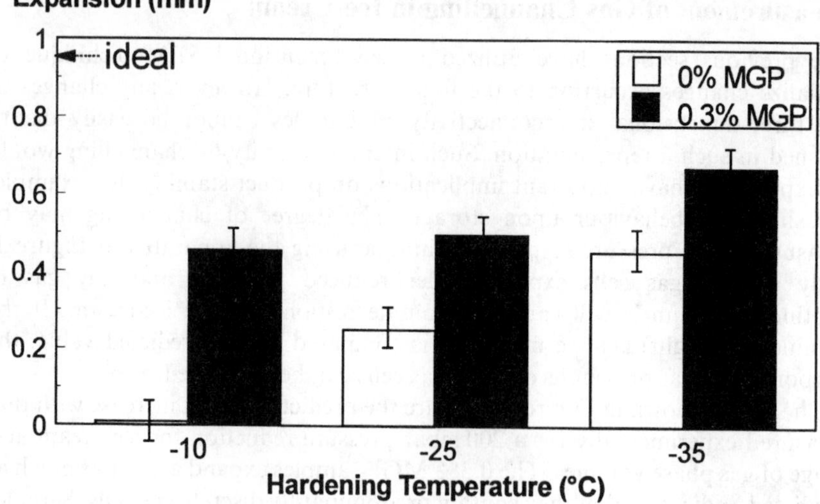

Figure 10 *Influence of hardening temperature on the pressure expansion of ice-cream. 'Ideal' refers to the expansion predicted for an ideal gas by Boyle's Law*

important role in stabilizing the gas phase (Figure 10). Less channelling was observed in samples which had been hardened at −35 °C compared to −10 °C. This result was as expected, since the longer the time taken to harden a sample, the more likely coalescence was to occur. As seen before, samples with emulsifier were less channelled due to the stabilizing effect of the de-emulsified fat.

Using the pressure expansion technique it was also possible to monitor changes in gas channelling as a function of the storage conditions. At low storage temperatures, the rate of channel formation is slow, but it may be accelerated by temperature increase or cycling. The temperature of ice-cream with 0.3% MGP was cycled weekly via the following procedure. The blocks of hardened ice-cream were laid on a tray without touching and stored in a −22 °C cold store for three days and a −10 °C store for four days. The degree of gas channelling was measured weekly (Figure 11). The fresh sample contained predominantly discrete gas cells. Upon temperature cycling the degree of expansion, and hence the proportion of gas present as discrete gas cells, was found to decrease. Although the changing gas structure could be clearly observed, there was no measured change in the gas phase volume for up to five weeks of storage. Hence, it was possible to measure the formation of the channelled structure before any volume loss (*i.e.* shrinkage) could be detected. Shrinkage was, however, observed upon extended storage, and after sixteen weeks storage the gas phase volume fraction had reduced from 0.5 to 0.47.

4 Conclusions

The gas phase in fresh ice-cream at −5 °C was found to be unstable and the small spherical bubbles coarsened rapidly at that temperature. Coalescence of

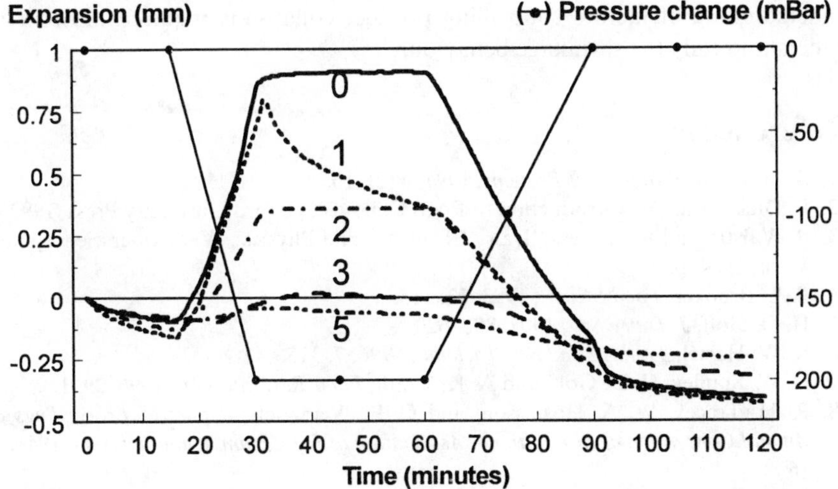

Figure 11 *Expansion response of temperature-cycled ice-cream (25 mm height) upon pressure reduction (•). Numbers indicate number of temperature cycles, each consisting of 3 days at −22 °C and 4 days at −10 °C*

bubbles under quiescent conditions is an important destabilizing mechanism for ice-cream. The tendency for film rupture increases when bubbles are close or touching for a long time or when they are pushed together by growing ice crystals. Key factors determining the extent of structural change are the product viscosity, the de-emulsified fat level, and the time taken to harden. Adsorption of de-emulsified fat particles at the air–water interface stabilized the foam by reducing drainage and hence coalescence. However, there is insufficient fat in ice-cream to form a network as seen in a whipped cream, and so rapid coarsening of the ice-cream foam is observed. Increasing the product viscosity by hardening reduces coalescence. However, in the time taken to harden, significant coarsening of the foam occurs, resulting in the formation of larger distorted and, in some cases, interlinked, gas cells. At intermediate levels of de-emulsified fat and fast hardening rates, there is a decrease in the rate of coalescence. High levels of de-emulsified fat promote gas coarsening. The extent of ice recrystallization during hardening is also dependent on the gas stability. The presence of a fine gas dispersion reduces the extent of ice–ice contact and hence ice accretion.

A pressure response technique has been developed to measure the degree of gas channelling within a product. The technique can monitor channel formation before any volume loss (*i.e.* shrinkage) can be detected. The rate of coalescence was found to be low at a storage temperature of −22 °C. However, upon temperature abuse, deterioration of the gas phase structure into interlinked gas channels during storage was observed. It is now possible when studying shrinkage to separate the two processes of coalescence and collapse. This is an important advance since valuable information can now be gained about the state of the gas phase before any measurement of volume loss. However, a better understanding of factors controlling product collapse is required in order to understand fully the shrinkage behaviour.

References

1. S. Turan, *IOP Industrial Proteins*, 1997, **4** (2), 10.
2. E. Dickinson, 'An Introduction to Food Colloids', Oxford University Press, 1992.
3. P. Walstra and R. Jenness, 'Dairy Chemistry and Physics', Wiley Interscience, New York, 1984, p. 279.
4. B. E. Brooker, *Food Struct.*, 1993, **12**, 115.
5. H. D. Goff, *J. Dairy Sci.*, 1997, **80**, 2620.
6. R. W. Hartel, *Trends Food Sci. Technol.*, 1996, **7**, 315.
7. D. W. Stanley, H. D. Goff, and A. K. Smith, *Food Res. Internat.*, 1996, **29**, 1.
8. R. H. Tracey, W. A. Hoskisson, and C. F. Weinreich, *Report of Proceedings of Annual Convention of International Association of Ice-Cream Manufacturers*, 1941, **2**, 16.
9. H. H. Sommer, 'Theory and Practice of Ice Cream Making', Sommer, Madison, WI, 1951.
10. U. K. Dubey and C. H. White, *J. Dairy Sci.*, 1997, **80**, 3439.
11. S. Turan, *Proc. Inter-Eis*, Zentralfachschule der Deutschen Süsswarenwirtschaft, Solingen, Germany, 1997, p. 199.
12. R. J. Ramsey, *Report of Proceedings of Annual Convention of International Association of Ice-Cream Manufacturers*, 1946, **2**, 58.
13. A. J. Lachmann and D. H. Volman, *Ice Cream Field*, 1950, **55** (4), 56.
14. H. D. Goff, W. Wiegersma, K. Meyer, and S. Crawford, *Canadian Dairy*, 1995, June/July, 12.
15. W. C. Cole, *Ice Cream Field*, 1939, October, 32.
16. A. Fink and J. E. Kessler, *Milchwissenschaft*, 1983, **38**, 330.
17. K. B. Caldwell, H. D. Goff, and D. W. Stanley, *Food Struct.*, 1992, **11**, 11.
18. B. Groh, T. Amend, and F. Dannenberg, *Proc. Inter-Eis*, Zentralfachschule der Deutschen Süsswarenwirtschaft, Germany, 1994, p. 24.
19. B. M. C. Pelan, K. M. Watts, I. J. Campbell, and A. Lips, 'Food Colloids: Proteins, Lipids and Polysaccharides', ed. E. Dickinson and B. Bergenståhl, Royal Society of Chemistry, Cambridge, 1997, p. 55.
20. E. Windhab and S. Bolliger, *European Dairy Magazine*, 1995, **1**, 28.

Effect of Pectinate on Properties of Oil-in-Water Emulsions Stabilized by α_{s1}-Casein and β-Casein

By Maria G. Semenova, Anna S. Antipova, Larisa E. Belyakova, Eric Dickinson,[1] Rupert Brown,[2] Edward G. Pelan,[2] and Ian T. Norton[2]

INSTITUTE OF BIOCHEMICAL PHYSICS, RUSSIAN ACADEMY OF SCIENCE, VAVILOV STR. 28, 117813 MOSCOW, RUSSIA
[1]PROCTER DEPARTMENT OF FOOD SCIENCE, UNIVERSITY OF LEEDS, LEEDS LS2 9JT, UK
[2]UNILEVER RESEARCH, COLWORTH HOUSE, SHARNBROOK, BEDFORD MK44 1LQ, UK

1 Introduction

During the manufacture of protein-stabilized oil-in-water food emulsions, polysaccharides are generally added to improve microstructure and enhance nutritional benefits. As substances with a rather high hydrophilicity, polysaccharides do not adsorb at the surface of emulsion droplets, but rather they influence the emulsion rheology and stability through interactions with protein or structure formation in the bulk aqueous phase.[1,2] Protein–polysaccharide interactions may influence the protein surface activity,[3] leading to changes both in the protein adsorbed layer at the oil–water interface,[4] and in the bulk emulsion,[5] and thus they can affect the emulsion stability.[6] It is known that low levels (below the critical (overlap) concentration) of polysaccharide can induce depletion flocculation of emulsion droplets. This flocculation is, in turn, closely related to enhanced emulsion creaming[7] in moderately dilute emulsions and a high low-stress bulk viscosity in concentrated emulsions.[8] Conversely, both flocculation and creaming can be strongly inhibited at polysaccharide concentrations well above the (overlap) concentration due to the immobilization of the dispersed particles in a gel-like polysaccharide network having a very high limiting low-stress shear viscosity.[6]

Under particular experimental conditions, where there is complex formation between protein and polysaccharide or surface activity of the polysaccharide, there may be co-adsorption of polysaccharide together with protein in the adsorbed layer around the emulsion droplets, and thus increased emulsion stabilization by enhanced steric and/or electrostatic mechanisms.[9] It may also be possible that the polysaccharide molecules can adsorb on several droplets simultaneously and cause flocculation by bridging.[10] Thus, to control emulsion structure and stability effectively in the presence of polysaccharide, we need to understand the role of protein–polysaccharide interactions, both during emulsion formation (adsorbed layer composition) and afterwards during storage (bulk composition).

This paper reports on the effect of adding high-methoxy pectinate to oil-in-water emulsions stabilized by one of the major individual components of bovine milk (α_{s1}-casein or β-casein). Our objective here is to relate the structure and stability properties of the α_{s1}-casein and β-casein containing emulsions to the observed nature and strength of the casein–pectinate interactions, both in the bulk aqueous phase and at the oil–water interface. The emulsion systems have been studied at two pH values (5.5 and 7.0) above the protein isoelectric point (pH 4.6) and over a range of ionic strength, *i.e.*, 0.01–0.20 M NaCl. The pectin–protein interactions have been measured in dilute solution using light scattering and phase diagram determinations under solution conditions identical to those of the emulsion aqueous phases. The emulsion composition and experimental procedures are identical to those in a previous study of protein-stabilized emulsions[11,12] in the absence of pectin.

2 Results and Discussion

Correlation of Emulsion Properties with Protein–Polysaccharide Interactions at pH 7.0

Emulsions prepared with both pure milk proteins at ionic strength 0.01 M were found to be of similar average droplet size ($d_{32} = 0.66 \pm 0.03\ \mu m$, $d_{43} = 0.98 \pm 0.05\ \mu m$). Figure 1 shows that increasing pectinate concentration in the aqueous phase during emulsification leads to a slight reduction in average droplet size for both (a) α_{s1}-casein and (b) β-casein concentrated emulsions (40 vol% oil, 2 wt% protein). The measured zeta potential ζ (not shown) in these emulsions shows no significant dependence on pectin presence, with the values of $|\zeta|$ for the highly charged α_{s1}-casein-coated droplets remaining substantially higher ($\zeta \approx -70$ mV $\pm$ 2 mV) than the values for droplets coated with β-casein ($\zeta \approx -40$ mV $\pm$ 2 mV). This result indicates no adsorption of the negatively charged pectinate on casein-coated emulsion droplets under these conditions (pH 7, I = 0.01 M).

To understand the molecular mechanism of the influence of pectinate on the emulsion properties, we have obtained the thermodynamic parameters of the casein–pectinate pair interactions. This thermodynamic parameter is a cross second virial coefficient (A_{23}) measuring the strength and nature of the interaction between pairs of different macromolecules in mixed biopolymer solutions.[13] It was determined from static light-scattering measurements on dilute mixed protein + polysaccharide solutions as a function of the solute

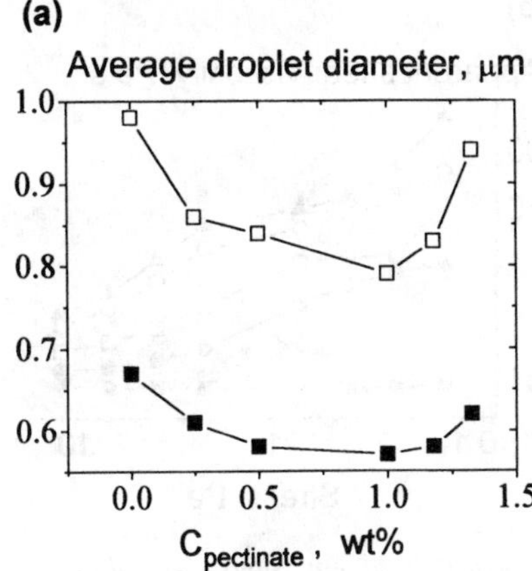

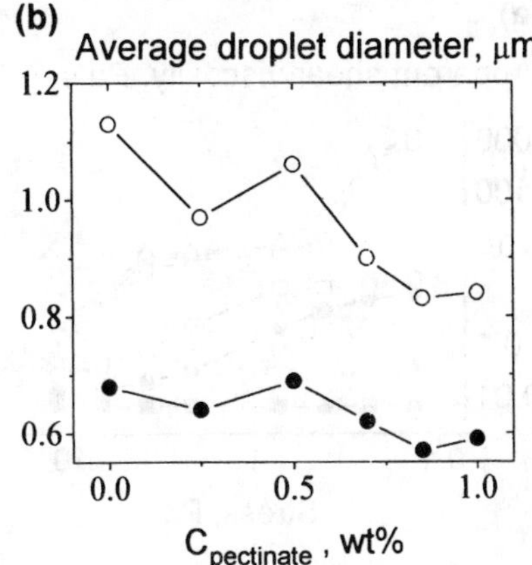

Figure 1 *Effect of polysaccharide on the initial droplet size of casein-stabilized emulsions (40 vol% oil, 2 wt% protein, pH 7.0, ionic strength 0.01 M) (a) Average droplet diameter is plotted against pectinate concentration* C_p *for* α_{s1}*-casein-stabilized emulsions:* ■, d_{32}; □, d_{43}. *(b) As (a), except for β-casein-stabilized emulsions:* ●, d_{32} ○, d_{43}

concentration and the scattering angle. We have found the value $A_{23} = 0.008 \times 10^{-4} \pm 0.001$ cm^3 mol g^{-2} for α_{s1}-casein–pectinate pair interactions and $A_{23} = 0.46 \times 10^{-4} \pm 0.05$ cm^3 mol g^{-2} for β-casein–pectinate pair interactions at pH 7 and ionic strength 0.01 M. These positive values of the cross second virial coefficient indicate net repulsive (thermodynamically unfavourable) interactions between both caseins and the pectinate in the aqueous phase. Thus, the measured decrease in average droplet diameter with increasing pectinate concentration can be attributed to a rise in protein loading at the droplet interface due to an increase in the protein thermodynamic activity in the bulk arising from the thermodynamically unfavourable interaction between protein and polysaccharide in accordance with the equation:[14]

$$\mu_2 = \mu_2^0 + RT(\ln(m_2/m^0) + A_{22}{*}m_2 + A_{23}{*}m_3). \quad (1)$$

Here μ_2 is the chemical potential of protein in the bulk, μ_i^0 and m_i are the standard chemical potential and concentration (molal scale) of the component i (i = 1–3; index 1 for water, 2 for protein, 3 for polysaccharide), respectively, A_{22}* is the second virial coefficient (molal scale) characterizing the pair interaction of the protein molecules in the bulk, A_{23}* is the cross second virial coefficient (molal scale) for the mixed polymer pair interaction, m^0 is the standard-state molality, R is the gas constant, and T is the absolute temperature.

Despite pectinate not binding to casein-coated droplets at pH 7, its presence in the emulsion can nevertheless influence the rheological and creaming behaviour. The concentrated emulsions stabilized with α_{s1}-casein or β-casein

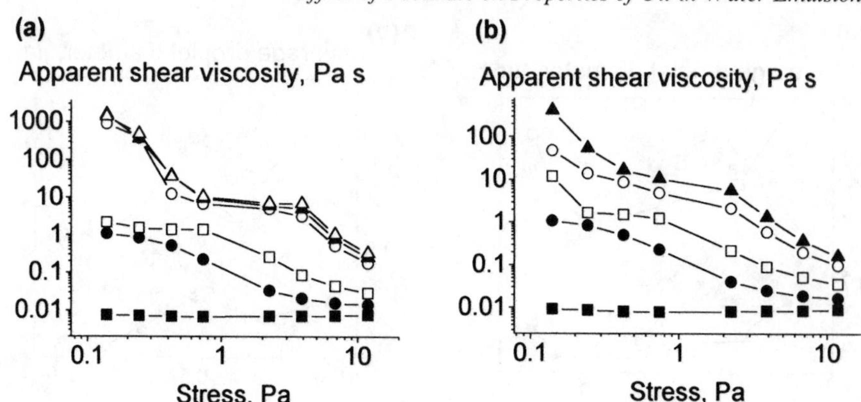

Figure 2 *Steady-state shear viscometry of casein-stabilized emulsions (40 vol% oil, 2 wt% protein, pH 7.0, ionic strength 0.01 M) containing various concentrations of pectinate. Apparent shear viscosity at 22 °C is plotted against stress: (a) α_{s1}-casein-stabilized emulsions; ■, no pectinate present; ●, 0.25 wt% pectinate; □, 0.5 wt% pectinate; ○, 1.0 wt%; ▲, 1.175 wt%; △, 1.325 wt% pectinate; (b) β-casein-stabilized emulsions; ■, no pectinate present; ●, 0.25 wt% pectinate; □, 0.5 wt% pectinate; ○, 0.85 wt% pectinate; ▲, 1.0 wt% pectinate*

(40 vol% oil, 2 wt% protein) with added pectinate are substantially shear-thinning, in contrast to the observed Newtonian behaviour of identical protein-stabilized emulsions made without pectinate. Figure 2 shows the dependence of the apparent viscosities on shear stress for emulsions stabilized by (a) α_{s1}-casein and (b) β-casein with different pectinate concentrations. The high values of the low-stress emulsion viscosity ($\sim 10^3$ Pa s) cannot be due to the viscous contribution from the polysaccharide in the continuous phase: a 2 wt% pectinate solution (pH 7, ionic strength 0.01 M) is a Newtonian liquid with a viscosity of 5×10^{-2} Pa s. We note that, whereas the original casein-stabilized emulsions are unflocculated, the ones with pectinate added are extensively flocculated. The highest increase in viscoelasticity was found for pectinate concentrations corresponding to phase separation (binodal position, see Figure 3) in the mixed aqueous solutions. The values of the complex shear modulus G^* at 1 Hz (Figure 4) show a substantial increase in G^* in the same concentration region for both casein + pectinate pairs. It is interesting to note that, in accordance with a more strongly thermodynamically unfavourable protein–polysaccharide interaction ($|A_{23}|$) and binodal position (see Figure 3), the values of G^* are several times higher for the β-casein-stabilized emulsion than for the α_{s1}-casein-stabilized ones at the same pectinate concentrations (>0.7 wt%). These observed rheological changes (*i.e.*, thickening) may suggest strengthening of the attraction forces between emulsion droplets driven by phase separation in the aqueous phase.

The effects of pure casein type and pH on the creaming stability of emulsions made with 11 vol% oil and 0.6 wt% α_{s1}-casein or β-casein + 0.28 wt% pectinate are shown in Figure 5. Whereas casein-stabilized emulsions give no discernible serum separation, those containing pectinate are very unstable with

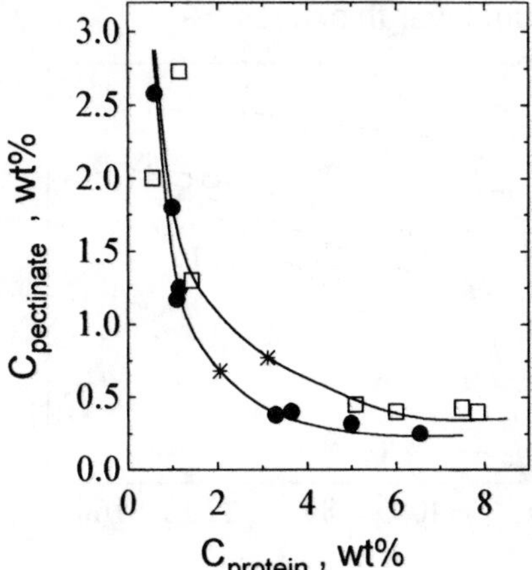

Figure 3 *Phase diagrams for β-casein + pectinate and α_{s1}-casein + pectinate mixed aqueous solutions at pH 7.0, I = 0.01 M: -●-, binodal line for β-casein + pectinate systems; -□-, binodal line for α_{s1}-casein + pectinate system; ★, critical points*

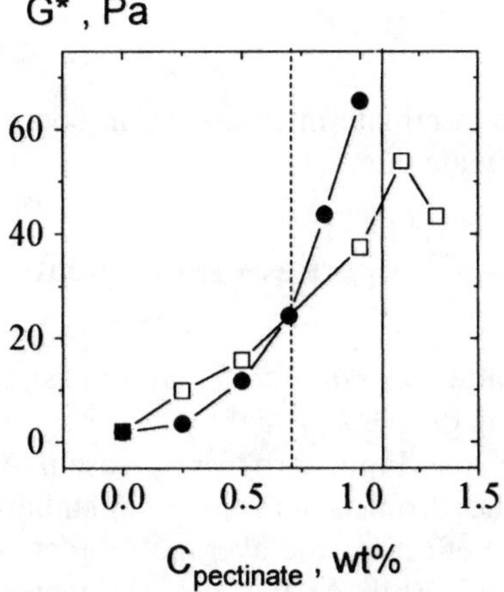

Figure 4 *Effect of polysaccharide on the viscoelasticity of casein-stabilized emulsions (40 vol% oil, 2 wt% protein, pH 7.0, ionic strength 0.01 M, 22 °C). Complex shear modulus* G* *at 1 Hz is plotted against pectinate concentration: ●, β-casein; □, α_{s1}-casein; lines indicating the pectinate concentration where phase separation occurs: ----, for β-casein + pectinate system; ——, for α_{s1}-casein + pectinate system*

respect to creaming. This behaviour, in combination with no observed incorporation of the hydrocolloid polymer into the casein adsorbed layer, is characteristic of a depletion flocculated system.[6,8] With addition of pectinate, the β-casein-stabilized emulsions cream more rapidly than their α_{s1}-casein-stabilized counterparts. This result is consistent with the greater strength of the

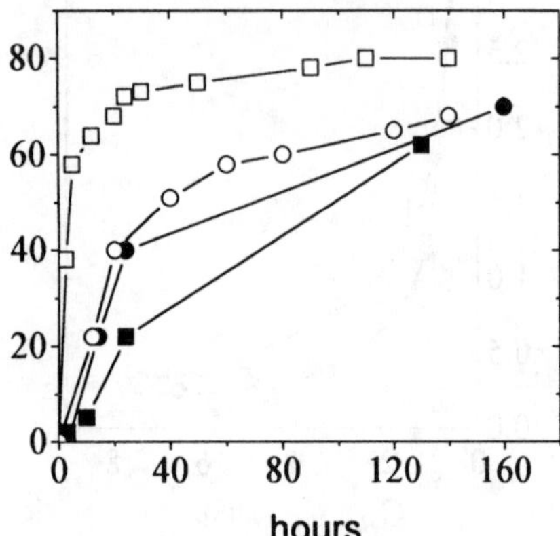

Figure 5 *Effect of polysaccharide on the creaming of casein-stabilized emulsions (11 vol% oil, 0.6 wt% protein, 0.28 wt% polysaccharide). The percentage serum separation is plotted against time of emulsion storage at 22 °C: ■, α_{s1}-casein, ●, β-casein, at pH 7.0, ionic strength 0.01 M; ○, β-casein, □, α_{s1}-casein, at pH 5.5, ionic strength of 0.01 M*

(more repulsive) β-casein–pectinate interactions, in comparison with the (less repulsive) α_{s1}-casein–pectinate interactions.

Correlation of Emulsion Properties with Protein–Polysaccharide Interactions at pH 5.5

Lowering the pH of the aqueous phase to 5.5 at ionic strength 0.01 M in both systems produces emulsion droplets of slightly larger average size as illustrated in Figure 6(a) for β-casein, and Figure 6(b) for α_{s1}-casein. Adding pectinate now leads to significantly smaller droplets in the β-casein-stabilized emulsion (Figure 6(a)) but it has negligible effect on the average droplet size in the α_{s1}-casein-stabilized emulsion [Figure 6(b)]. At this pH, the increasingly negative zeta potential values as a function of pectinate concentration in Figure 7 strongly suggest adsorption of pectinate onto the droplet surface. A determination of the free (unadsorbed) pectinate concentration in the serum phases of these emulsions also suggests that a small amount of pectinate (0.1 and 0.2 ± 0.05 mg m^{-2} for α_{s1}-casein and β-casein, respectively) is associated with the protein adsorbed layer at the surface of the droplets. Tensiometry data also confirm some adsorption of high-methoxy pectinate at the oil–water interface (the interfacial pressure is 11 mN m^{-1} after 10 hours, for a pectinate concentration of 0.1 wt%). Surface activity of high-methoxy pectinate has been reported previously.[15] Under these conditions (pH 5.5, ionic strength 0.01 M) the protein–polysaccharide pair interactions are repulsive (thermodynamically unfavourable) for the β-casein–pectinate pair (A_{23} = 2.6 ± 0.3 × 10^{-4} cm^3 mol g^{-2}),

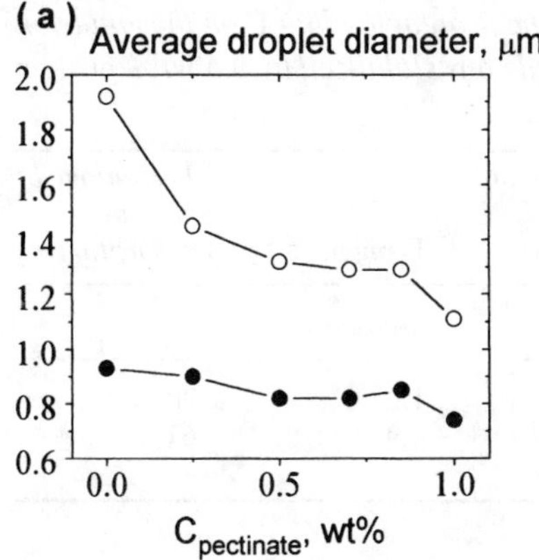

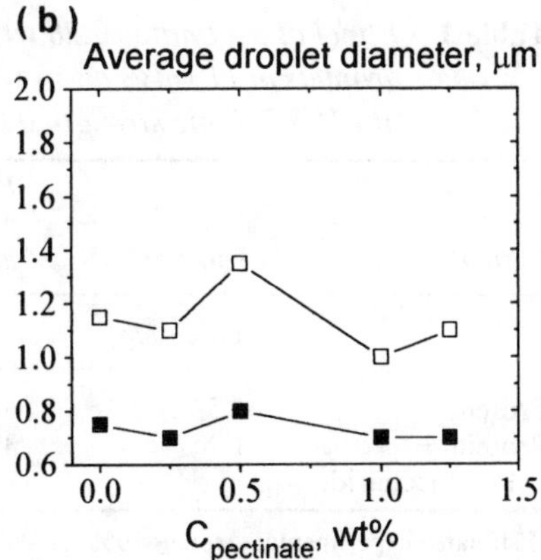

Figure 6 *Effect of polysaccharide on the initial droplet size of casein-stabilized emulsions (40 vol% oil, 2 wt% protein, pH 5.5, ionic strength 0.01 M). (a) Average droplet diameter is plotted against pectinate concentration C_p for β-casein-stabilized emulsions:* ●, d_{32}, ○, d_{43}. *(b) As (a) except for α_{s1}-casein-stabilized emulsions:* ■, d_{32}; □, d_{43}

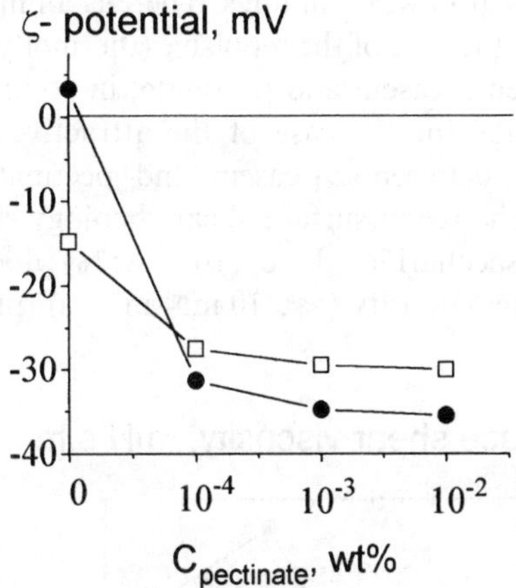

Figure 7 *Effect of polysaccharide on the zeta potential of emulsion droplets stabilized by caseins. Zeta potential ζ is plotted against pectinate concentration:* ●, *β-casein;* □, *α_{s1}-casein*

but are attractive (thermodynamically favourable) for the α_{s1}-casein–pectinate pair ($A_{23} = -7.6 \pm 0.8 \times 10^{-4}\,cm^3\,mol\,g^{-2}$).

So, on the basis of all the available information, we can propose that the decrease in emulsion droplet size with increasing pectinate concentration could be a result of two factors. Firstly, in the α_{s1}-casein-stabilized emulsions, it could be due to additional stabilization (electrostatic and/or steric) of the emulsion droplets caused by associative adsorption of the high-methoxy pectinate onto the interface. Secondly, for the case of β-casein-stabilized emulsions, it may be

Table 1 *Effect of pectinate (0.28 wt%) on the protein loading Γ on the emulsion droplets in 11 vol% oil-in-water emulsions stabilized by 0.6 wt% protein at pH 5.5, ionic strength 0.01 M*

System	Γ/*mg* m^{-2} [a]	*Percentage of protein adsorption*	Γ/*mg* m^{-2} [a]	*Percentage of protein adsorption*
	α_{s1}-*casein*		β-*casein*	
Protein	1.5	30	1.2	20
Protein + Polysaccharide	1.7	37	4.9	67

[a]Estimated experimental error ± 10%

due to increased protein loading due to the increase in protein thermodynamic activity in the aqueous medium driven by the repulsive (thermodynamically unfavourable) protein–polysaccharide interaction [see equation (1)].

The significance for emulsion formation of the character of the protein–polysaccharide interactions, namely either net repulsive or net attractive, is further illustrated by the following findings. There is an increase in the protein surface concentration in the case of the repulsive (thermodynamically unfavourable) interaction between β-casein and pectinate, in contrast to the effectively constant surface coverage for the case of the attractive (thermodynamically favourable) interaction between α_{s1}-casein and pectinate (see Table 1). A separate finding is the change in surface shear rheology shown in Figure 8. A bulk solution of polysaccharide alone (10^{-3} wt%) does not lead to any measurable surface shear viscosity ($\ll$ 10 mN m^{-1} s) (pH 5.5, ionic strength

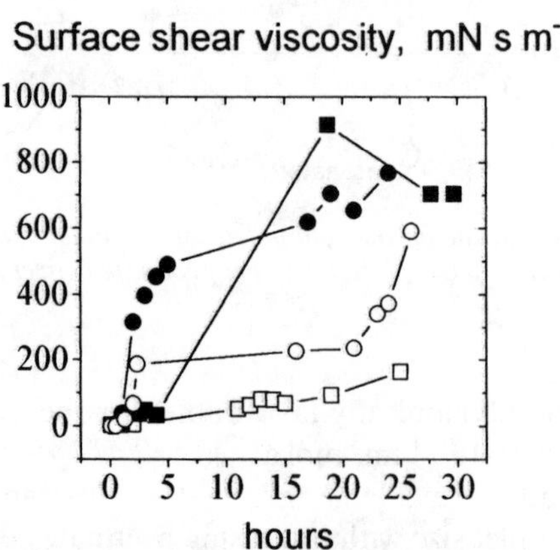

Figure 8 *Surface shear viscosity of adsorbed casein layers at the oil–water interface (10^{-3} wt% protein, pH 5.5, ionic strength 0.01 M, 25 °C) in the presence (filled symbols) and absence (open symbols) of pectin (10^{-3} wt%). Apparent surface viscosity η is plotted against time: □, ■, α_{s1}-casein; ○, ●, β-casein*

0.01 M). However, when mixed 1:1 with casein in dilute solution, the presence of pectin led to a substantial increase in the protein surface shear viscosity measured after several hours. This behaviour is characteristic of the strengthening of the intermolecular interactions in the protein adsorbed layer.[16] So, for the case of α_{s1}-casein, the interfacial shear viscosity after 24 hours increases fivefold (values of 820 ± 80 and 160 ± 20 mN m^{-1} s with and without pectin, respectively). For β-casein, it increases by only 20–30% (values of 750 ± 75 and 590 ± 60 mN m^{-1} s with and without pectinate, respectively). For α_{s1}-casein, these results may be caused by the rather strong attractive protein–polysaccharide interaction. In the case of the β-casein adsorbed layer, the results could be attributed to increased protein loading due to a rise in protein thermodynamic activity in the aqueous medium driven by repulsive (thermodynamically unfavourable) protein–polysaccharide interaction [see equation (1)]. These findings would seem to suggest that, under these conditions (pH 5.5, ionic strength 0.01 M), α_{s1}-casein is able to form a mechanically stronger structure within the adsorbed layer in the presence of pectinate. Again these shear viscosity results seem to fit in with the measured strength and nature of the protein–polysaccharide interaction.

The adsorption of pectinate to both kinds of protein-stabilized droplets and the specific character of the protein–polysaccharide interaction at pH 5.5 and ionic strength 0.01 M have a significant influence on the rheological and creaming behaviour of the emulsion systems. Figure 9 shows a strong increase in the limiting zero-shear-rate viscosity of the α_{s1}-casein-stabilized emulsions with increasing pectinate concentration. Under similar conditions a much smaller increase in the absolute value of the limiting zero-shear-rate viscosity (inset to Figure 9) was observed for β-casein-stabilized emulsions with added pectinate. The rather high values of the low-stress emulsion viscosity (10–10^3 Pa s) cannot be due to the viscous contribution from the polysaccharide in the continuous phase: a 2.5 wt% pectinate solution (pH 5.5, ionic strength 0.01 M) is a Newtonian liquid with a viscosity of 1.5×10^{-1} Pa s. Dynamic viscometry, Figure 10, shows an increase in the complex shear modulus G^* of the emulsions stabilized by both the milk protein fractions with increasing pectinate concentration. This increase is more pronounced for the α_s-casein system. The creaming results in Figure 5 are indicative of the appearance of an extensive serum layer in 11 vol% emulsions containing 0.28 wt% pectinate, similar to what was observed at pH 7.0 for the case of β-casein-stabilized emulsions and much more strongly for α_{s1}-casein-stabilized emulsions. We note that, whereas the original casein-stabilized emulsions are unflocculated, the ones with pectinate added are extensively flocculated.

The observed differences in both rheological and creaming behaviour of the casein-stabilized emulsions in the presence of pectinate could be attributed to the essential difference in both the character and the strength of the protein–polysaccharide interactions in these systems. Thus, for example, the more strongly attractive interaction between α_{s1}-casein and pectinate can induce stronger droplet flocculation, probably by a bridging mechanism. Conversely,

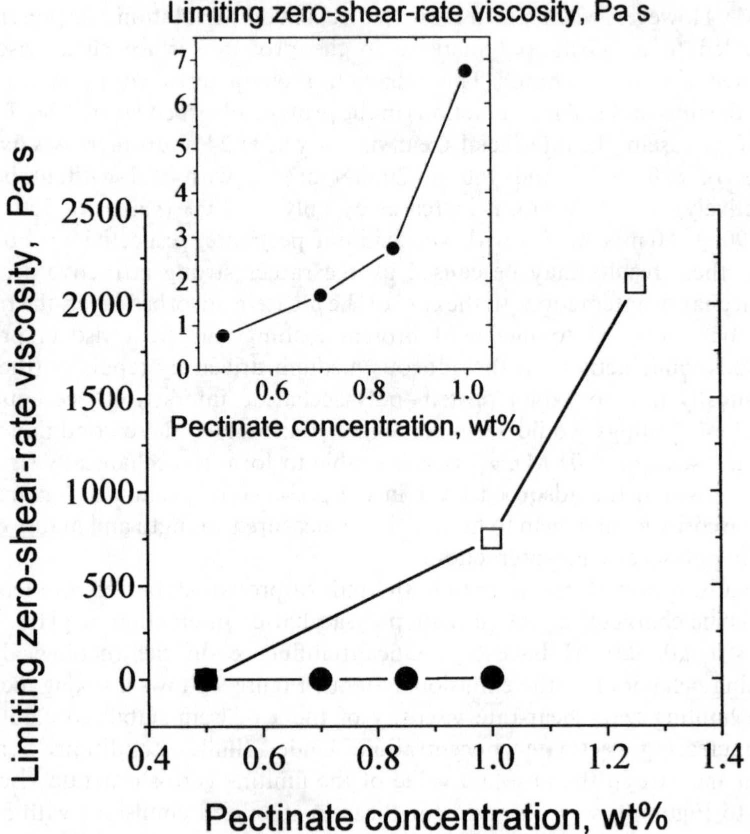

Figure 9 *Steady-state shear viscometry of casein-stabilized emulsions (40 vol% oil, 2 wt% protein, pH 5.5, ionic strength 0.01 M) containing various concentrations of pectin. Limiting zero-shear-rate viscosity at 22 °C is plotted against pectinate concentration: □, α_{s1}-casein; ●, β-casein*

the weaker repulsive interaction between β-casein and pectinate can cause weaker droplet flocculation, possibly by a depletion mechanism.

The effect of increasing ionic strength is to reduce the value of $|\zeta|$ (due to charge screening) for both types of casein-coated droplets as well as to change the character (both nature and strength) of the protein–pectinate interactions in the emulsion systems studied. For example, at pH 5.5 and ionic strength 0.15 M, the zeta potential values were, respectively, found to be -15 ± 2 mV and 0 ± 2 mV for β-casein emulsions with or without pectinate. Zeta potentials for α_{s1}-casein emulsions with or without pectinate at ionic strength 0.2 M were found to be $\zeta = -16$ mV $\pm$ 2 mV and $\zeta = -5$ mV $\pm$ 2 mV, respectively. These data clearly suggest pectinate adsorption onto the casein-coated droplets over the whole range of ionic strength studied at this pH.

Table 2 shows the change in the nature and strength of the protein–pectinate interactions with increase in ionic strength in aqueous solution. There is a direct

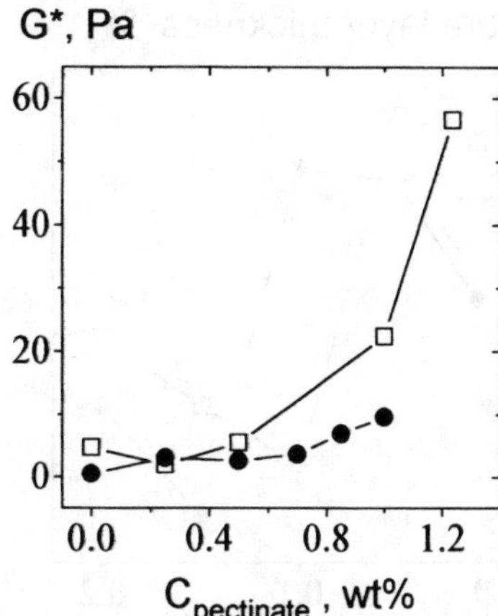

Figure 10 *Effect of polysaccharide on the viscoelasticity of casein-stabilized emulsions (40 vol% oil, 2 wt% protein, pH 5.5, ionic strength 0.01 M, 22 °C). Complex shear modulus* G* *at 1 Hz is plotted against pectinate concentration:* ●, *β-casein;* □, *α_{s1}-casein*

correlation between the ionic strength dependence of the character of the protein–pectinate interactions and the viscoelastic and the creaming behaviour of the corresponding emulsions. This is shown most particularly in the creaming behaviour of the 11 vol% emulsions (0.6 wt% protein, 0.28 wt% polysaccharide). Figure 11 indicates that the change in the creaming rate is in accordance with the expected alteration of the character of the protein–pectinate interactions. Comparison of the creaming behaviour of the emulsions stabilized by the protein + polysaccharide mixtures with that for those stabilized by α_{s1}-casein alone (see Figure 12) or β-casein alone (no serum layer formation) provides additional evidence of the importance of the influence of the nature and strength of the protein–polysaccharide interactions on the emulsion stability properties.

Table 2 *Cross second virial coefficient A_{23} for pair interactions α_{s1}-casein + pectinate or β-casein + pectinate at pH 5.5, 22 °C, as a function of ionic strength*

Ionic strength/M	$A_{23} \times 10^4$/*cm*^{3} *mol g*$^{-2}$ α_{s1}-*casein + pectinate*	$A_{23} \times 10^4$/*cm*^{3} *mol g*$^{-2}$ β-*casein + pectinate*
0.01	−7.6	2.6
0.05	−9.9	4.6
0.10	4.8	0.3
0.15	0.8	−20.0
0.20	2.2	

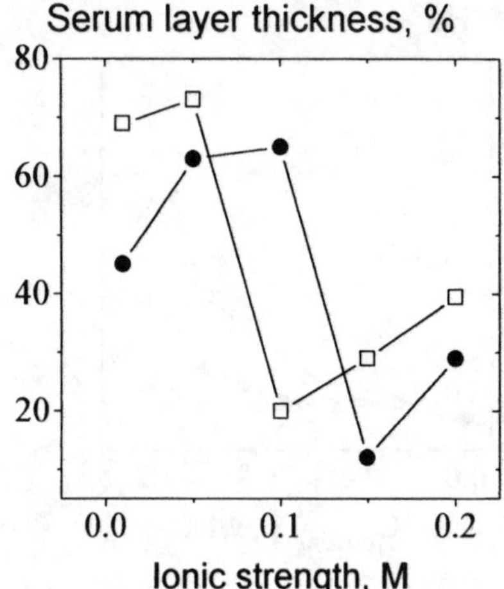

Figure 11 *Effect of added salt on the creaming of casein-stabilized emulsions (11 vol% oil, 0.6 wt% protein, 0.28 wt% polysaccharide, pH 5.5). The percentage serum separation is plotted against time of emulsion storage at 22 °C:* □, α_{s1}*-casein,* ●, *β-casein*

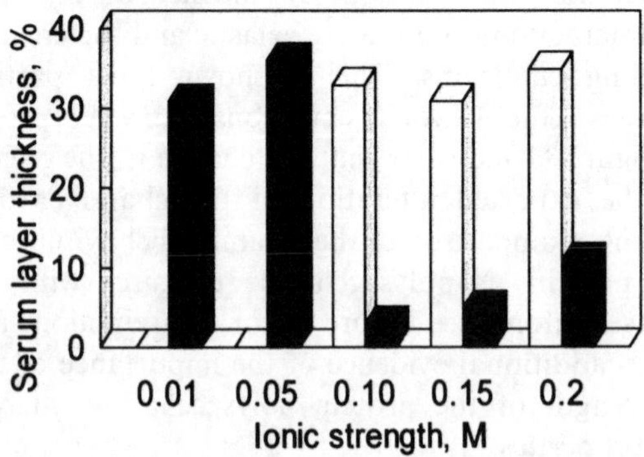

Figure 12 *Effect of added salt on the creaming of* α_{s1}*-casein-stabilized emulsions (11 vol% oil, 0.6 wt% protein, pH 5.5, 22 °C). The percentage serum separation is plotted against ionic strength in the presence (filled symbols) and absence (open symbols) of 0.28 wt% polysaccharide*

3 Conclusions

The results indicate that the emulsion rheology and stability are consistent with the thermodynamics of the casein–pectinate interactions over a wide polymer concentration range. Thus, the relation of emulsion properties to the nature and strength of the interaction between different biopolymer emulsion components can really be a useful approach to getting a better understanding of the mechanisms of emulsion formation and stabilization at the molecular level.

Acknowledgement

M.S, A.A., and L.B. are grateful to Unilever Research (Colworth Laboratory) for the financial support of this research.

References

1. E. Dickinson, in 'Biolpolymer Mixtures', ed. S. E. Harding, S. E. Hill, and J. R. Mitchell, Nottingham University Press, Nottingham, 1995, p. 349.
2. E. Dickinson and D. J. McClements, 'Advances in Food Colloids', Blackie, Glasgow, 1995, chap. 3.
3. G. E. Pavlovskaya, M. G. Semenova, E. N. Thzapkina, and V. B. Tolstoguzov, *Food Hydrocolloids*, 1993, **7**, 1.
4. E. N. Tsapkina, M. G. Semenova, G. E. Pavlovskaya, A. L. Leontiev, and V. B. Tolstoguzov, *Food Hydrocolloids*, 1992, **6**, 237.
5. K. Pawlowsky and E. Dickinson, in 'Food Colloids: Proteins, Lipids and Polysaccharides', ed. E. Dickinson and B. Bergenståhl, Royal Society of Chemistry, Cambridge, 1997, p. 258.
6. E. Dickinson, *ACS Symp. Ser.*, 1996, **650**, 197.
7. L. Heeney, M.Sc. Thesis, University of Leeds, 1995.
8. E. Dickinson, J. Ma, and M. J. W. Povey, *Food Hydrocolloids*, 1994, **8**, 481.
9. Y. K. Leong, P. J. Scales, T. W. Healy, and D. V. Boger, *Colloids Surf. A*, 1995, **95**, 43.
10. E. Dickinson and V. B. Galazka, *Food Hydrocolloids*, 1991, **5**, 281.
11. E. Dickinson, M. G. Semenova, and A. S. Antipova, *Food Hydrocolloids*, 1998, **12**, 227.
12. E. Dickinson, M. G. Semenova, A. S. Antipova, and E. Pelan, *Food Hydrocolloids*, 1998, **12**, 425.
13. P. Kratochvill and L. O. Sudelof, *Acta Pharm. Suec.*, 1986, **23**, 31.
14. E. Edmond and A. G. Ogston, *Biochem J.*, 1968, **109**, 569.
15. G. Berth, H. Anger, I. G. Plashina, E. E. Braudo, and V. B. Tolstoguzov, *Carbohydr. Polym.*, 1982, **2**, 1.
16. E. Dickinson and S. Euston, in 'Food Polymers, Gels and Colloids', ed. E. Dickinson, Royal Society of Chemistry, Cambridge, 1991, p. 132.

Brownian Dynamics Simulation of Colloidal Aggregation and Gelation

By Michel Mellema, Joost H. J. van Opheusden,[1] and Ton van Vliet

DEPARTMENT OF FOOD TECHNOLOGY AND NUTRITIONAL SCIENCES, FOOD PHYSICS GROUP, WAGENINGEN AGRICULTURAL UNIVERSITY, BOMENWEG 2, 6703 HD WAGENINGEN, THE NETHERLANDS
[1]DEPARTMENT OF AGRICULTURAL, ENVIRONMENTAL AND SYSTEMS TECHNOLOGY, APPLIED PHYSICS GROUP, WAGENINGEN AGRICULTURAL UNIVERSITY, BOMENWEG 4, 6703 HD WAGENINGEN, THE NETHERLANDS

1 Introduction

In a destabilized dispersion of colloidal or Brownian particles, clusters may grow such that they finally occupy the whole system. Usually such a system is soft and elastic, in which case it is called a particle gel. In recent years, Brownian dynamics (BD) simulation models have been tested to study particle gel formation.[1–5] For the present study, the simulation parameters were chosen in such a way as to provide particle aggregation that gives gelation in a manner akin to the formation of, *e.g.*, casein particle gels. An understanding of the relationship between the particle interactions and the structure of the final gel matrix is essential in order to be able to produce, for instance, casein particle gels (like yoghurt and curd) with controlled physical properties.[6,7]

Casein particle gels are formed by aggregation of the protein particles present in milk; these are the so-called casein micelles.[8,9] Native casein micelles have 'hairs' on the surface that mainly consist of κ-casein and provide steric stabilization. An early step in the production of cheese involves adding a rennet enzyme that splits off the protruding hydrophilic part of κ-casein, while in the production of yoghurt lowering of the pH is involved. In both these ways, the κ-casein layer thickness is decreased and colloidal instability is induced. There are indications of involvement of α_{s1}-casein and calcium bridging in the attraction between casein micelles.[10] The experimental bonding rates are surprisingly low[11] suggesting that remains of the layer or electrostatic forces may act as a repulsive barrier at intermediate distances.

In view of uncertainties about the details of the interactions, the potential in our simulations is not intended to represent the real system perfectly. Only the minimal features necessary for the formation of the type of structures observed in casein particle gels were incorporated. Confocal microscopy studies on casein gels reveals a cloudy matrix of protein particle strings and clusters. Typically, the aggregates formed have a relatively high fractal dimensionality D_f of about 2.1–2.3.[12,13] During (long-term) ageing of casein gels the value of D_f may increase, and it can be accompanied by expulsion of liquid (*i.e. syneresis*) and a decrease in the modulus of elasticity. The rate at which these processes occur may vary greatly. However, they are always present because a casein gel is not a stable system. It will eventually phase separate; if not occurring immediately due to the presence of long-range attractive forces during aggregation,[2] phase separation will occur at a later stage due to rearrangements.[14] In our simulations, we prevent quick local compaction by incorporating (a) slow aggregation and (b) irreversible, flexible bonding. Slow aggregation is obtained by incorporating a repulsive barrier at intermediate distances.

Low bonding rates due to repulsive interactions usually lead to the formation of structures with relatively high D_f,[1–4] which is what we also observed for casein gels. Another factor is that in simulations it is often found that D_f depends strongly on the volume fraction φ. If the bonding rates are much slower than the diffusion rate, this effect diminishes. Probably, the formation of depletion zones around the clusters is reduced.

An irreversible bonding scheme represents a strong attraction upon contact. This allows for a large influence of the particle kinetics on the structure. Reversible bonding allows rearrangements leading to a situation of phase separation. In our model, the bonds between the particles are flexible,[1] which allows for some rearrangements in a more restricted way (no phase separation).

The net rate of particle aggregation has a large effect on the final gel properties.[1,12,15,16] This study goes beyond that in showing that (a) the aggregation rate, (b) the rearrangement rate and (c) the fractal gel properties can all be predicted from the interaction details using the theoretical framework of Smoluchowski[17] and Fuchs.[18]

2 Brownian Dynamics Model

By definition the solvent in BD is considered as a continuum. In addition, the temperature T is constant since the average resultant force on each particle is always kept zero (all the energy is assumed to be completely dissipated by friction). In our BD numerical simulations we keep track of the position of 1000 hard-core spherical particles of unit radius in three dimensions. All parameters corresponding to sizes or distances are normalized to the radius of one spherical particle, and all parameters corresponding to energies are normalized to units of kT. The cubic box has edges set equal to 50 corresponding to a volume fraction φ of *ca.* 3.4% (the effect of varying φ will be discussed in a later paper[19]). Features explicitly included in our model are flexible bonds, rotational diffusion, and periodic boundary conditions.

The simulation model is based on Newton's equation of motion for a macroscopic particle with a fluctuating random force added to account for the thermal collisions of the solvent molecules with the particle. Such a stochastic equation of motion is called a Langevin equation.[1–3,20]

$$F_{\mathrm{res}} = m\frac{\mathrm{d}^2}{\mathrm{d}t^2}r_i = \sum_j F_{ij}(r_{ij}) + R_i + H_i. \quad (1)$$

Here F_{ij} is the net force of interaction between a pair of particles i and j, R_i is the random (Brownian) force with a Gaussian distribution, H_i is the Stokes friction force acting on particle i, r_i is the position of particle i, and r_{ij} is the relative position of particle i with respect to particle j ($= (r_i(t) - r_j(t))$, where t is the time).

The friction force H_i is proportional to the velocity, *i.e.*

$$H_i = \frac{\mathrm{d}r_i}{\mathrm{d}t}\gamma, \quad (2)$$

where $\gamma = 6\pi\eta_0$ is the Stokes friction coefficient and η_0 is the solvent shear viscosity. Many-body hydrodynamic interactions are neglected.

Equation (1) is solved numerically to extract the movement of the particles, using a constant timestep Δt. We choose this Brownian timestep to be much larger than the relaxation time of the fluid, but also small enough to ensure that the interaction forces do not change significantly during one timestep. The former assumption allows us to neglect the second-order term in equation (1). Using the Euler forward method[20] to solve the remaining first-order differential equation, we arrive at:

$$\Delta r_i(t + \Delta t) = \frac{\Delta t}{\gamma}\left(\sum_j F_{ij}(t) + R_i(t)\right). \quad (3)$$

The effect of the force R_i is a translational displacement which is tuned to obey Einstein's law for an isolated particle. For instance in the x-direction this gives us

$$\Delta x_i^R(t + \Delta t) = N_s\sqrt{6D_c\Delta t}, \quad (4)$$

where $D_c = \gamma^{-1}$ is the diffusion coefficient, which is normalized to unity. The parameter N_s is the random seed number, and has a uniform distribution on the interval $[-1,1]$. The root-mean-square average is given by:

$$\sigma^2 = \frac{1}{2}\int_{-1}^{1} x^2\mathrm{d}x = \frac{1}{3}. \quad (5)$$

The quantity $\sqrt{(2D_c\Delta t)}$ is the dimensionless r.m.s. displacement in the absence of interactions. If the number of particles is N, in one timestep $3N$ numbers are drawn to ensure uncorrelated distributions for $F_{ij} = 0$. The number of steps (Δt or $\sqrt{(2D_c\Delta t)}$) by which the simulation has passed, is given by $N_{\Delta t}$. The parameter $N_{\Delta t}$ is a direct measure of time.

In addition to translational diffusion, the individual particles also undergo rotational diffusion according to the corresponding Langevin equation for the rotational degrees of freedom. The rotational motion is governed by a diffusion coefficient D_R (particle radius = 1, $kT = 1$):

$$D_R = \frac{3D_c}{4}. \tag{6}$$

What is crucial for particle aggregation models to be able to imitate casein aggregation is the opportunity for the clusters to deform and/or rearrange. We incorporate this feature by introducing a certain bond flexibility. Once a bond is formed it does not break. The points at which the bonds are attached to the particle surfaces are fixed, and the angles between two (or more) bond attachment points on the same particle do not change. Through rotational and translational motion of the particles, the aggregated clusters can change conformation. The relative particle motion is restricted such that the surface-to-surface bond length does not exceed the maximum specified bond length $D_{\text{string}} = 0.1$.

3 Particle Interactions

In the literature,[21] two limiting starting points for the modelling of particle aggregation are described: diffusion limited aggregation (DLA) and reaction limited aggregation (RLA). Pure DLA is the process whereby each particle encounter leads to bonding. For RLA an additional repulsive barrier has to be crossed prior to aggregation. This causes a delay in the aggregation process and the structures formed are usually more compact. For reaction limited cluster–cluster aggregation (RLCCA)[15,22] inter-clusters bonds are allowed to be formed, which leads to more compact structures that have distinct fractal features. Our simulations are close to RLCCA, due to the presence of intermediate/long-range repulsions (no barrier to *particle contact*) and irreversible bonding.

The particle pair interactions (in the algorithm represented by term F_{ij}) are described using a potential $u(r)$. The force F is constant at a centre-to-centre distance $r = [2.1, D]$:

$$u(r) = \begin{cases} 0; & r \geq D \\ F(r - D); & D_{\text{bond}} < r < D \\ 0; & 2 < r < D_{\text{bond}} \end{cases} \tag{7}$$

where D_{bond} is the bonding distance (always set to 2.1) and D is the maximum interaction distance (ranging from 2.1 to 3.3). We also introduce the parameter H, which equals the maximum value of the potential $u(r)$ at a distance $r = D_{\text{bond}}$. The parameter H ranges in value from 0 to ~9. The interactions are repulsive if F is negative and u and H are positive. For all the simulations we fix $\sqrt{(2D_c\Delta t)}$ at 0.003, which is small enough for the steepest potentials.

In RLA a parameter P_{bond} is introduced to represent the probability that a bond is formed upon contact of the hard spheres. Only for very small values of P_{bond} does one have real RLA. The parameter P_{bond} is always set to unity in our simulations. Instead of slowing down aggregation with a low bonding probability, we use a repulsive force of finite extent. This has a similar, but subtly different, effect. In that sense our simulations are of RLA type, albeit with still a rather high effective bonding probability. The interactions are schematically presented in Figure 1a. We show hard sphere particles diffusing translationally and rotationally. Interesting to notice is the difference between the effects of P_{bond} and u at the level of particle encounters, which is shown schematically in Figure 1b. If one particle encounters a doublet, the effects of u will add up. In contrast, the effect of P_{bond} is separate for both particles in the doublet. For the same effective 'bonding possibility', a high value of u is expected to lead to more stringy structures than a low value of P_{bond}.

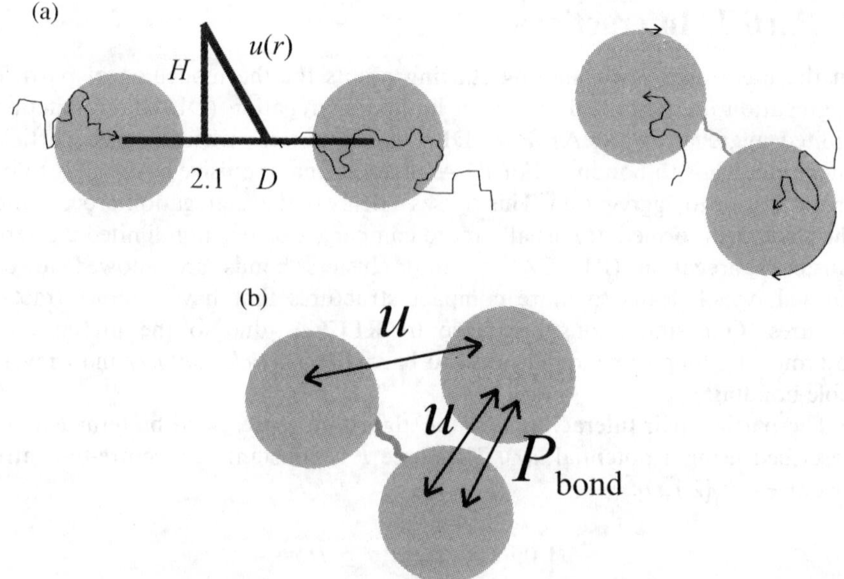

Figure 1 *(a) Schematic representation of the close encounter (left) and the resulting bonding (right) of two colloidal particles. (b) Representation of the close encounter of a doublet and a single particle. Arrows illustrate the difference in the effect of* P_{bond} *and* $u(r)$

4 Fuchs' Theory

We introduce Smoluchowski's classical concept[17] of the stability ratio, W_s, which measures the effectiveness of the potential barrier in preventing colloidal particles from aggregating:

$$W_s = \frac{\textit{average time for bonding; with repulsion}}{\textit{average time for bonding; with } u(r) = 0}. \quad (8)$$

In an actual simulation 'experiment', the stability ratio scales like

$$W_s = \frac{\left(\frac{\delta N_{\text{bonds}}}{\delta N_{\Delta t}}\right)_{u>0}}{\left(\frac{\delta N_{\text{bonds}}}{\delta N_{\Delta t}}\right)_{u=0}}, \quad (9)$$

where $\delta N_{\Delta t}$ is the 'time' elapsed since the initial random configuration. The reciprocal of W_s ($= 1/W_s$) is a measure of the stickiness of the particles. For our simulations, the theory only applies to the first few aggregation steps, *i.e.* a few times the time required for the number of aggregates, N_{agg}, to be reduced to half of the initial value. In this approach, a single particle is considered an aggregate of radius 1. At times beyond this 'Smoluchowski-regime' the shapes and dimensions of the aggregates formed influence the rate of ongoing aggregation. Actually, the Smoluchowski theory deals more precisely with coalescence rather than with aggregation. This is the reason for the inapplicability of the theory to the later stages of aggregation.

Fuchs[18] modified Smoluchowski's theory for the case of diffusion in a force field. The Fuchs ratio W_{F} for a single pair of particles having an interaction potential $u(r)$ is given by:

$$W_{\text{F}} = 2a \int_{2a}^{\infty} \exp\left(\frac{u(r)}{kT}\right) \frac{\mathrm{d}r}{r^2}. \quad (10)$$

Rewriting equation (10) incorporating the potentials used here leads to:

$$W_{\text{F}} = 2 \int_{D_{\text{bond}}}^{D} \exp\left(H\left[\frac{D-r}{D-D_{\text{bond}}}\right]\right) \frac{\mathrm{d}r}{r^2} + \frac{2}{D} + 1 - \frac{2}{D_{\text{bond}}}. \quad (11)$$

We can now approximate the integral in equation (11) numerically for any given values of H and D. We use this 'theoretical' W_{F} to compare with the 'experimental' W_s.

The Fuchs approach describes the aggregation ratio for hard-core spheres in suspension. It is specifically only applicable to the early aggregation stage. There may be other more appropriate parameters describing the rate at different later stages during the aggregation process. For comparison, we include the Boltzmann factor e^H and the quantity $(D - 2.1) \times H$ (the surface area of the potential) as measures of the aggregation retardation in the discussion of our results.

5 Power-law Behaviour

A particle aggregate can be considered fractal if the geometry looks the same over a reasonably large range of magnifications. This is called 'scale-invariance'. For an ideal fractal aggregate, the following scaling relation can be written[2,3]

$$\varphi \propto r^{D_f - 3}. \tag{12}$$

Here φ is the density or volume fraction inside the aggregate of size r, and D_f is the fractal dimensionality. For a particle aggregate, D_f is usually in the interval 1.7–2.4. A 3D homogeneous object would give $D_f = 3$, which means that φ is independent of the cluster size.

It should be noted that the fractal scaling relations apply not only to *deterministic* fractals, *i.e.* fractals which can be described using a constant algorithm. Such very regular structures are not found in real life. Particle gels can have *stochastic* fractal features, which are a product of the random Brownian mechanism of the aggregation.

The density distribution within an object, or an image of the object, can be probed by evaluating a pair correlation function $g_2(r)$ or density correlation function $C(r)$. Usually, the latter is used to analyse images where the particles are hard to recognize (*e.g.*, in confocal images). In computer simulations all the particle positions are known precisely, and so we can define $g_2(r)$. However, in estimating an effective D_f from the simulations, it is more convenient to smooth out the short-range oscillations in $g_2(r)$ by integration, leading to

$$n \propto \tilde{n}\left(\frac{r}{2a}\right)^{D_f} \qquad (2a \le r \le \xi) \tag{13}$$

where n is the average number of particles within range r of another particle, and ξ is the correlation length. The lower cut-off length is chosen here to be $2a$, in order for the prefactor $\tilde{n}$ to become equal to the coordination number. Note that the scaling relation in equation (13) is a fit to a linear regime in a plot of $\log(n)$ *versus* $\log(r)$.

It is important to realize that a gel matrix is only fractal over a limited regime of length scales.[23] The parameter ξ is the upper cut-off length of the fractal regime (or, alternatively, the lower cut-off length of the homogeneous regime). In a dispersion of separate particle clusters, the value of ξ corresponds to the average cluster separation. In a gel matrix ξ corresponds to the average radius of the gel pores. For a model situation, at length scales above ξ, we have $D_f = 3$. There is also a lower cut-off length r_0. The parameter r_0 is expected to be a little larger than $2a$. The values of both r_0 and $\tilde{n}$ give information on the degree of compactness of the gel matrix at small length scales. (All the parameters mentioned above are defined later in Figure 6.)

The uses of the fractal approximation (or power-law approach) to casein micelle aggregation and particle gels were introduced by, respectively, Horne[24] and Bremer *et al.*[12,25] The added value is that the density scaling in the fractal regime determines interesting practical parameters, like the elasticity modulus, the fracture strain and the permeability coefficient.[12,26,27] It was first proposed by Bos and van Opheusden[3] that not only D_f but also $\tilde{n}$ and r_0, are important parameters in experimental and simulational studies concerning the fractal structure of aggregating systems. Probably ξ is also equally important, for instance in determining large-deformation rheology[26] or the permeability properties of particle gels. We note that not all these fractal parameters are fully independent of each other, and additional information may still be needed. For instance, we need to quantify the curvature of particle strings, which cannot be detected using a fractal approach.

6 Results and Discussion

We have performed simulations for various values of the potential height H and distance D. We start with 1000 particles randomly distributed within the (periodic) cubic simulation box. The number of steps $N_{\Delta t}$ in the Smoluchowski regime was on average 1000. The $N_{\Delta t}$ for gelation ranged from 10^4 ($W_s \sim 1$) up to $> 5 \times 10^5$ ($W_s \sim 20$). Gelation, or the 'gel point', was defined arbitrarily as the moment when the number of aggregates N_{agg} had decreased to 10. We experienced that at this point usually more than 90% of the particles are in one or two large aggregates. Moreover, it required an excessive amount of time to reach $N_{agg} = 1$.

In Table 1 we give some examples of sets of potential parameters used in the simulations. It demonstrates that quantities like $(D - 2.1) \times H$, W_F, and e^H are not equivalent. For instance, we see that two simulations can be done at the same value of the ratio $(D - 2.1) \times H$, while $1/W_F$ is different and *vice versa*. Also shown are the ranges over which these quantities have been studied. The actual number of simulation runs was much more than the four examples given in Table 1.

Table 1 *Examples of sets of potential parameters used in the simulations*

Example	$D-2$	H	W_F	$(D - 2.1) \times H$	e^H
Range studied for					
kinetics	0–1.3	0–9	1–20	0–7	1–8000
power-law behaviour	0–1	0–5	1–6	0–5	1–150
simulation (a)	0.6	4.4	4.5	2.2	82
simulation (b)	0.5	5.6	9.1	2.2	270
simulation (c)	0.3	5.6	5	1.1	270
simulation (d)	0.5	4.8	5	1.9	122

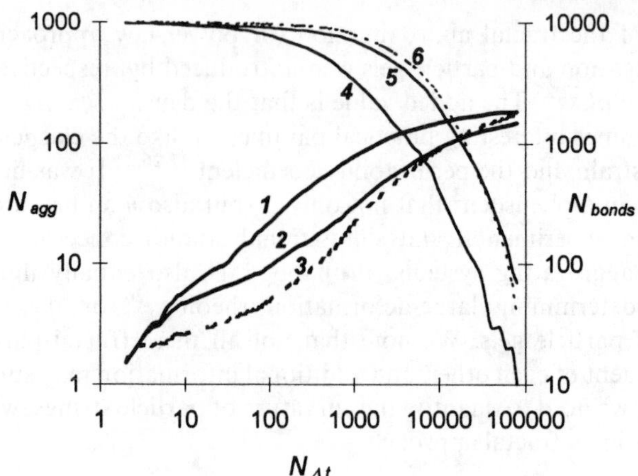

Figure 2 *The quantities* N_{agg} *and* N_{bonds} *as a function of* $N_{\Delta t}$ *for three simulated systems: (1)* N_{agg}; $W_F = 1$; *(2)* N_{agg}; $W_F = 2.4$, $D = 3$, $H = 2.7$; *(3)* N_{agg}; $W_F = 2.4$, $D = 2.4$, $H = 3.9$; *(4)* N_{bonds}; $W_F = 1$; *(5)* N_{bonds}; $W_F = 2.4$, $D = 3$, $H = 2.7$; *(6)* N_{bonds}; $W_F = 2.4$, $D = 2.4$, $H = 3.9$

Aggregation Kinetics

In any simulation the number of bonds (N_{bonds}) increases and the number of aggregates (N_{agg}) decreases with $N_{\Delta t}$. For several simulations this is shown in Figure 2. It is clear that the incorporation of large repulsive forces reduces the initial rate of bonding. In the very early stages, the bonding rate is influenced by the specific initial random configuration. After $N_{\Delta t} = 1000$ the shape and dimensions of the aggregates formed start to influence the bonding rate. As the simulation proceeds further, the value of N_{bonds} increases much more than the value of N_{agg} decreases. This can be taken as an indication of the presence of rearrangements.

From the increase in N_{bonds} with $N_{\Delta t}$ (up to $N_{\Delta t} = 1000$), we can derive a value for the Smoluchowski stability ratio W_{s} using equation (9). Each value of W_{s} is derived from one simulation with $u \neq 0$ and one using the same set of parameters with $u = 0$ (for normalization). We can approximate W_{F} by incorporating u in equation (11). We have performed simulations at four chosen values of W_{F}, but for different slopes of the potential. A weak, long-range potential can have the same value for W_{F}, as a strong, short-range potential. Hence $(D - 2)/H$ may serve to discriminate between different barriers with equal effective repulsion. The reciprocal of both these values is plotted in Figure 3, along with the ratio $(D - 2)/H$. From this graph we see that W_{F} predicts very well the aggregation rate as given by W_{s}. For $D < 2.2$, W_{s} is found to be usually smaller than expected based on the value of W_{F}. This is due to the short time spent by the particle in the interaction range. This simulation artefact could be corrected for by using smaller timesteps, but this would increase the total simulation time significantly.

We have also studied the influence of the interaction quantities e^H and

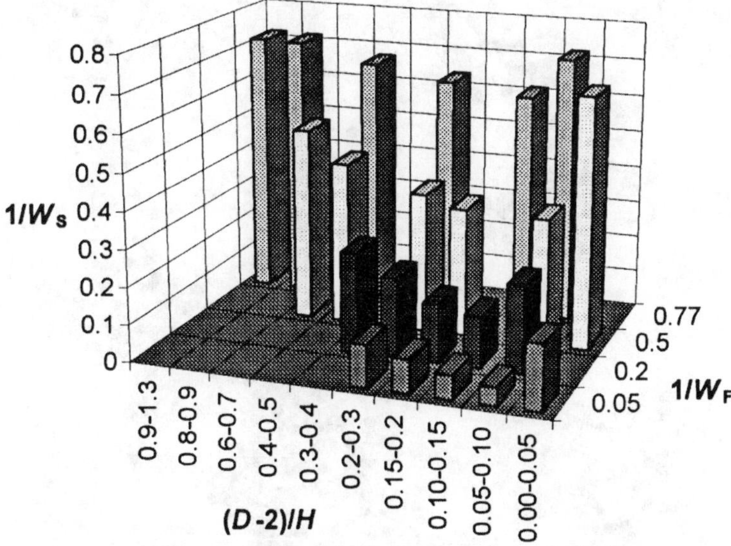

Figure 3 *The quantity* W_S *as a function of* W_F *and* $(D - 2)/H$ *for various simulated systems*

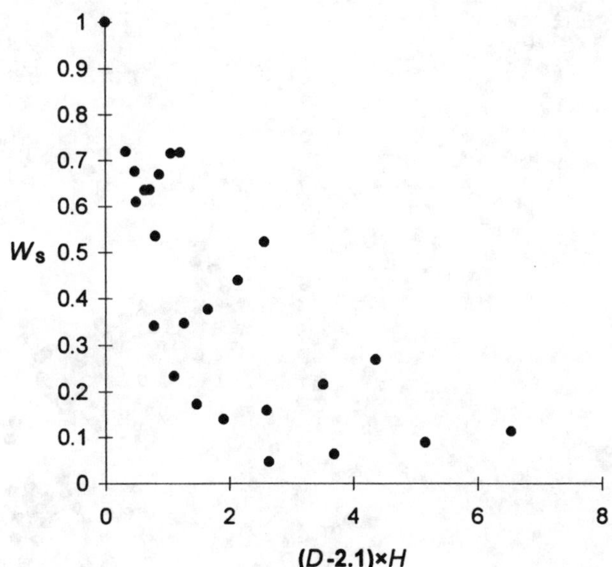

Figure 4 *The quantity* W_S *as a function of* $(D - 2.1) \times H$ *for various systems with different* $(D - 2)/H$

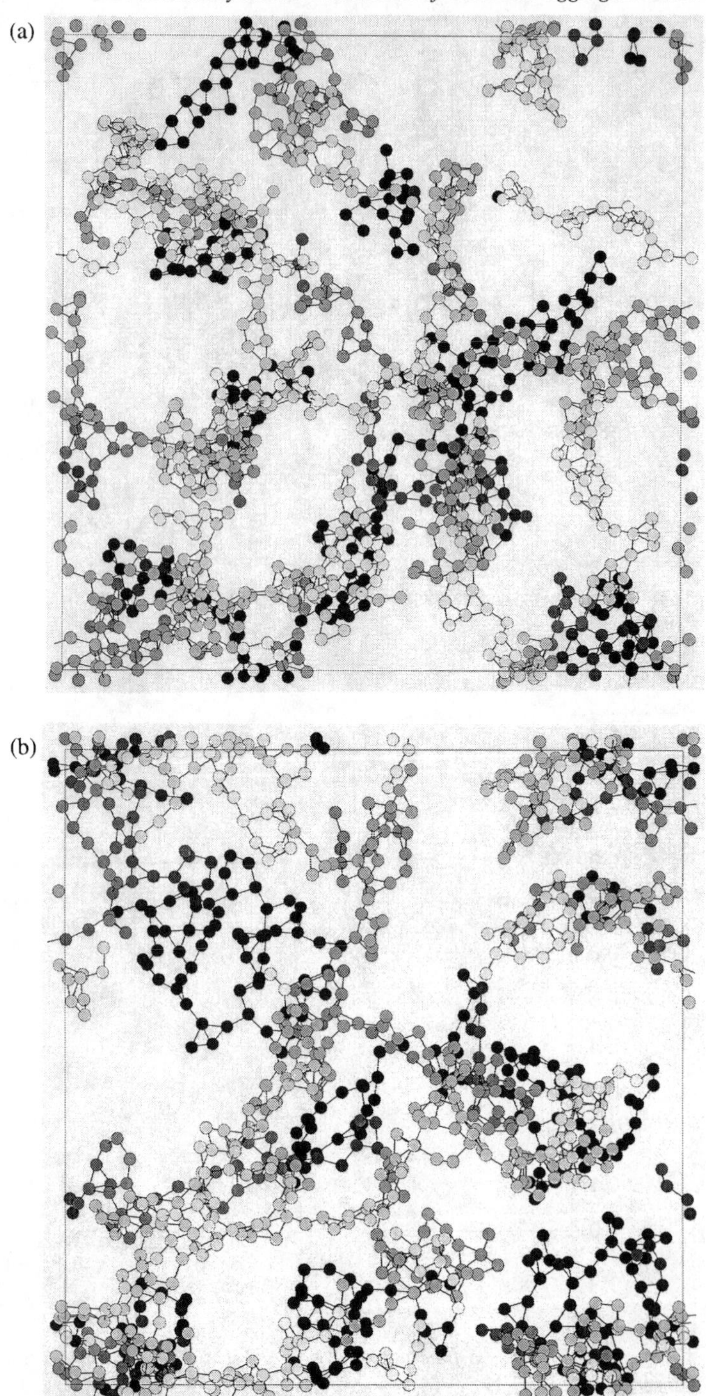

Figure 5 *Images of simulated gels (a)* $W_F = 1$ *at* $N_{agg} = 10$. *(b)* $W_F = 2.4$; $D = 3$; $H = 2.7$ *at* $N_{agg} = 10$. *Shading is used as indication of depth*

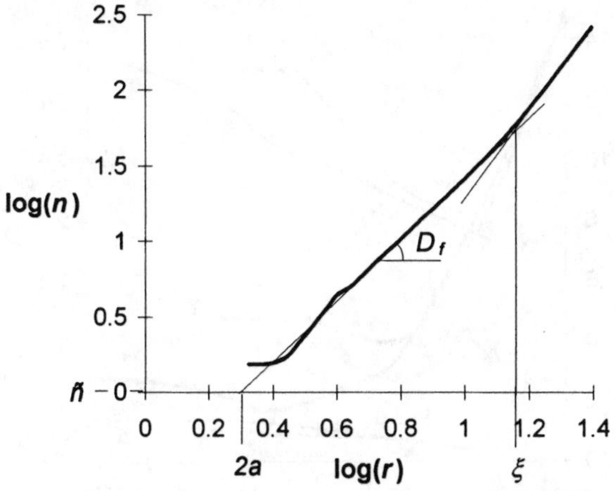

Figure 6 *Log–log plot of number of particles* n *as a function of length scale* r *of the system in Figure 5b. The quantities* D_f, ñ, ξ *and* 2a *are defined in the text*

$(D-2.1) \times H$. We have found that e^H is especially poor for predicting the actual bonding rate as given by W_s. Also for $(D - 2.1) \times H$ there is poor correlation with W_s (see Figure 4).

Power-law Behaviour

Two examples of particle configurations generated using the BD model are shown in Figure 5a ($W_F = 1$) and Figure 5b ($W_F = 2.4$). For the sake of clarity in these images, the particle radii have been reduced to half of their original values. In addition, the bonds are shown as strings of exaggerated length.

We can derive fractal parameters for the system at any time during the simulation. As an example, for the system in Figure 5b we have calculated $n(r)$ according to equation (13). The result is shown in Figure 6. We can clearly distinguish the fractal region at which $D_f = 1.8$ from the long-range region for which $D_f = 3$. The parameter ξ denotes the position of the transition between the linearized fractal and the homogeneous regimes.

As the aggregation proceeds, the apparent value of D_f decreases from 3 to its lowest value at around $N_{agg} = 10$. This is shown in Figure 7. In the time spent between $N_{agg} = 10$ and $N_{agg} = 1$, D_f has a tendency to increase, especially at low W_F. At high W_F, there is also a tendency for D_f to increase as the simulation continues beyond the point when $N_{agg} = 1$. In Figure 7 the correlation length ξ and the coordination number $\tilde{n}$ are also plotted. Both these parameters increase with decreasing N_{agg}. The factor $\tilde{n}$ seems to level off, whereas ξ continues to increase significantly. Probably most of the early-stage rearrangements are taking place over large length scales. The continued increase of the gel pore size ξ, along with a slowly increasing D_f, gives the first indication of micro-syneresis.

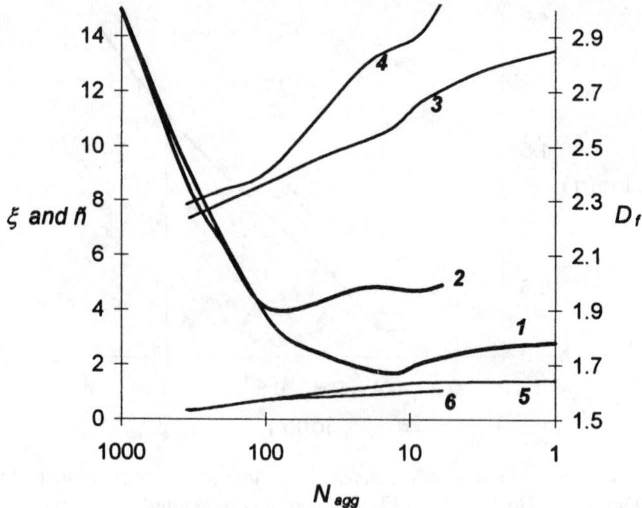

Figure 7 *Quantities* D_f, ξ *and* ñ *as a function of* N_{agg} *for two simulated systems: (1)* D_f; $W_F = 1$; *(2)* D_f; $W_F = 2.4$; *(3)* ξ; $W_F = 1$; *(4)* ξ; $W_F = 2.4$; *(5)* ñ; $W_F = 1$; *(6)* ñ; $W_F = 2.4$

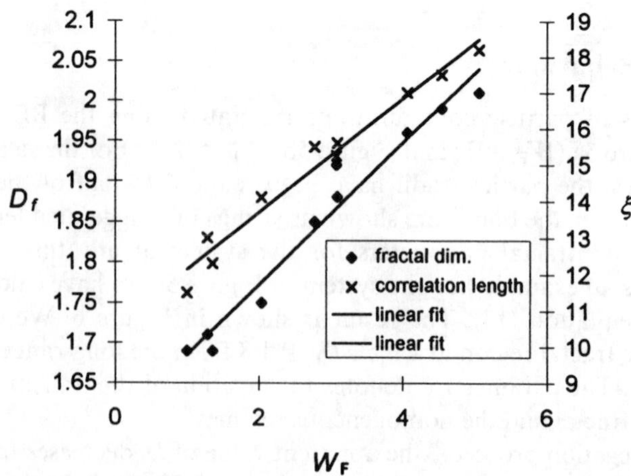

Figure 8 *Quantities* D_f *and* ξ *as a function of* W_F *for several simulated systems with different* $(D - 2)/H$

Figures 8, 9 and 10 show the influence of, respectively, W_F, $(D - 2.1) \times H$, and e^H on the values of D_f and ξ derived for several simulations at $N_{agg} = 10$. We can see clearly that only W_F has a predictive value for both fractal parameters.

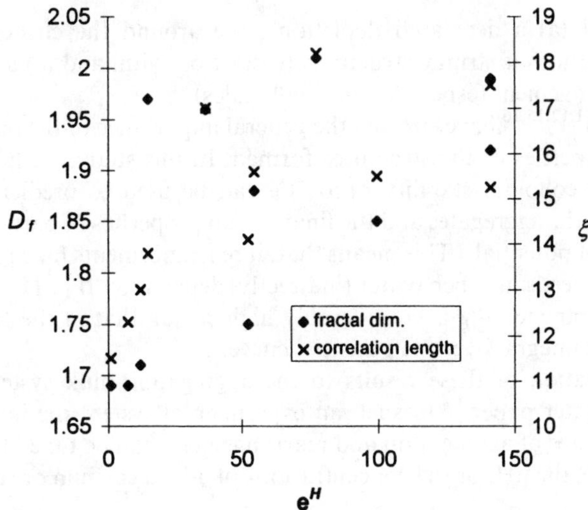

Figure 9 *Quantities* D_f *and* ξ *as a function of* e^H *for several simulated systems with different* $(D - 2)/H$

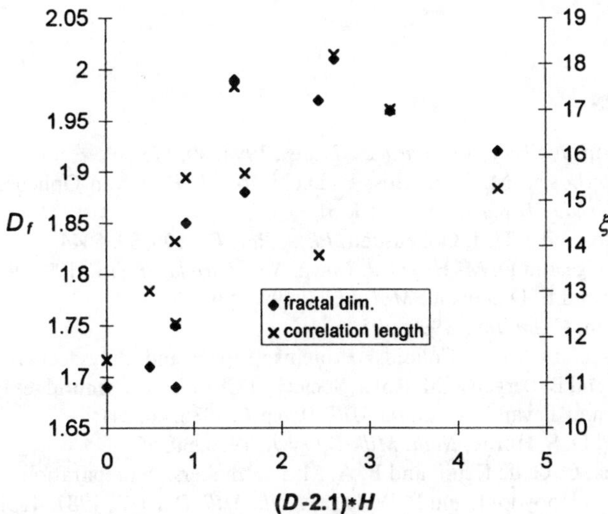

Figure 10 *Quantities* D_f *and* ξ *as a function of* $(D - 2.1) \times H$ *for several simulated systems with different* $(D - 2)/H$

7 Concluding Remarks

The existence of a gel is generally determined by a balance of cross-linking and phase separation. In our simulations, a higher repulsive barrier provides a delayed aggregation, which results in the formation of gels with a higher fractal dimensionality D_f and larger fractal scaling regimes. This is probably due to a combination of (a) a decreased depletion zone around the clusters, (b) an increased formation of stringy structures to start off with, and (c) a decreased extent of rearrangement (especially on small scales).

Other studies[1,12,15,16] have stressed the general importance of the net bonding rate on the properties of the structures formed. In this study, we have shown that the Fuchs colloidal stability ratio W_F can be used to predict the early bonding rate of the aggregates and the final fractal properties of particle gels for various forms of potentials. This means that, if rearrangements take place as the aggregation proceeds, this behaviour (indirectly) depends on W_F. The predictive value of the parameter W_F is considerably higher than that of the Boltzmann factor e^H or the integral over the potential curve.

The extrapolation of these results to the aggregating milk system will be discussed in a later paper.[19] In such an experimental system the interactions, and hence the rates of aggregation and rearrangement, can be tuned by altering the temperature, the pH, or the concentrations of added calcium or rennet.

Acknowledgement

The authors thank the European Union for financial support (FAIR-CT96-1216).

References

1. E. Dickinson, *J. Chem. Soc. Faraday Trans.*, 1994, **90**, 173.
2. B. H. Bijsterbosch, M. T. A. Bos, E. Dickinson, J. H. J. van Opheusden, and P. Walstra, *Faraday Discuss.*, 1995, **101**, 51.
3. M. T. A. Bos and J. H. J. Opheusden, *Phys. Rev. E*, 1996, **53**, 5044.
4. J. F. M. Lodge and D. M. Heyes, *J. Chem. Soc. Faraday Trans.*, 1997, **93**, 437.
5. M. Whittle and E. Dickinson, *Mol. Phys.*, 1997, **90**, 739.
6. E. Dickinson, *Chem. Ind.*, 1990, 595.
7. E. Dickinson, in 'Food Colloids: Proteins, Lipids and Polysaccharides', ed. E. Dickinson and B. Bergenståhl, Royal Society of Chemistry, Cambridge, 1997, p. 107.
8. P. Walstra and T. van Vliet, *Neth. Milk Dairy J.*, 1986, **40**, 241.
9. C. Holt and D. S. Horne, *Neth. Milk Dairy J.*, 1996, **50**, 85.
10. M. Mellema, C. G. de Kruif, and F. A. M. Leermakers, in preparation.
11. A. C. M. van Hooydonk and P. Walstra, *Neth. Milk Dairy J.*, 1987, **41**, 19.
12. L. G. B. Bremer, B. H. Bijsterbosch, P. Walstra, and T. van Vliet, *Adv. Colloid Interface Sci.*, 1993, **46**, 117.
13. M. T. A. Bos, Ph.D. Thesis, Wageningen Agricultural University, The Netherlands, 1997.
14. T. van Vliet, J. A. Lucey, K. Grolle, and P. Walstra, in 'Food Colloids: Proteins, Lipids and Polysaccharides', ed. E. Dickinson and B. Bergenståhl, Royal Society of Chemistry, Cambridge, 1997, p. 335.
15. P. Meakin, *J. Colloid Interface Sci.*, 1989, **134**, 235.
16. M. Verheul, S. P. F. M. Roefs, M. Mellema, and C. G. de Kruif, *Langmuir*, 1998, submitted.
17. M. Smoluchowski, *Physik Z.*, 1916, **17**, 557; *Z. Physik Chem.*, 1917, **92**, 129.

18. N. Fuchs, *Z. Physik*, 1934, **89**, 736.
19. M. Mellema, J. H. J. van Opheusden, and T. van Vliet, 1998, in preparation.
20. M. P. Allen and D. J. Tildesley, 'Computer Simulation of Liquids', Oxford University Press, Oxford, 1987.
21. P. Meakin, *Adv. Colloid Interface Sci.*, 1988, **28**, 249.
22. M. D. Haw, M. Sievwright, W. C. K. Poon, and P. N. Pusey, *Adv. Colloid Interface Sci.*, 1995, **62**, 1.
23. D. Avnir, O. Biham, D. Lidar, and O. Malcai, *Science*, 1998, **279**, 39.
24. D. S. Horne, *Faraday Disc. Chem. Soc.*, 1987, **83**, 259.
25. L. G. B. Bremer, T. van Vliet, and P. Walstra, *J. Chem. Soc. Faraday Trans.*, 1989, **85**, 3359.
26. T. van Vliet and P. Walstra, *Faraday Discuss.*, 1995, **101**, 359.
27. T. van Vliet, this volume, p. 307.

Fluid–Fluid Interfaces

Dilational Rheology of Proteins Adsorbed at Fluid Interfaces

By E. H. Lucassen-Reynders* and J. Benjamins

UNILEVER RESEARCH VLAARDINGEN, OLIVIER VAN NOORTLAAN 120, 3133 AT VLAARDINGEN, THE NETHERLANDS

1 Introduction

The essential stabilizing function of proteins and other surface-active molecules during emulsification and foaming is that they enable the interface to resist tangential stresses from the adjoining flowing liquids.[1] An interface covered with surface-active material is a two-dimensional body with its own rheological properties, which provides the liquid film separating emulsion drops or foam bubbles with a mechanism for dynamic stabilization. Rheological interfacial parameters can be defined for both compressional deformation and shearing motion in the interface, and both types of parameters can be measured at small periodic deformations[2,3] as well as under large continuous expansion or shear.[4,5] Surface shear viscosity has enjoyed the greater attention of experimentalists working with food proteins.[5–9] Their studies have shown that surface shear viscosities of protein layers are far higher than those of small molecules, and that they keep increasing considerably with increasing age of the interface. Surface rheological experiments with proteins at small deformations, however, have revealed that the viscoelasticity in shear is smaller than that in compression/deformation.[10] While surface shear viscosity may contribute appreciably to the long-term stability of emulsions and foams, it can be expected to be less relevant to their stability in the production stage for two reasons. Firstly, the dominant type of deformation that interfaces undergo during emulsification and foaming is expansion and, to a lesser extent, compression rather than shear. Secondly, the new interface which is continuously formed during the dispersion process, over timescales which may be as low as 1 ms or less, cannot build up the high shear viscosities found for aged interfaces. For short-term stability, therefore, we consider interfacial rheology in compression/expansion to be more relevant.

In compression/expansion, the surface dilational modulus is defined by the

*Present address: Mathenesselaan 11, 2343 HA Oegstgeest, The Netherlands

expression originally proposed by Gibbs[11] for the surface elasticity of a soap-stabilized liquid film as the increase in surface tension for a small increase in area of a surface element:

$$\varepsilon = \frac{d\gamma}{d\ln A}. \tag{1}$$

Here γ is the surface tension and A is the area of the surface element. In the simplest case, the modulus is a pure elasticity with a limiting value ε_0 to be deduced from the surface equation of state, *i.e.*, from the equilibrium relationship between surface tension and surfactant adsorption, Γ:

$$\varepsilon_0 = \left(\frac{-d\gamma}{d\ln\Gamma}\right)_{eq}. \tag{2}$$

This limiting value is reached only if there is no exchange of surfactant with the adjoining bulk solution (*i.e.*, $\Gamma \times A$ is constant), and if, moreover, the surface tension after a deformation adjusts instantaneously to its equilibrium value for the new adsorption. Deviations from this simple limit occur when relaxation processes in or near the surface affect either γ or Γ within the time of the measurement. In such cases, the measured modulus ε is a surface viscoelasticity, with an elastic part accounting for the recoverable energy stored in the interface and a viscous contribution reflecting the loss of energy through any relaxation processes occurring at or near the surface.

This article aims to survey the experimental rheological data available for adsorbed protein layers under compression/dilation and, where possible, to relate these data to the molecular properties of the protein through independently measured equilibrium adsorption properties.

2 Experimental Methods

Convenient techniques for measuring surface dilational moduli are derived from the longitudinal wave method.[2,12] In this method, the surface is subjected to small periodic compressions and expansions at a given frequency, usually by a barrier oscillating in the plane of the surface; the response of the surface tension is monitored by a probe, *e.g.*, a Wilhelmy plate. In such experiments the viscoelastic modulus ε is a complex number, with a real part ε' (the storage modulus) equal to the elasticity, ε_d, and the imaginary part ε'' (the loss modulus) given by the product of the viscosity, η_d, and the imposed angular frequency, ω, of the area variations:

$$\varepsilon = \varepsilon' + i\varepsilon'' = \varepsilon_d + i\omega\eta_d. \tag{3}$$

The viscous contribution to the modulus ε is reflected in a phase difference ϕ between stress ($d\gamma$) and strain (dA), which means that the elastic and viscous contributions are given by:

$$\varepsilon' = |\varepsilon| \cos \phi; \quad \varepsilon'' = |\varepsilon| \sin \phi. \tag{4}$$

Thus, dilational elasticity and viscosity can be determined separately from measured values of the absolute value of the modulus, $|\varepsilon|$, and the phase angle, ϕ.

In the conventional form of the method, the oscillating barrier produces a surface deformation which is uni-directional, *i.e.*, non-isotropic, and the Wilhelmy plate is positioned some distance away from the barrier. Problems can arise when (i) the surface has appreciable resistance to shearing motion and/or (ii) one of the bulk phases is so highly viscous that the area variations generated by the barrier are substantially damped at the position of the probe. The former point is relevant to proteinaceous surfaces and the latter to interfaces with viscous oils, *e.g.*, vegetable oil; both interfere with the proper measurement of ε. Two modifications have therefore been applied to the conventional technique: (i) at the air–water surface, surface shear effects have been eliminated by the use of a square-band barrier,[13] and (ii) interfaces with highly viscous oil have been studied with a dynamic drop tensiometer,[14] in which wave propagation effects are minimized by measuring ΔA and $\Delta\gamma$ on one single small area. At the air–water surface, protein adsorption has been measured by ellipsometry,[13] under the same experimental conditions as for the surface dilational moduli.

3 Experimental Results for Adsorbed Proteins

Air–Water Surface

Figures 1–4 present a selection of results obtained with flexible proteins (*i.e.*, β-casein, κ-casein, whole casein) and globular proteins (*i.e.*, bovine serum albumin and ovalbumin).[15] These results were found to agree within experimental error with other published data under comparable conditions.[16–20]

Figure 1 illustrates the slow equilibration of very dilute solutions, with the adsorption increasing immediately after the start of the experiment, but the surface pressure Π and the modulus ε reaching measurable values only after an 'induction time'. Such induction times generally indicate a severely non-ideal surface equation of state,[21] as will be demonstrated later. Surface pressure and modulus depend on both concentration and time, but these two variables merely serve to determine the changing values of the adsorption: for each protein, all measurements at different bulk concentrations and different surface ages were found approximately to coincide on a single curve of the modulus as a function of adsorption. For some of the proteins, Figure 2 gives examples of the Π *versus* Γ relationship obtained from experiments at different concentrations during equilibration for 21 h, which is considered to result in near-equilibrium.

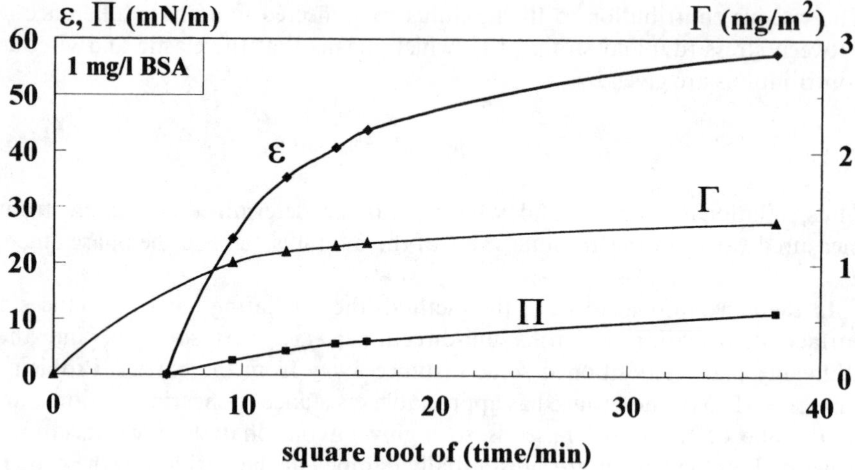

Figure 1 *Surface dilational modulus ε, adsorption Γ and surface pressure Π as a function of adsorption time for bovine serum albumin (BSA). Frequency = 0.84 rad s^{-1}; pH = 6.7; ΔA/A = 0.07*

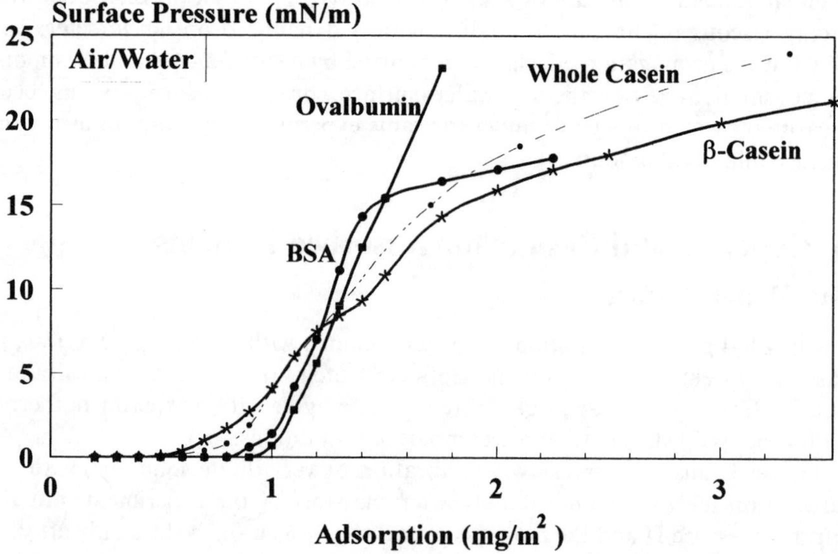

Figure 2 *Surface pressure Π as a function of adsorption Γ for β-casein, whole casein, BSA, and ovalbumin at pH = 6.7 and 20 °C*

This means that equilibrium *in* the surface (*i.e.*, in the relationship between Π and Γ) is established much faster than that between surface and bulk solution. As a result, as illustrated in Figure 3, the modulus ε at each given frequency is a unique function of the surface pressure for different concentrations and surface ages. At the highest frequency, purely elastic behaviour was found for all

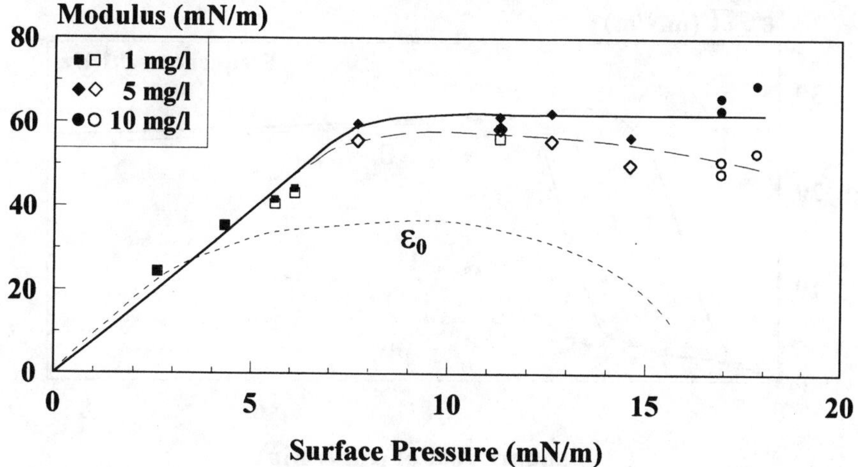

Figure 3 *Surface dilational modulus ε as a function of surface pressure Π for BSA at pH = 6.7; ΔA/A = 0.07. Closed symbols: ω = 0.84 rad/s; open symbols: ω = 0.084 rad/s. Drawn line: results at highest frequency. Dashed line: limiting modulus, ε_0, from Π vs Γ curve in Figure 2*

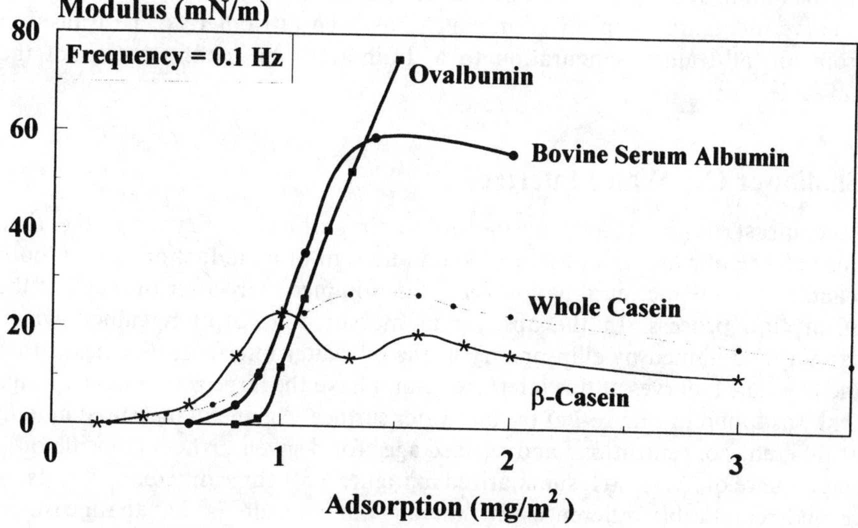

Figure 4 *Surface dilational modulus ε as a function of adsorption Γ for whole casein, β-casein, BSA and ovalbumin*

proteins; at the lower frequencies viscous phase angles of at most 20° were found for most proteins, with an exceptionally high value of 42° for whole casein at high concentration and low frequency, *i.e.*, corresponding to viscous contributions equal to 36% and 90%, respectively, of the elastic contribution to the modulus. Modulus *versus* adsorption curves are summarized in Figure 4, where

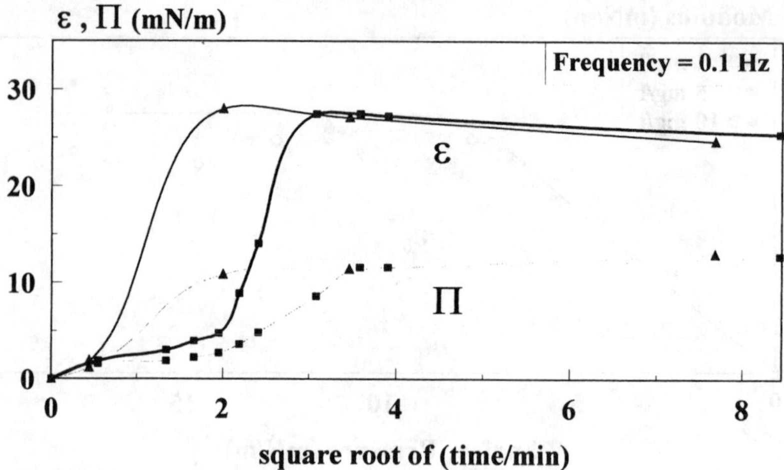

Figure 5 *Time dependence of the modulus, |ε|, and the interfacial pressure Π at the sunflower oil–water interface for two concentrations of BSA:*[14] ■, *5 mg l^{-1};* ▲, *10 mg l^{-1}*

the maximum at 1 mg m^{-2} of β-casein reflects the inflexion point in the Π *versus* Γ curve in Figure 2, an inflexion which has been attributed to the transition from an 'all-train' configuration to a 'train-and-loop' conformation of the molecule.[16]

Sunflower Oil–Water Interface

In contrast to the results at the air–water surface in Figure 1, the time dependence of ε and Π in Figure 5 shows no sign of an induction period: both ε and Π assume measurably non-zero values immediately after the start of the adsorption process. In the absence of measured adsorption values, which cannot be obtained by ellipsometry at the oil–water interface, this means that the Π *versus* Γ curves at this interface cannot have the large zero pressure range that was found in Figure 2 at the air–water surface. Again, all measured moduli at different concentrations and surface ages for a given protein coincide on a single curve of ε *versus* Π, summarized in Figure 6 for three different proteins. A second remarkable difference with the air–water results is that the measured moduli are generally much lower at the oil–water interface, as illustrated in Figure 7 for whole casein.

4 Interpretation of Results

Effect of Surface Equation of State

In the pure elasticity range, *i.e.*, at not too high a surface pressure, the dynamic behaviour is dominated by the surface equation of state, and the modulus ε

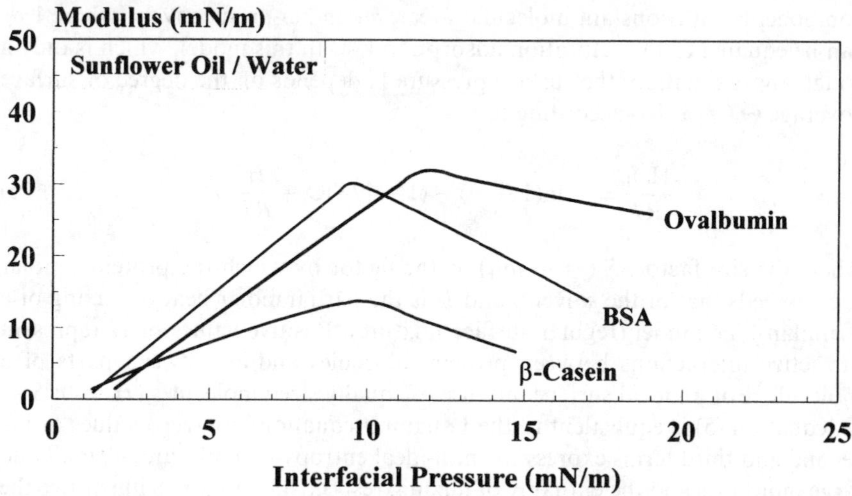

Figure 6 *Modulus* ε versus *interfacial pressure* Π *for flexible and compact proteins at the sunflower oil–water interface at 0.1 Hz*[14]

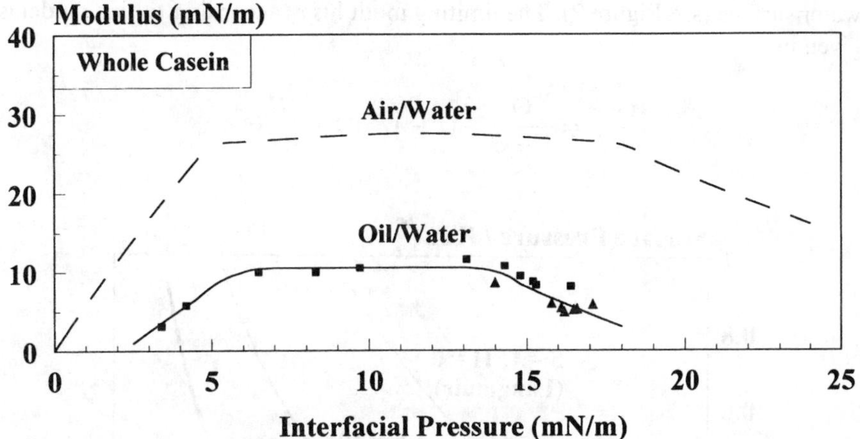

Figure 7 *Modulus* ε versus *interfacial pressure* Π *for whole casein at the air–water and sunflower oil–water interfaces at 0.1 Hz. Dashed line: data from Figure 4;* ■, *10 mg* l^{-1}; ▲, *100 mg* l^{-1} [14]

should equal the limiting value ε_0 as defined in equation (2). This is found to be reasonably accurate up to Π values of 15 mN m^{-1} for the caseins, and over a smaller range of Π for the globular proteins (BSA and ovalbumin). Existing analytical equations of state are unable to describe all aspects of protein behaviour, but a simple treatment accounts for the main features of the measured moduli. The two-dimensional model proposed[15,22] considers both entropy and enthalpy contributions to first order, for solvent and protein components with constant molecular areas, ω_1 and ω_2, respectively, where $1/\omega_i$ can be equated to the saturation adsorption Γ_i^∞. In this model, which is only a rough approximation, the surface pressure Π depends on the degree of surface coverage Θ ($= \omega_2\Gamma_2$) according to

$$\frac{\Pi\omega_1}{RT} = -\ln(1-\Theta) - (1 - 1/S)\Theta - \frac{H}{RT}\Theta^2, \tag{5}$$

where the size factor, S ($= \omega_2/\omega_1$), is the factor by which the protein's molar area exceeds that of the solvent, and H is the partial molar heat of mixing of a Frumkin-type model (regular surface mixture). Positive values of H represent attractive interactions between protein molecules and/or between parts of a molecule. For an ideal surface mixture of equally sized molecules ($H = 0$; $S = 1$), equation (5) is equivalent to the Langmuir equation; non-zero values of the second and third terms express the non-ideal entropy of a mixture of small and large molecules and the enthalpy of mixing, respectively. Figure 8 illustrates the pressure *versus* adsorption isotherms for the three cases representative of (i) ideal mixing, (ii) non-ideal entropy, and (iii) non-ideal entropy combined with heat of mixing. The combination of entropy and enthalpy factors, in particular, results in very low pressures at low surface coverage, as is often found at the air–water surface (see Figure 2). The limiting modulus ε_0 according to this model is given by

$$\frac{\varepsilon_0\omega_1}{RT} = \frac{\Theta}{1-\Theta} - (1 - 1/S)\Theta - \frac{2H}{RT}\Theta^2. \tag{6}$$

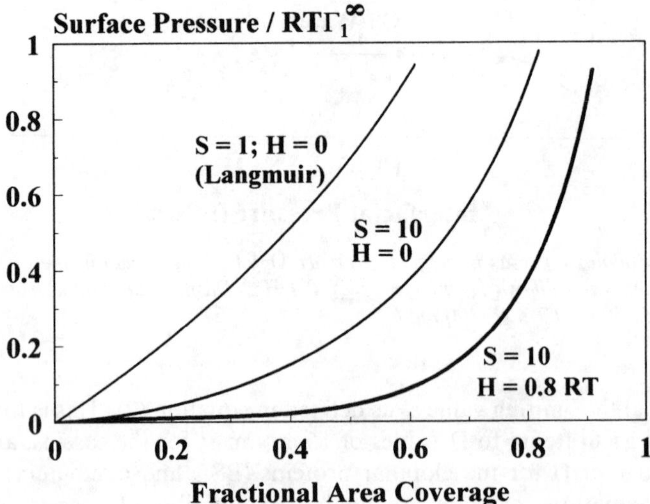

Figure 8 *Effects of non-ideal entropy and enthalpy terms on surface pressure* versus *surface coverage according to equation (5). See text for definitions of quantities* S *and* H

The effects of non-ideal entropy and enthalpy on the ε_0 *versus* Π relationship are shown in Figure 9. Interestingly, it is only the combination of entropy and enthalpy factors that produces the steep linear ascent of the modulus at moderately high surface coverage observed experimentally. Figure 10 illustrates that equations (5) and (6) describe quite well a large range of experimental data

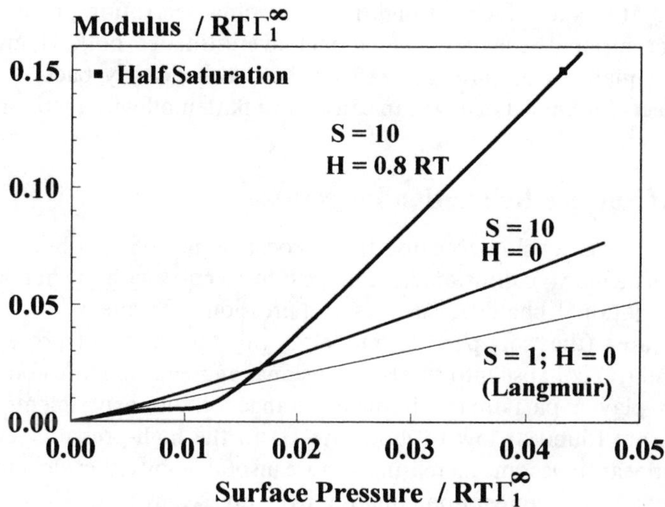

Figure 9 *Effects of non-ideal entropy and enthalpy terms on limiting modulus* versus *surface pressure according to equation (6). See text for definitions of quantities* S *and* H

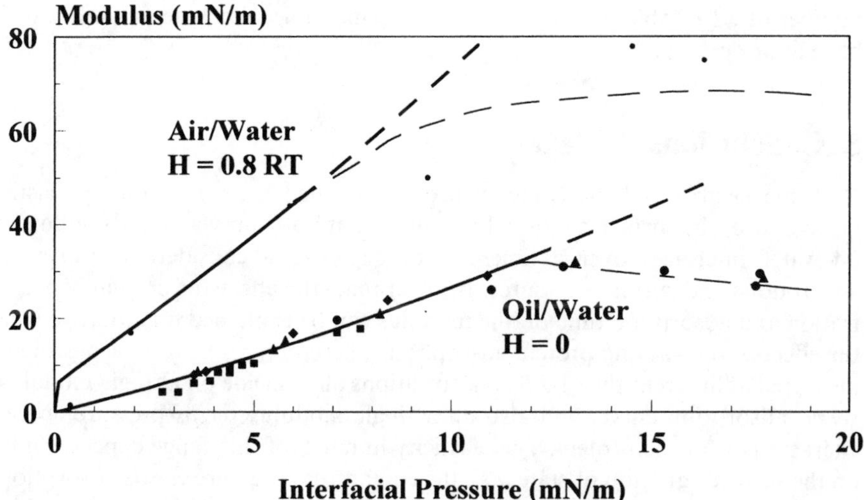

Figure 10 *Comparison of experiment and theory for ovalbumin at air–water and sunflower oil–water interfaces. Points and thin lines: experiments at 0.1 Hz. Bold lines: theory of equations (5) and (6) for* ω_1 = *0.1* nm^2 *per molecule and* ω_2 = *10* nm^2 *per molecule*

for ovalbumin at the two interfaces for non-optimized values of ω_1 and ω_2, assuming $S = 100$. The data at the air–water surface require a high value of the enthalpy term H, which, at the chosen value for the size factor S, produces phase separation in the surface and a very steep increase of ε_0 at near-zero values of Π.[23] At the sunflower oil–water interface, however, the data do not require any enthalpy contribution. This explains the apparent absence of any induction time in Figure 5 at this interface. At high Π, where slow relaxation processes cause viscoelastic surface behaviour, the model overestimates both Π and ε_0, as analytical equations of state generally do. In such densely packed surfaces, several effects (discussed below) can cause Π to flatten off with increasing Γ.

Effects of Surface Relaxation Processes

Processes sufficiently slow to cause the viscoelastic behaviour observed at low frequencies include (i) diffusional exchange of molecules with the bulk solution, (ii) conformational changes, such as compression of protein into a smaller molecular area (the 'soft particle' model[24]), (iii) formation of protein aggregates,[25] and (iv) collapse into three-dimensional aggregates. Diffusion does not appear to play a part in the frequency range of our experiments as pure elasticities are found at low to moderate Π. In the high-pressure region, the proteins appear to become increasingly more insoluble, by either slow molecular reconformation or aggregation/collapse, over timescales from 10^2 to 10^5 s. In the elastic region, reconformation processes such as unfolding are fast, *i.e.*, they are essentially completed in seconds. In close-packed layers, such processes may proceed at a much slower rate. Current modelling of such relaxation phenomena is still in the state of curve fitting in terms of simple first-order kinetics with a number of adjustable parameters; more specific molecular kinetic models will have to be developed.

5 Conclusions

Both the flexible and the compact proteins studied here show purely elastic behaviour of the surface dilational modulus, ε, at both air–water and sunflower oil–water interfaces over a timescale of 1 s, over a considerable range of adsorptions and surface pressures. In this range, the effects of protein concentration and adsorption time on the modulus can be explained quantitatively as the effect of the varying protein adsorption. Therefore, results for each protein measured at different times and concentrations all coincide on a single modulus *versus* adsorption curve, and also on a single modulus *versus* pressure curve, characteristic of the protein. The elasticity in much of this range depends only on the surface equation of state, *i.e.*, the equilibrium pressure *versus* adsorption relationship of the protein. Common features distinguishing this relationship from that for simple small molecules at the air–water surface are (i) extremely low values of the surface pressure and the modulus in a range of adsorptions up to roughly 1 mg m^{-2}, followed by (ii) a sudden and steep increase of both surface pressure and modulus with increasing adsorption at higher adsorptions (up to roughly 2 mg m^{-2}) and, finally, (iii) a flattening off of both modulus and pressure at the highest adsorptions.

At low and moderate adsorptions, the modulus is found to be purely elastic over timescales from 1 to 100 s, implying that relaxation processes do not play any part over these timescales. This leads to the two-fold conclusion that diffusional interchange with the solution does not take place (because this mechanism requires longer times), and that any relaxation processes *in* the adsorbed layer take place at shorter times. In this elastic range, differences between individual proteins are related to different degrees of severe non-ideality of the surface equation of state, caused by molecular reconformations and interactions. An adequate description of the characteristic features of the elastic region is obtained from a simple equation-of-state model combining non-ideal entropic and enthalpic effects. The measured steep increase of the elasticity points to an overriding influence of attractive molecular interactions. Such non-ideality is shown to be much less pronounced at the vegetable oil–water interface than at the air–water surface, resulting in appreciably lower moduli at the former interface. Less severe non-ideality of the oil–water interface is also indicated by the far shorter observed 'induction' times. As a result of the intramolecular and intermolecular interactions, the modulus at both interfaces generally increases with decreasing flexibility of the molecules.

At high adsorptions, where the protein layers are close-packed and visco-elastic, existing equations of state have to be extended in order to account for the measured moduli. Relaxation processes in this region cause the adsorbed protein to become increasingly more insoluble, by slow molecular reconformation and/or aggregation phenomena, over time scales from 10^2 to 10^5 s, *i.e.*, much slower than for the lower adsorptions of the elastic region. Specific molecular kinetic models will have to be developed for a satisfactory description of such slow relaxation processes.

References

1. M. van den Tempel, in 3rd International Congress of Surface Active Agents, Cologne, 1960, Vol. II, p. 345.
2. J. Lucassen and M. van den Tempel, *J. Chem. Sci.*, 1972, **27**, 1283.
3. J. A. de Feijter and J. Benjamins, *J. Colloid Interface Sci.*, 1979, **70**, 375.
4. F. van Voorst Vader, T. F. Erkens, and M. van den Tempel, *Trans. Faraday Soc.*, 1964, **60**, 1170.
5. E. Dickinson, B. S. Murray, and G. Stainsby, *J. Chem. Soc. Faraday Trans. I*, 1988, **84**, 871.
6. D. E. Graham and M. C. Phillips, *J. Colloid Interface Sci.*, 1980, **76**, 240.
7. A. Martinez-Mendoza and P. Sherman, *J. Dispersion Sci. Technol.*, 1990, **11**, 347.
8. V. Kiosseoglou, *J. Dispersion Sci. Technol.*, 1992, **13**, 135.
9. B. S. Murray and E. Dickinson, *Food Sci. Technol. Int.* (Japan), 1996, **2**, 31.
10. J. Benjamins and F. van Voorst Vader, *Colloids Surf.*, 1992, **65**, 161.
11. J. W. Gibbs, 'Collected Works', Dover Publications, New York, 1928, Vol. 1, p. 302.
12. J. Lucassen and M. van den Tempel, *J. Colloid Interface Sci.*, 1972, **41**, 491.

13. J. Benjamins, J. A. de Feijter, M. T. A. Evans, D. E. Graham, and M. C. Phillips, *Faraday Discuss. Chem. Soc.*, 1975, **59**, 218.
14. J. Benjamins, A. Cagna, and E. H. Lucassen-Reynders, *Colloids Surf.*, 1996, **114**, 245.
15. J. Benjamins and E. H. Lucassen-Reynders, in 'Proteins at Liquid Interfaces', ed. D. Möbius and R. Miller, Elsevier, Amsterdam, 1998, p. 341.
16. D. E. Graham and M. C. Phillips, *J. Colloid Interface Sci.*, 1980, **76**, 227.
17. C-S. Gau, H. Yu, and G. Zografi, *J. Colloid Interface Sci.*, 1994, **162**, 214.
18. G. A. van Aken and M. T. E. Merks, *Progr. Colloid Polym. Sci.*, 1994, **97**, 281.
19. R. Douillard, M. Daoud, J. Lefebvre, C. Minier, G. Lecannu, and J. Coutret, *J. Colloid Interface Sci.*, 1994, **163**, 277.
20. G. A. van Aken and M. T. E. Merks, *Voedingsmiddelen Technologie*, 1994, **27**, 11.
21. M. van den Tempel and E. H. Lucassen-Reynders, *Adv. Colloid Interface Sci.*, 1983, **18**, 281.
22. E. H. Lucassen-Reynders, *Colloids Surf.*, 1994, **91**, 79.
23. E. H. Lucassen-Reynders and J. Benjamins, to be published.
24. J. A. de Feijter and J. Benjamins, *J. Colloid Interface Sci.*, 1982, **90**, 289.
25. A. Dussaud and M. Vignes-Adler, *J. Colloid Interface Sci.*, 1994, **167**, 256.

Dynamic Properties of Protein + Surfactant Mixtures at the Air–Liquid Interface

By Reinhard Miller, Valentin B. Fainerman,[1] Alexander V. Makievski,[1] Dmitri O. Grigoriev,[2] Peter Wilde,[3] and Jürgen Krägel

MAX PLANCK INSTITUTE OF COLLOIDS AND INTERFACES, RUDOWER CHAUSSEE 5, BERLIN-ADLERSHOF, D-12489, GERMANY
[1]INSTITUTE OF TECHNICAL ECOLOGY, 25, BULV. SHEVCHENKO, DONETSK, 340017, UKRAINE
[2]INSTITUTE OF CHEMISTRY, ST. PETERSBURG STATE UNIVERSITY, UNIV. PR. 2, 1989084 ST. PETERSBURG, RUSSIA
[3]INSTITUTE OF FOOD RESEARCH, NORWICH RESEARCH PARK, COLNEY, NORWICH, NR4 7UA, UK

1 Introduction

In practice adsorption layers of mixed protein + surfactant systems are more relevant rather than pure protein layers. There is a large variety of interfacial properties which can be tuned by changing the composition of the system or the bulk properties, such as pH, temperature, or ionic strength. This peculiarity is caused by the interplay between the two components which are very different in their nature, size, and flexibility. Most important in this respect is that the interaction between protein and surfactant leads to the formation of new compounds, the interfacial behaviour of which can be much different from that of the single components.

Mixed protein + surfactant adsorption layers are vital for a number of technologies, *e.g.*, in coating, pharmacy, oil recovery, food technology, as discussed in detail in various books.[1–3] In a number of papers dealing with mixed systems, however, often technical surfactants have been used, the quantitative characterization of which is impossible. Also, when surfactants of different nature are mixed with proteins, for easy comparison it is recommended to select samples of comparable surface activity. For example, the non-ionic Tween 20, which is often used in fundamental food studies, and the ionic surfactant sodium dodecyl sulfate (SDS) have surface activities that differ by more than three orders of magnitude.

Studies of mixed protein + surfactant systems have been performed by many authors using various experimental techniques,[4–13] *e.g.*, the dynamic surface tension behaviour of protein + surfactant mixtures,[4–6] the importance of protein + surfactants layers for the stability of emulsions[7–9] or foams,[10,11] and the surface rheological investigations on mixed protein + surfactants.[12,13] A theoretical description of such mixed protein + surfactants systems, however, does not exist. Only qualitative and semi-quantitative analysis is available so far, assuming that the compounds act independently.

In the present paper the adsorption kinetics and the surface rheological behaviour of model food proteins (β-lactoglobulin and β-casein) mixed with surfactants (sodium tetradecyl sulfate, cetyl trimethyl ammonium bromide, and decyl dimethyl phosphine oxide) are presented. The adsorption behaviour of the pure proteins has been described recently.[14] The adsorption kinetics and the adsorbed layer state are determined via measurements of surface tension (using drop and bubble techniques) and adsorption layer thickness (using ellipsometry) as a function of time and concentration. Interfacial shear rheology was studied using a torsion pendulum rheometer.

2 Materials

The sodium dodecyl sulfate (SDS) and sodium tetradecyl sulfate (STS) were obtained from Dr. G. Czichocki (MPI, KGF, Berlin) to a high degree of purity. Cetyl trimethyl ammonium bromide (CTAB) was purchased from FLUKA and re-crystallized several times before use. The decyl dimethyl phosphine oxide (C_{10}DMPO) was purchased from Gamma-LAB Dr. Schano, Berlin, Germany. The sample has a special degree of purity suitable for interfacial studies. High purity polyoxyethylene 20 sorbitan monolaurate (Tween 20) was purchased from Pierce Chemical Co. (Surfact-Amp 20, Prod. No. 28320).

The model proteins β-casein (C6905 Lot 12H9550, molecular weight 2.4×10^4 daltons), β-lactoglobulin (L-0130, Lot 91H7005, molecular weight 18.4×10^3 daltons) from bovine milk were purchased from Sigma (Germany). The β-casein was used without further purification, while the β-lactoglobulin sample was purified according to the procedure given by Clark *et al.*[15] For comparison, also studies with human serum albumin (HSA) have been performed. This sample was purchased from Sigma (Germany, A8301, Lot 94H8270, molecular weight 6.9×10^4 daltons). The phosphate buffer solution used was prepared from the appropriate stock solutions of Na_2HPO_4 and NaH_2PO_4. The surface tension of the buffer solution at pH 7 was 72.5 mN m^{-1}.

3 Methods

The adsorption kinetics and the adsorbed layer state were determined via measurements of surface tension using axisymmetric drop shape analysis (ADSA).[16] The drop was formed at the tip of a PTFE capillary in a water-saturated atmosphere. The volume of the drops was *ca.* 0.03 cm^3. The accuracy of the ADSA method in these studies was ± 0.2 mN m^{-1}. Surface dilational viscosity and elasticity can be determined from pendant drop transient relaxation experiments.[17] In this case the drop volume, and hence the adsorbed layer on its surface, was increased by small increments of the order of less than 10% and the surface tension change was monitored as a function of time. Further experimental details can be found elsewhere.[18]

The adsorption layer thickness was measured as a function of time and concentration using the multifunctional optical instrument 'Multiscope' (OPTREL GmbH, Dr. H. Motschmann) in the ellipsometric mode. Ellipsometry is a conventional method for investigating thin films; it yields information about the thickness, refractive index and adsorbed amount of an adsorbed or a spread monolayer at the interface between two different media. The common principle of ellipsometry is to measure the change of the state of polarization of monochromatic parallel light upon reflection at the interface. These changes are connected with the two measured ellipsometric angles, Δ and Ψ. The relation between these parameters and the optical properties of the incident light, immersion medium, adsorbed layer, and reflecting substrate is given by the following equation:

$$[\tan\Psi \exp(i\Delta')]/[\tan\Psi \exp(i\Delta)] = [1 + i4\pi d_1 \cos\phi_0 \times \sin^2\phi_0 n_2^2 M]/[\lambda(n_2^2 - n_0^2)(n_0^2 \sin^2\phi_0 - n_2^2\cos^2\phi_0)]. \tag{1}$$

Here d_1 is the film thickness, n_0, n_1 and n_2 are the refractive indices of air, monolayer and water, respectively, λ is the wavelength of light, ϕ_0 the angle of incidence of light, and the quantity M is defined by

$$M = n_0^2 + n_2^2 - (n_1^2 + n_0^2 n_2^2/n_1^2). \tag{2}$$

Ψ and Δ are the angles measured for the clean water surface while Ψ' and Δ' are the angles for the monolayer covered surface. Equation (1) cannot be solved analytically but it can be solved numerically. In this manner the thickness and the refractive index of the adsorbed layer can be determined. If the refractive index of the interfacial layer is assumed to be a linear function of the solute concentration the adsorption Γ can be calculated from the equation

$$\Gamma = \frac{d_1(n_1 - n_2)}{a}; \quad a = \mathrm{d}n/\mathrm{d}c, \tag{3}$$

where a is the refractive-index increment of the solute.

The evaluation procedure to derive the monolayer characteristics according to equation (1) was performed by means of the evaluation and simulation programme developed by H. Motschmann. This software, using the Jones matrix formalism,[19] solves equation (1) numerically and yields the values of the thickness and refractive index of the monolayer. The adsorption values were calculated from equation (3) taking into account the a values given elsewhere.[20]

The interfacial shear rheology was studied using a torsion pendulum rheometer as described elsewhere.[21] The small deflection angle of 1–3 degrees used in the measurements allows studies with a minimum disturbance of the surface layer structure. From the damped oscillation behaviour of the ring, both the shear viscosity η_s and the shear elasticity can be deduced. There is no direct relationship between shear rheological and thermodynamic quantities of an adsorbed layer. However, the shear rheology allows us to estimate the strength of structures formed in the interfacial layers. This is of particular interest for mixed layers, as the rheological parameters can give evidence about the presence of different components and their concentration.

4 Results

The set of surfactants was specially selected such that the surface activity was in a certain narrow range, and three of them were therefore directly comparable: the non-ionic C_{10}DMPO, the cationic CTAB, and the anionic surfactant STS. Figure 1 shows the surface tension isotherms of the pure proteins. These proteins have a surface activity which is essentially of the same order of magnitude. Figure 2 shows the surface tension isotherms of the pure surfactants. It is evident that the three surfactants CTAB, STS and C_{10}DMPO are quite similar in their surface activity. The use of STS instead of the often used SDS is preferable as its surface activity, as given by the position of the isotherm, is very close to that of CTAB or C_{10}DMPO. The non-ionic surfactant Tween 20, as one can see, is of extremely high surface activity; it forms micelles at bulk

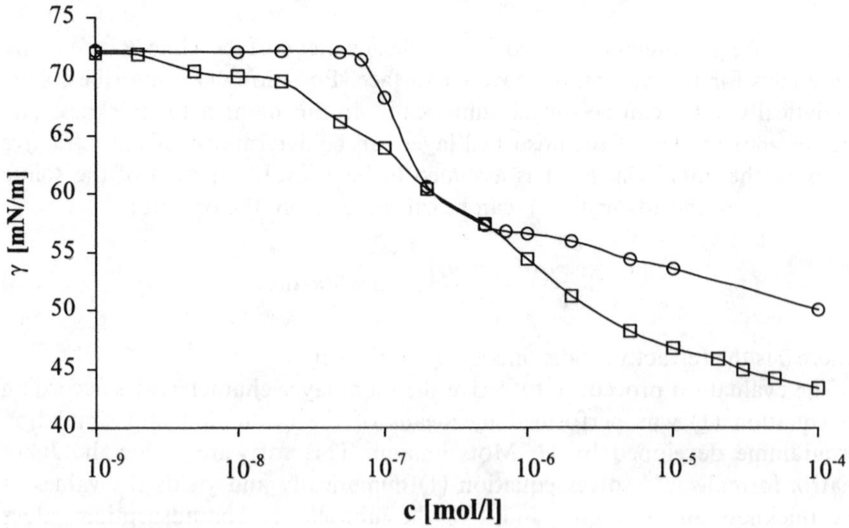

Figure 1 *Adsorption isotherms of β-lactoglobulin (□) and β-casein (○) at the water–air interface at 25 °C*

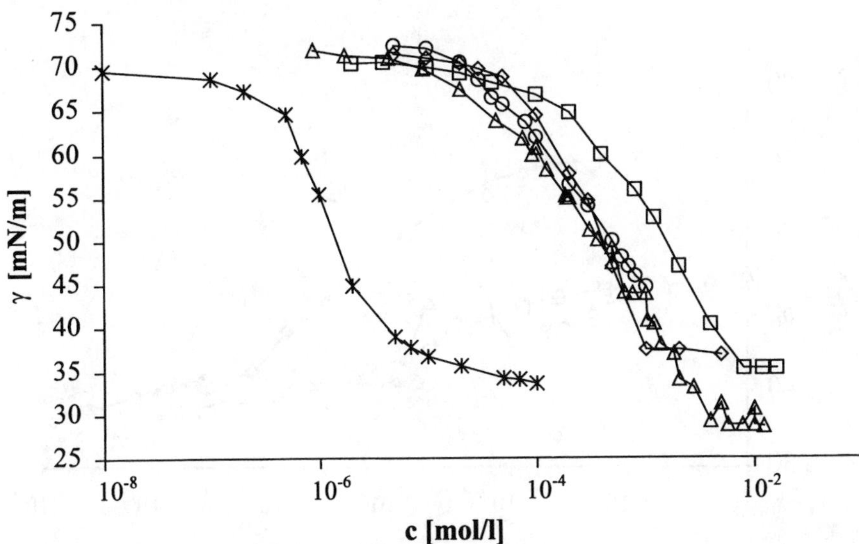

Figure 2 *Adsorption isotherms of the studied surfactants at the water–air interface at 25 °C: SDS (□), STS (○), CTAB (◇), Tween 20 (★), and $C_{10}DMPO$ (△)*

surfactant concentration about three orders of magnitude below that for the three selected surfactants chosen for this study.

The surface tension isotherms of the mixed protein + surfactant systems are given in Figure 3. For both β-lactoglobulin and β-casein, the mixtures with CTAB are shown for two constant protein concentrations. Some examples of dynamic surface tensions of mixed protein + surfactant systems are given in Figure 4. At a constant protein content of 10^{-6} mol l^{-1} increasing amounts of surfactant were found to lead to significant changes in the adsorption kinetics.

The adsorption layer thicknesses for the two proteins have been measured by ellipsometry. As an example, the isotherm of the adsorbed layer thickness is shown in Figure 5a, while the adsorbed amount as a function of concentration is given in Figure 5b. The presence of increasing amounts of surfactant influences the adsorbed layer thickness of the mixed system as compared with that for the protein alone. In Figure 6 the change in thickness is shown for β-casein at $c = 10^{-6}$ mol l^{-1} mixed with increasing amounts of $C_{10}DMPO$.

In order to verify ideas about the structure of the mixed protein + surfactant layers, interfacial shear rheology studies of the two proteins have been performed. Figure 7 demonstrates the change of surface shear viscosity η_s with time at different β-casein concentrations. The shear rheological behaviour of mixed β-casein + CTAB systems at constant contents of protein and decreasing protein + surfactant concentration ratio is given in Figure 8, while the corresponding results for mixtures of β-lactoglobulin + $C_{10}DMPO$ are summarized in Figure 9.

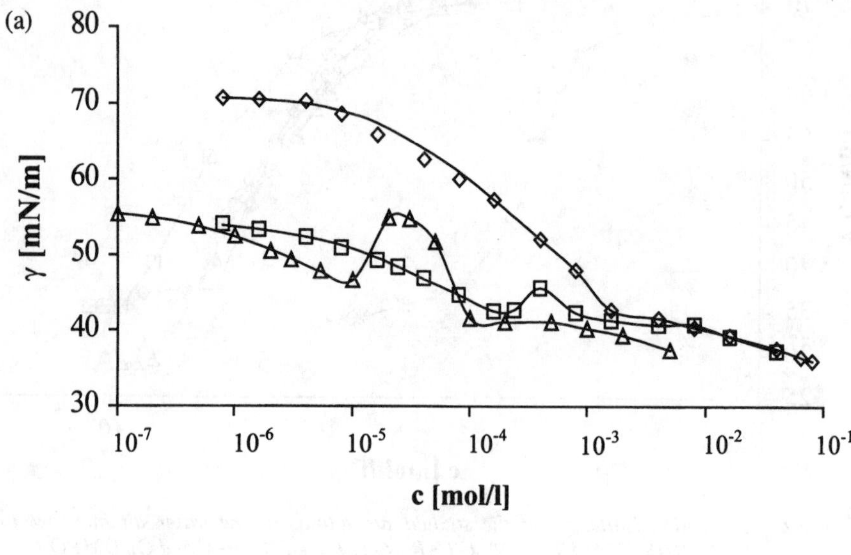

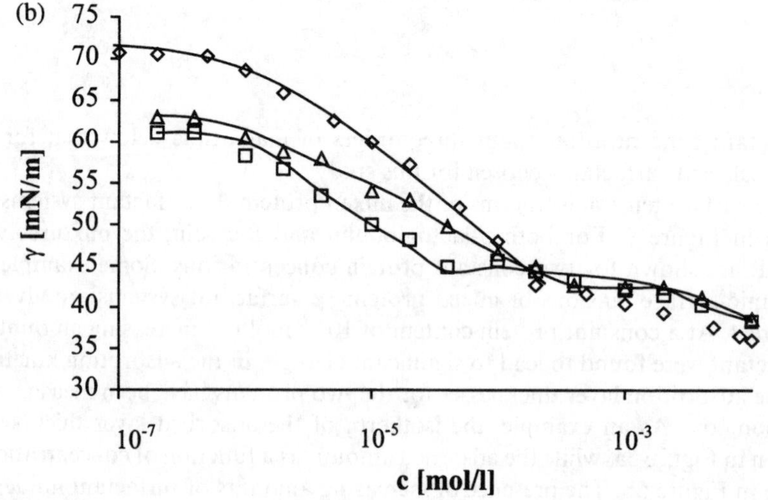

Figure 3 *Surface tension isotherms of CTAB (◇) and mixed protein + CTAB systems at two constant protein concentrations,* 10^{-6} *mol* l^{-1} *(□) and* 10^{-7} *mol* l^{-1} *(△); (a) β-casein, (b) β-lactoglobulin*

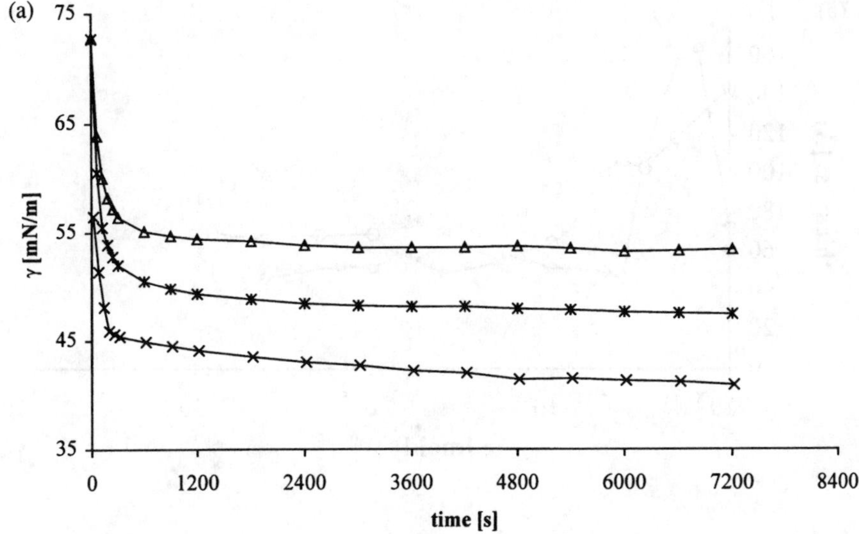

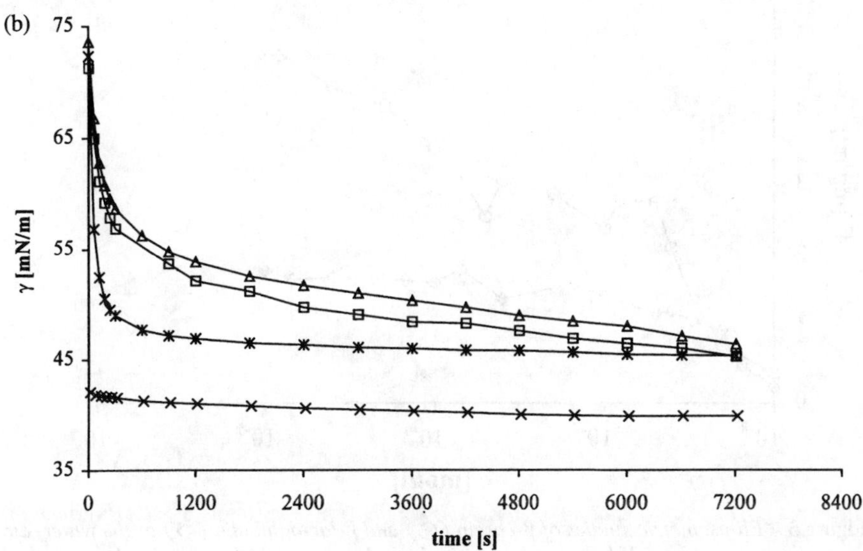

Figure 4 *Dynamic surface tension of the water–air interface at 25 °C for (a) β-casein + CTAB mixtures and (b) β-lactoglobulin + CTAB mixtures at constant protein concentration of 10^{-6} mol l^{-1} and different CTAB concentrations of 10^{-7} mol l^{-1} (△), 5 × 10^{-6} mol l^{-1} (□), 10^{-5} mol l^{-1} (★), and 10^{-3} mol l^{-1} (×)*

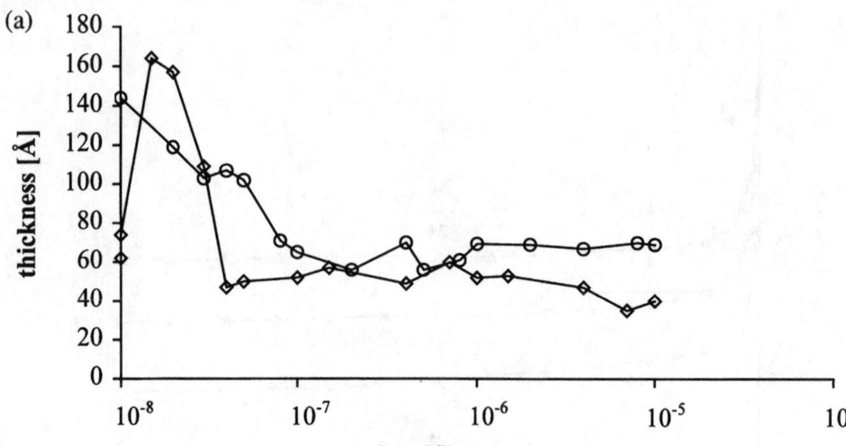

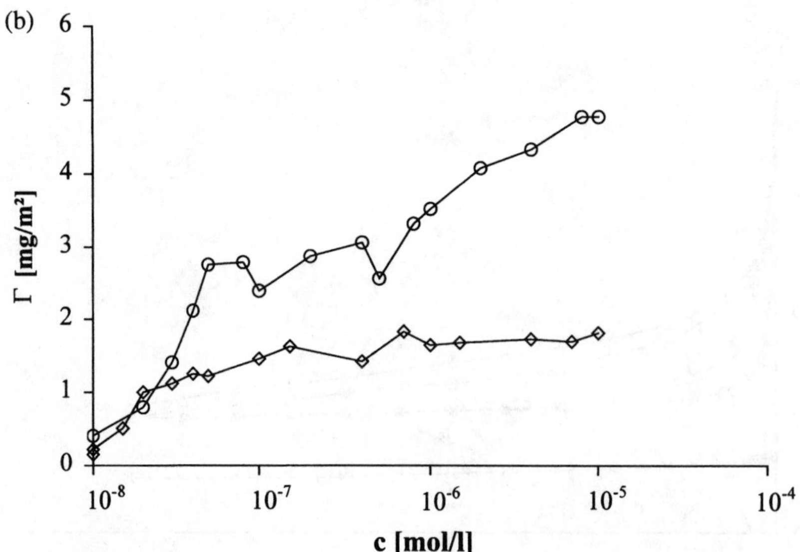

Figure 5 *Ellipsometric studies of β-casein (○) and β-lactoglobulin (◇) at the water–air interface at 25 °C showing (a) adsorption layer thickness and (b) adsorbed amount Γ*

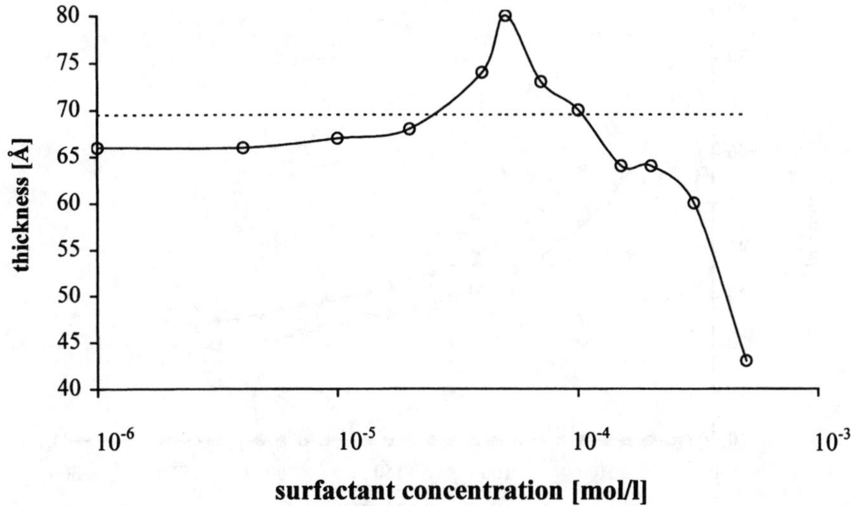

Figure 6 *Adsorption layer thickness of the β-casein + $C_{10}DMPO$ mixed system at a protein concentration of $c = 10^{-6}$ mol l^{-1} at 25 °C. The dashed line indicates the thickness of the pure β-casein layer*

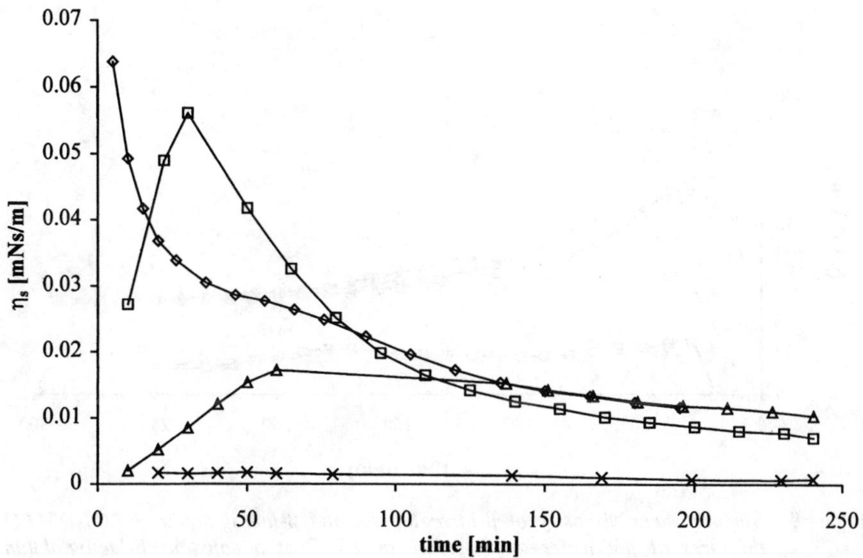

Figure 7 *Surface shear viscosity of β-casein at the water–air interface at 25 °C at different concentration of 10^{-8} mol l^{-1} (×), 10^{-7} mol l^{-1} (△), 10^{-6} mol l^{-1} (□), and 10^{-5} mol l^{-1} (◇)*

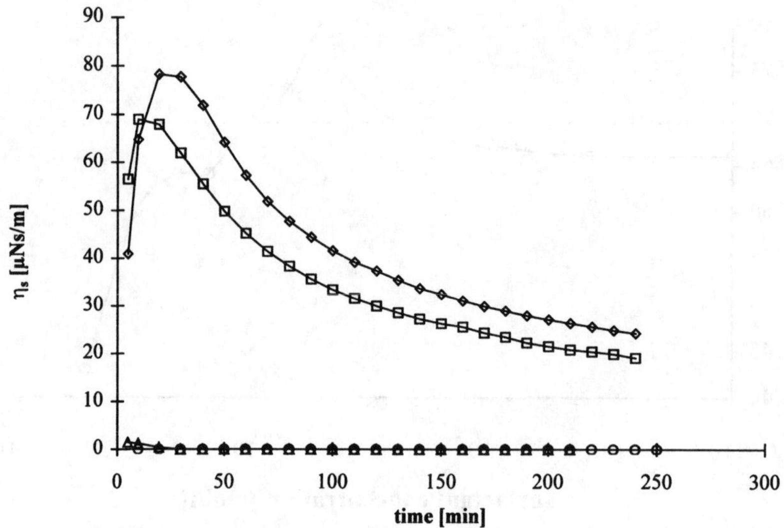

Figure 8 *Surface shear viscosity of β-casein and β-casein + CTAB mixtures at the water–air interface at 25 °C at a constant β-casein concentration of 10^{-6} mol l^{-1} (◇) and different concentrations of CTAB. Molar ratio (β-casein:CTAB): 10 (□), 0.1 (△) and 0.001 (○)*

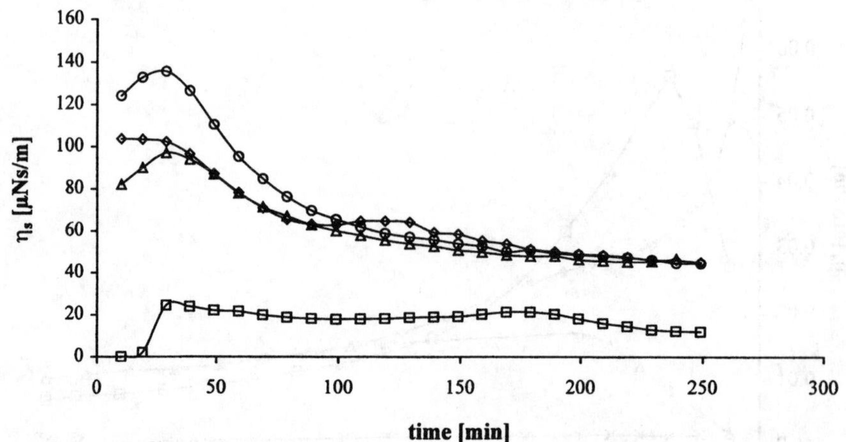

Figure 9 *Surface shear viscosity of β-lactoglobulin and β-lactoglobulin + $C_{10}DMPO$ mixtures at the water–air interface at 25 °C at a constant β-lactoglobulin concentration of 10^{-6} mol l^{-1} (◇) and different concentrations of $C_{10}DMPO$. Molar ratio (β-lactoglobulin:$C_{10}DMPO$): 0.1 (○), 0.06 (△) and 0.03 (□)*

5 Discussion

Adsorption Behaviour of the Pure Proteins

As shown elsewhere,[14] the $\Pi(c)$ isotherms of the two proteins can be described by a theoretical expression of the form[22]

$$\Pi = -\frac{RT}{\omega_\Sigma}[\ln(1-\Gamma_\Sigma\omega_\Sigma) - a_{el}\Gamma_\Sigma^2\omega_\Sigma^2], \tag{4}$$

$$b_1 c = \frac{\Gamma_1\omega_\Sigma}{(1-\Gamma_\Sigma\omega_\Sigma)^{\omega_{min}/\omega_\Sigma}}, \tag{5}$$

where R and T are the gas constant and absolute temperature, respectively, b_1 is a constant characterizing the surface activity of the protein, and a_{el} is a parameter representing the electrical charge effect. The total adsorption in these equations can be expressed via the adsorption in state 1, Γ_1, and the minimum partial molar area ω_{min}:

$$\Gamma_\Sigma = \Gamma_1 \sum_{i=1}^{n} i^\alpha \exp\frac{\omega_{min}(i-1)}{\omega_\Sigma}\exp\left[-\frac{(i-1)\Pi\omega_{min}}{RT}\right]. \tag{6}$$

The mean partial molar area of all states, ω_Σ, and the adsorption in any i^{th} state Γ_i, can be expressed by

$$\omega_\Sigma = \omega_{min}\frac{\sum_{i=1}^{n} i^{(\alpha+1)}\exp\left(-\frac{i\Pi\omega_{min}}{RT}\right)}{\sum_{i=1}^{n} i^{\alpha}\exp\left(-\frac{i\Pi\omega_{min}}{RT}\right)}, \quad \Gamma_i = \Gamma_\Sigma\frac{i^\alpha\exp\left[-\frac{(i-1)\Pi\omega_{min}}{RT}\right]}{\sum_{i=1}^{n} i^\alpha\exp\left[-\frac{(i-1)\Pi\omega_{min}}{RT}\right]}. \tag{7}$$

The maximum molar area ω_{max} in this model is defined by $\omega_{max} = n\omega_{min}$, and it represents the most extended conformation of the protein molecule at the interface.

The results of fitting the model to the surface tension data for the two model proteins of Figure 1 are summarized in Table 1. They show that, although the initial area per molecule (ω_{max}) is much larger for the casein, the final molar area ω_{min} for both proteins is comparable. Due to the compact structure of β-lactoglobulin, the final thickness is lower than that of β-casein (*cf.* Figures 5a and 5b). Also the adsorbed amount Γ is about two times larger for β-casein than for β-lactoglobulin. The comparatively large thickness of the measured β-lactoglobulin adsorbed layer at low bulk concentration (>150 Å) is difficult to understand (Figure 5a).

Table 1 *Adsorption parameters of the model proteins β-lactoglobulin and β-casein*

Protein	$\omega_{max}/(nm^2)$	$\omega_{min}/(nm^2)$	a_{el}
β-Casein	100	7	80–150
β-Lactoglobulin	15–20	8	40

Adsorption Behaviour of Mixed Protein + Surfactant Systems

The isotherms of the three selected surfactants have been measured in the presence of different amounts of the two proteins. As an example, Figure 3 gives the results for CTAB at two protein concentrations, 10^{-7} mol l^{-1} and 10^{-6} mol l^{-1}. A decrease of the surface tension γ becomes remarkable already at a concentration of the ionic surfactant (CTAB or STS) where no change in γ is observed in the absence of protein. This indicates that the protein–surfactant complexes formed are highly surface-active. With increasing surfactant concentration the isotherms of the mixed systems approach those of the pure surfactant with a slight shift to higher concentrations (the shift is caused by loss of surfactant due to binding to the protein). This means that the complexes are continuously replaced at the surface by surfactant molecules, although the complexes are expected to be much more surface-active than the protein alone. This effect can be explained either by a change in the type of interaction with the surfactant (hydrophobic rather than ionic) or by aggregate formation of the complexes.

This picture of the mixed layers is supported by the dynamic surface tension data given in Figure 4. The adsorption dynamics changes from that of a pure protein (at low surfactant concentration) to the behaviour of a pure surfactant. A detailed analysis of the adsorption kinetics has been given elsewhere.[23] To determine the nature and strength of surfactant binding to the protein, dynamic surface tension data are not suitable. This information can possibly be obtained from relaxation studies at higher frequencies, such as oscillating bubble or capillary wave studies. Such investigations are underway.

The behaviour for both proteins and both ionic surfactants (CTAB and STS) is quite similar. In contrast, the adsorption of protein + non-ionic surfactant mixtures seems to be purely competitive. Only when the adsorption of C_{10}DMPO increases significantly (*i.e.* γ decreases remarkably) is the protein repelled from the surface layer. Finally, the isotherm of the mixture exactly coincides with that of the pure surfactant; *i.e.*, no significant amount of surfactant is bound to the protein.

Adsorption Layer Thicknesses of Pure and Mixed Systems

Ellipsometric measurements of the thickness of surfactant adsorbed layers is rather difficult and inaccurate when the molecules are too short. For the samples used in the present study, exact data could not be obtained. The experiments for pure protein adsorbed layers, however, as shown in Figure 5, do give some insight into the interfacial molecular arrangement. The final thickness of the β-lactoglobulin adsorption layer is approximately the length of the molecule (56 Å).[24] The thickness of the β-casein layer is about twice as large. Measurements on the effect of added ionic surfactants on the layer thickness are still under way and results are not available yet.

Measurements of the adsorbed layer thickness of the mixed protein + non-ionic surfactant systems support the model of a pure competitive adsorption. At a given protein concentration, the interfacial layer thickness remains constant until a replacement of the protein due to increasing adsorption of the non-ionic surfactant sets in. This is demonstrated for the system β-casein + C_{10}DMPO in Figure 6. When the surfactant concentration reaches the critical micelle concentration (CMC) the adsorbed layer thickness decreases significantly, approaching a value close to zero. This value corresponds to the surfactant adsorbed layer, the thickness of which is too small to be measured accurately by ellipsometry.

Surface Shear Viscosity of Pure and Mixed Systems

The measured surface shear viscosities of the pure proteins show a general type of behaviour as given in Figure 7 for the case of β-casein. At a low bulk concentration of 10^{-8} mol l^{-1}, no measurable viscosity was obtained in the time period of 3 hours. With increasing concentration the viscosity η_s increases, eventually passing through a maximum and then levelling off. For the β-lactoglobulin a similar behaviour is obtained. However, although the surface tension isotherms (Figure 1) document a similar surface activity, the viscosity of the β-lactoglobulin adsorbed layer almost diminishes at a concentration of 10^{-5} mol l^{-1}, while that of the β-casein at the same concentration passes through comparatively high values (about 60 μN m^{-1} s) and then levels off at 10 μN m^{-1} s. The β-lactoglobulin layer seems to be well ordered and the adsorbed molecules do not resist against shear deformation above 10^{-5} mol l^{-1}. In contrast, the casein molecules are still in a conformation in a rather packed interfacial layer which allows strong mutual interaction.

The surface shear viscosity of β-casein layers behaves similarly when ionic surfactant is added (Figure 8). Keeping the protein concentration constant, the addition of increasing amounts of CTAB leads to a viscosity which passes through a maximum with time. This effect may be due to displacement of the protein–surfactant complexes by the unbound surfactant molecules. Increasing surfactant concentration leads to the behaviour of a pure surfactant adsorption layer. This, however, seems to be very unrealistic. Inspection of the isotherms of the pure and mixed systems in Figure 3a shows that the adsorption of CTAB only starts at a bulk concentration of 10^{-5} mol l^{-1}. The adsorbed layer is for sure formed of protein–surfactant complexes. At a protein–surfactant ratio of 0.1, the conformation of these complexes seems to become highly surface-active and very compact, so that the applied shear stress does not feel a remarkable resistance. The general behaviour of the mixtures of β-lactoglobulin with ionic surfactants is similar. It has to be noted here, however, that the surface properties of β-lactoglobulin depend very much on the sample origin. The effect of sample purification has been discussed recently elsewhere.[15] Depending on the lot preparation number of the commercial sample number and the method of purification, the surface shear viscosity was found to change significantly. This particular behaviour is discussed further elsewhere.[25]

The rheological behaviour of the mixed protein + non-ionic surfactant systems is very different from the systems containing the ionics. It strongly supports the idea of a competitive adsorption between the protein and surfactant; however, a certain degree of hydrophobic interaction seems to exist. The surface shear viscosity of β-lactoglobulin + C_{10}DMPO mixtures is shown in Figure 9. As one can see, a change in the η_s values sets in when the surfactant starts to adsorb. With increasing surfactant concentration, the surface viscosity slightly increases, then decreases, and reaches values close to zero at the CMC, which is characteristic for a pure surfactant layer.

6 Conclusions

The equilibrium state of the pure protein adsorbed layers is described by a thermodynamic model. Their adsorption kinetics is described by a diffusion theory based on the respective thermodynamic equations of state. The surface tension data correlate well with data from adsorbed layer thickness and surface rheological measurements.

For the mixed systems, quantitative theoretical models do not exist, and thus the thermodynamic and kinetic data can be discussed only qualitatively. The assumption that there are different types of interaction between protein and surfactant molecules, and a specific surface activity and conformation of the protein–surfactant complexes, is in agreement with the adsorption and rheological data obtained.

The complexes between proteins and ionic surfactants can have a higher surface activity than the pure protein, as evident from the more than additive surface pressure increase. The presence of ionic surfactant seems also to modify the structure of the adsorbed protein. Thus, even β-casein becomes rather more compact at the interface at a certain STS or CTAB concentration.

In the presence of non-ionic surfactant, the adsorption layer is mainly formed as a result of competitive adsorption between the compounds. This behaviour is supported by all the quantities measured: surface tension isotherm, adsorbed layer thickness, and surface shear viscosity.

Acknowledgements

The work was financially supported by a project of the European Community (INCO ERB-IC15-CT96-0809), the DFG (Mi418/9-1, Mi418/7-11), and the Fonds der Chemischen Industrie (RM 400429). A research fellowship from the

Alexander von Humboldt Stiftung (A.V.M., BB-1034737) is also gratefully acknowledged.

References

1. 'Food Colloids and Polymers: Stability and Mechanical Properties', ed. E. Dickinson and P. Walstra, Royal Society of Chemistry, Cambridge, 1993.
2. 'Studies of Interface Science', ed. D. Möbius and R. Miller, Elsevier, Amsterdam, vol. 7, 1998.
3. 'Emulsions: Fundamentals and Applications in the Petroleum Industry', ed. L. L. Schramm, ACS Advances in Chemistry Series No. 231, American Chemical Society, Washington, D.C., 1992.
4. J. A. de Feijter and J. Benjamins, in 'Food Emulsions and Foams', ed. E. Dickinson, Royal Society of Chemistry, London, 1987, p. 72.
5. P. Chen, Z. Policova, S. S. Susnar, C. R. Pace-Asciak, P. M. Demin, and A. W. Neumann, *Colloids Surf. A*, 1996, **114**, 99.
6. R. Wüstneck, J. Krägel, R. Miller, P. J. Wilde, and D. C. Clark, *Colloids Surf. A*, 1996, **114**, 255.
7. J. Chen and E. Dickinson, *Colloids Surf. A*, 1995, **100**, 267.
8. E. Dickinson, R. K. Owusu, and A. Williams, *J. Chem. Soc. Faraday Trans.*, 1993, **89**, 865.
9. M. Cornec, A. R. Mackie, P. J. Wilde, and D. C. Clark, *Colloids Surf. A*, 1996, **114**, 237.
10. D. C. Clark, J. Mingins, F. S. Sloan, L. J. Smith, and D. R. Wilson, in 'Food Emulsions and Foams', ed. E. Dickinson, Royal Society of Chemistry, London, 1987, p. 110.
11. M. Coke, P. J. Wilde, E. J. Russel, and D. C. Clark, *J. Colloid Interface Sci.*, 1990, **138**, 489.
12. R. Wüstneck, J. Krägel, R. Miller, P. J. Wilde, and D. C. Clark, *Colloids Surf. A*, 1996, **114**, 255.
13. D. K. Sarker, P. J. Wilde, and D. C. Clark, *J. Agric. Food Chem.*, 1995, **43**, 295.
14. A. V. Makievski, R. Wüstneck, D. O. Grigoriev, J. Krägel, and D. V. Trukhin, *Colloids Surf. A*, in press.
15. D. C. Clark, F. Husband, P. J. Wilde, M. Cornec, R. Miller, J. Krägel, and R. Wüstneck, *J. Chem. Soc. Faraday Trans.*, 1995, **91**, 1991.
16. P. Chen, D. Y. Kwok, R. M. Prokop, O. I. del Rio, S. S. Susnar, and A. W. Neumann, in 'Studies of Interface Science', ed. D. Möbius and R. Miller, Elsevier, Amsterdam, 1998, vol. 6, p. 61.
17. R. Miller, R. Sedev, K.-H. Schano, Ch. Ng, and A. W. Neumann, *Colloids Surf.*, 1993, **69**, 209.
18. A. V. Makievski, R. Miller, V. B. Fainerman, J. Krägel, and R. Wüstneck, this volume, p. 269.
19. R. M. A. Azzam and N. M. Bashara, 'Ellipsometry and Polarized Light', North Holland, Amsterdam, 1979.
20. J. A. de Feijter, J. Benjamins, and F. A. Veer, *Biopolymers*, 1978, **17**, 1759.
21. J. Krägel, S. Siegel, R. Miller, M. Born, and K.-H. Schano, *Colloids Surf. A*, 1994, **91**, 169.
22. V. B. Fainerman and R. Miller, in 'Studies of Interface Science', ed. D. Möbius and R. Miller, Elsevier, Amsterdam, 1998, vol. 7, p. 51.

23. R. Wüstneck, J. Krägel, R. Miller, V. B. Fainerman, P. J. Wilde, D. K. Sarker, and D. C. Clark, *Food Hydrocolloids*, 1996, **10**, 395.
24. S. Brownlow, J. H. Morais-Cabral and L. Sawyer, Protein Data Bank, www.pdb.bnl.gov/pdb-bin/pdbids, 1BEB.
25. J. Krägel, R. Wüstneck, F. Husband, P. J. Wilde, A. V. Makievski, D. O. Grigoriev, and J. B. Li, *Colloids Surf. B*, in press.

Comparison of the Dynamic Behaviour of Protein Films at Air–Water and Oil–Water Interfaces

By Brent S. Murray, Merete Færgemand, Marion Trotereau, and Ann Ventura

FOOD COLLOIDS GROUP, PROCTER DEPARTMENT OF FOOD SCIENCE, UNIVERSITY OF LEEDS, LEEDS LS2 9JT, UK

1 Introduction

The dynamic behaviour of protein films is recognized as being of importance to the formation and stability of the wide variety of food colloids in which proteins occur.[1,2] By dynamic behaviour is meant the response of the film to the changing stresses and strains imposed upon the film as consequence of flow within the colloidal system. The study of such changes is described as interfacial (or surface) rheology. However, the exchange of surface-active material between the interface and the bulk phases, and the close proximity of particle surfaces in concentrated systems, means that interfacial rheology is also intimately connected with the overall composition and bulk rheology of the material. The interfacial rheology of a protein film reflects the complex three-dimensional structure which the constituent molecules possess. On adsorption at an interface this structure changes in a complex way, with existing bonds being broken and new bonds being formed, both within and between molecules, during the formation of a structurally coherent film.

Proteins are very surface-active materials, in that they will readily adsorb at most types of interface, *e.g.*, air–water (A–W), oil–water (O–W) or solid–water, *etc.*, often with a high affinity.[3] Most of the detailed experimental studies of the adsorption of proteins have been carried out at the interface between air and water, or at the interface between water and well characterized solids. The latter have typically been suitably transparent materials (sometimes after surface chemistry modification) for the application of various types of surface spectroscopy,[4,5] using techniques which in some instances (*e.g.*, ellipsometry) are applicable also at the A–W interface. Whilst such studies[4–6] have provided a wealth of important information on the formation and structure of adsorbed layers at such surfaces, the technologically more important interface between water and oil phases has been relatively neglected. The presence of the second bulk (oil) phase, which is also fluid and has a refractive index closer to that of the aqueous phase, often inhibits the application of techniques developed for the A–W or solid–water interfaces. The purpose of this paper is to draw attention to interfacial rheology measurements on proteins at O–W interfaces in order to highlight the fact that much more work needs to be done in this area. This is because the behaviour at an O–W interface is not necessarily the same as that at an A–W (or solid–water) interface, due to the different state of unfolding of the protein molecules at the different surfaces.

Dilatational and Shear Interfacial Rheology

Two types of interfacial rheology[2] may be distinguished—dilatational rheology, where the strain is the fractional area change of the interface, and shear rheology, where the strain is the gradient of displacement within the interface (with no area change). It is important to distinguish between the two types because, if an area change occurs, the adsorption equilibrium between the interface and the bulk becomes disturbed, and the interfacial composition may vary during the deformation. Dilatational rheological behaviour is thus associated in some way with the rate of adsorption/desorption of the surface-active agent. Since the formation and collapse of emulsion droplets or foam bubbles involves an increase and decrease of interfacial area, respectively, attention tends to be focused on dilatational rheology rather than shear rheology.[1] In practice, however, all interfacial disturbances will involve a combination of shear and dilatational deformation, and the complex network-like nature of most protein films means that the shear and dilatational moduli are necessarily related.[1,7] Consequently, emulsion and foam stability has been correlated with both shear and dilatational moduli.[1,8]

Previous Studies at O–W and A–W Interfaces

As mentioned above, there have been relatively few measurements of protein structure and dynamics at O–W interfaces—including measurements of dilatational rheology. Shear rheological measurements are somewhat easier to perform, and they have been shown[9] to be highly sensitive to changes in solution conditions and interfacial composition. Shear measurements are difficult to model in terms of film structure, however. The most comprehensive attempt to measure the dynamic behaviour of pure proteins at both A–W and O–W interfaces is the study by Graham and Phillips.[10,11] Radiolabelled proteins were used, however, which can modify the surface activity,[12,13] so that some of the measurements may not be directly related to the behaviour of the native proteins. Also, different oils (toluene doped with radiation scintillator, paraffin oil or dodecane) were used for different O–W measurements. Noting these deficiencies, the picture which emerges from the studies of Graham and Phillips is a complex one—some proteins (*e.g.*, lysozyme) appear to be more unfolded at the O–W interface than at the A–W interface, whilst other proteins (*e.g.*, β-casein) show the opposite behaviour. Similarly, depending on the bulk protein concentration, C_b, or the interfacial protein concentration, Γ, the interfacial moduli sometimes appear to be much higher at the O–W interface than at the A–W interface, whilst at other times they are lower. More recent measurements have indicated that the relative magnitudes of the moduli at the two types of interface depend on the bulk protein concentration and also the oil type.[14,15] It is of interest to establish whether the apparent differences at O–W interfaces are a result of differences in the inherent solvency of the oils, the presence of other surface-active materials in the oils, or the result of differing measurement techniques (probing different rates and extent of film deformation).

2 Materials and Methods

Imidazole, potassium dihydrogen phosphate, disodium hydrogen phosphate, sodium chloride, concentrated nitric acid, hydrochloric acid, granulated activated charcoal (product no. 33034) were all *AnalaR* grade reagents from BDH Merck. Bovine serum albumin (product code A-7638, lot no. 14H9348), β-lactoglobulin (product code L-0131, lot no. 91H7005), lysozyme (chicken egg white, product code L6876, lot number 111H7010), research grade hexa-oxyethylene glycol *n*-dodecyl ether ($C_{12}E_6$) and *n*-tetradecane (99%) were from Sigma Chemicals (Poole, UK). The microbial Ca^{2+}-independent transglutaminase (TGase) derived from *Streptoverticillium mobaraense* was obtained from Ajinomoto Co. (Japan) and purified as described previously.[16] Commercial whey protein, whey protein isolate PSDI-2400, was from MD Foods Ingredients (Videbaek, Denmark). Commercial sodium caseinate was from DMV (Veghel, Netherlands). Purified β-casein was obtained from the Hannah Research Institute (Ayr, UK). Sunflower oil was purchased from a local supermarket (Morrisson's). All water used was from a Millipore Alpha-Q purification system with a surface tension of 72.0 mN m^{-1} at 25 °C.

Surface measurements were made on a Langmuir trough apparatus that was described in detail previously.[17] The barrier containing the interfacial film consists of a continuous, rhombus-shaped piece of Teflon which can flex at the corners to change the film area. Interfacial pressure (π) versus area per molecule (A) measurements were made at the O–W and A–W interfaces as described elsewhere.[17] 'Spread monolayers' were formed by spreading aqueous protein solution at the interface. 'Adsorbed monolayers' were formed via the following procedure. The trough was filled with protein solution and left for a certain time to allow protein to adsorb at the interface. Most of the protein solution was then slowly run out of the trough, but leaving enough to ensure that the interfacial film remained enclosed within the Teflon barrier. The protein solution removed was then slowly replaced with buffer to dilute down the original subphase. The subphase was gently mixed with a long glass rod, beneath the barrier. This procedure was repeated several times to give a subphase of negligible protein concentration (typically 10^{-7} wt% protein) with an adsorbed film of protein at the A–W interface. All monolayers were compressed at speeds slow enough that no hysteresis was observed on immediate re-compression of the newly formed monolayer. Dilatational rheological measurements were made in two different ways. (i) The interface was subjected to a sudden (1 second), step-change in area and the resultant change in π *versus* time curve was Fourier transformed[18] to obtain the dynamic storage ε' and loss moduli ε'' as a function of frequency. (ii) The interfacial area, A_0, was subjected to a sinusoidal variation (amplitude ΔA_0) at fixed frequency, f, and the resultant sinusoidal variation in π monitored. The dynamic elastic modulus, ε_0, was then calculated from

$$\varepsilon_0 = \frac{\Delta\pi}{\Delta A_0/A_0}, \tag{1}$$

where $\Delta\pi$ is the average amplitude in π measured. All protein solutions were buffered with either 0.02 mol dm^{-3} imidazole or phosphate, adjusted to pH 7.0 by addition of hydrochloric acid. Charcoal treatment was used to further purify the pure β-lactoglobulin, as described previously.[19] The temperature was 25 $\pm$ 2 °C, unless stated otherwise. Oils were purified by passing through a column of freshly activated alumina.

3 Results and Discussion

π–A isotherms

Figure 1(a) shows π–A isotherms for spread films of β-casein and β-lactoglobulin at the n-tetradecane–water and A–W interfaces. Figure 1(b) shows the behaviour of the commercial sodium caseinate and the commercial whey protein at the same interfaces. All proteins were initially more expanded at the O–W interface, though it is seen also that all proteins were more compressible at the O–W interface: the O–W isotherms crossed over the A–W isotherms at higher π, to give lower A for the same π. A similar picture has been observed for the globular protein BSA. The A–W isotherms for β-casein and β-lactoglobulin are identical to those obtained by Mitchell *et al.*[19] some time ago, but considerably less expanded than the 'adsorbed' π–A isotherms reported by Graham and Phillips for the A–W interface. The latter authors constructed[10] adsorbed π–A isotherms for β-casein, BSA and lysozyme from the equilibrium adsorbed Γ and π determined for the radiolabelled proteins at O–W and A–W interfaces. As such, agreement with spread monolayers might only be expected at high A (low Γ) where A is relatively independent of the bulk concentration from which the protein has adsorbed.[20] The adsorbed isotherms obtained by Graham and Phillips for β-casein at the paraffin–water interface were also quite different from those measured here at the n-tetradecane interface. As far as the authors are aware there are no other O–W isotherms to compare with for these proteins.

Taken on their own, the spread monolayer results indicate that proteins with both a globular structure (*e.g.*, BSA, β-lactoglobulin, lysozyme, *etc.*) and a more

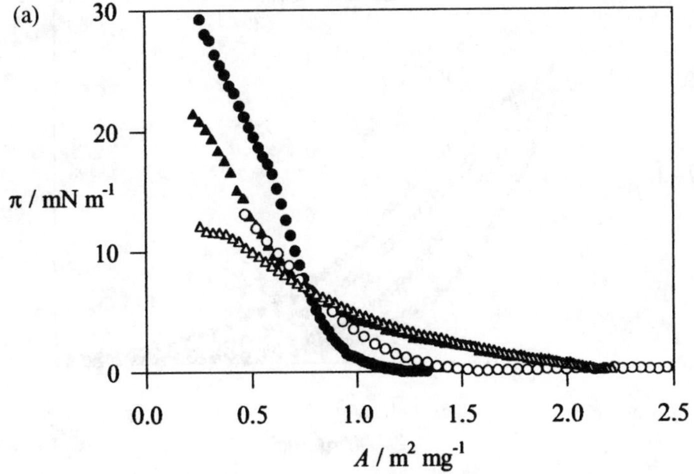

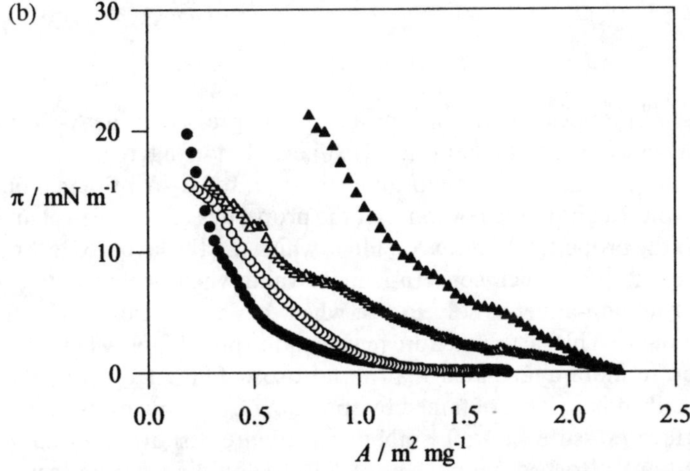

Figure 1 *(a) π–A isotherms at the A–W interface for spread films of β-casein (○) and β-lactoglobulin (●), and at the* n*-tetradecane–water interface for β-casein (△) and β-lactoglobulin (▲). (b) π–A isotherms at the A–W interface for spread films of sodium caseinate (○) and whey protein (●), and at the* n*-tetradecane–water interface for sodium caseinate (△) and whey protein (▲)*

random structure (*e.g.*, the caseins) are more unfolded at the O–W interface than at the A–W interface. Previous work[17,19] has also indicated that A–W protein monolayers slowly become more expanded with time, but that this more expanded state is lost on compression. More recent work[21] has confirmed this behaviour, as indicated in Figure 2. Spread films of β-lactoglobulin which were left for a long time before compression slowly became more expanded. But on repeated compression the isotherm shifted back towards the isotherm for the freshly formed monolayer. This effect was not observed at the O–W interface,

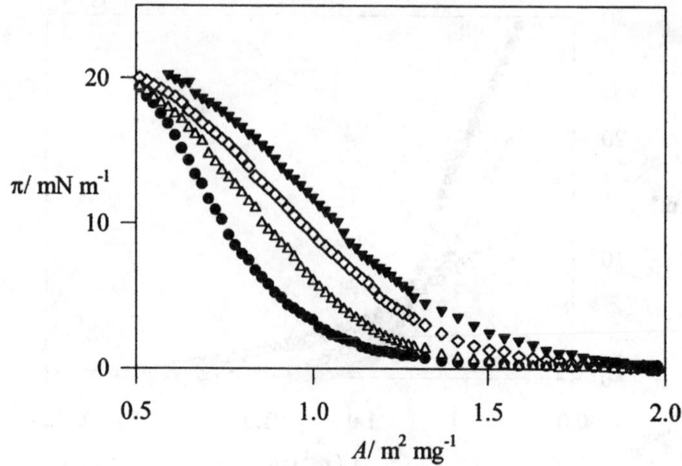

Figure 2 *π–A isotherms at the A–W interface for spread films of β-lactoglobulin, for different film ages: 0 h (●), 24 h (◇) and 68 h (▼). Also shown is the isotherm after 3 successive compressions of the 24 h film (△)*

however.[17–19] It appears that, when protein is compressed at the A–W interface, it tends to persist in a state that is less flexible and more aggregated.

Observation of such non-equilibrium behaviour for A–W protein monolayers prompts concern about the relevance of the properties of *spread* protein films in relation to the properties of *adsorbed* films, which are the layers of technological importance in foods. Such concern is more valid when proteins are spread as solids or from non-aqueous solvents, or when they are spread rapidly to a high initial π (low A). This last procedure may lead to spread films where the protein molecules are more aggregated than in adsorbed films. However, the monolayers described here were obtained by spreading from aqueous solution to low initial surface pressure ($\pi < 0.1$ mN m^{-1}). Figure 3 shows π–A curves for a spread and an 'adsorbed' monolayer of β-lactoglobulin (formed from adsorption at 2×10^{-4} wt% protein). Since the surface concentration is actually unknown for the adsorbed monolayer, the A values for the adsorbed monolayer were derived by assuming that the A values are equal for both types of film at $\pi = 25$ mN m^{-1}. The agreement between the two curves is reasonably close, considering that the adsorbed monolayer was 12–14 h old, so that it may have been slightly more expanded due to the ageing effect described above. Thus, the real π–A isotherms, and hence the structures of the monolayers formed in the two very different ways, must be close: spread monolayers are identical to films formed by adsorption, at least for adsorption from low bulk concentrations. Figure 3 also shows a π–A isotherm for an adsorbed film of lysozyme, formed (and calculated) as for the β-lactoglobulin film. This illustrates the usefulness of the adsorbed monolayer technique, since lysozyme cannot be reliably spread under these conditions,[13] because of its tendency to desorb on spreading due to its high charge at neutral pH.

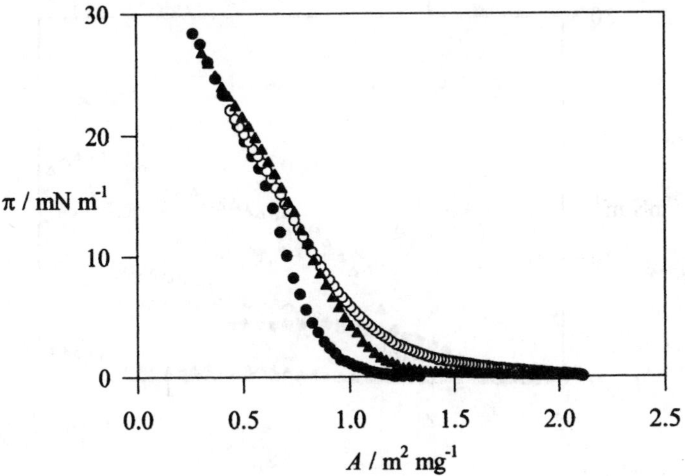

Figure 3 *π–A isotherm at the A–W interface for a spread monolayer of β-lactoglobulin (●), an 'adsorbed' monolayer of β-lactoglobulin (○) and an 'adsorbed' monolayer of lysozyme (▲). The* A *values for the adsorbed monolayers were derived by assuming the* A *values were equal to the value for the spread film at* $\pi = 25\ mN\ m^{-1}$

Dilatational Interfacial Rheology and Enzymic Cross-linking

Figure 4 shows the dilatational rheology of adsorbed β-lactoglobulin and sodium caseinate films measured at the A–W interface and at the O–W interface, measured via the step area change technique. Consistent with other measurements made under similar conditions,[22] higher moduli were observed for β-lactoglobulin at the O–W interface (where the oil was *n*-tetradecane) than at the A–W interface. If proteins are more unfolded at the O–W interface, then this might be expected to lead to greater intermolecular interaction and higher film elasticity and viscosity. Williams and Prins[14] showed a similar trend for the same value of C_b at the paraffin–water interface (f = 0.1 Hz and $\Delta A_0/A_0$ = 3%). At lower and higher concentrations they observed that the difference between the A–W and O–W interfaces decreases. At low C_b the moduli at both interfaces must necessarily decrease to zero, as Γ tends to zero. At high C_b and such low ΔA_0, one might expect moduli to be dominated by the extent of protein adsorption from the bulk, so that the type of interface becomes less important (providing adsorption barriers do not vary significantly amongst the set of interfaces). In agreement with the results of Cagna *et al.*[15] (also at f = 0.1 Hz and $\Delta A_0/A_0$ = 3%), the moduli for caseinate were lower at the O–W interface than at the A–W interface. Possibly this is due to the even greater unfolded state at the O–W interface, which makes the protein extremely flexible and mobile, giving the low modular behaviour characteristic of low-molecular-weight surfactants.

In order for a globular protein such as β-lactoglobulin to become intermolecularly cross-linked by the enzyme transglutaminase, the protein must be unfolded to some extent, to give the enzyme access the reacting amino acid

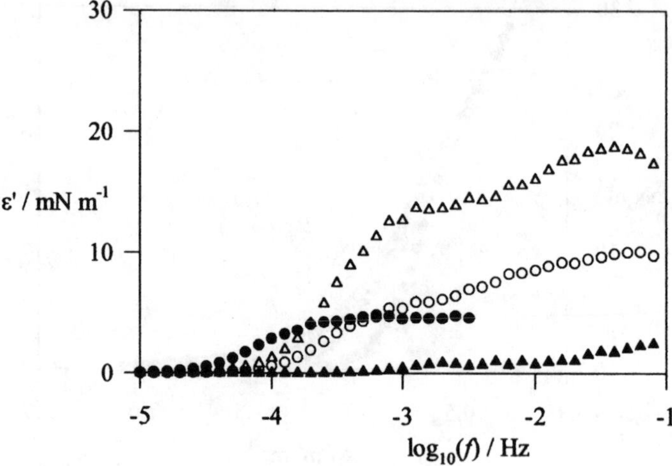

Figure 4 *Dynamic dilatational storage moduli, ε', at the A–W interface for adsorbed films of sodium caseinate (●) and β-lactoglobulin (○) and at the O–W interface for sodium caseinate (▲) and β-lactoglobulin (△). Measurements made via the step area technique, at 40 °C, for bulk protein concentrations 10^{-3} wt%, with oil phase* n-*tetradecane*

residues (GLN and LYS). In the bulk this normally requires heating the protein, but on adsorption there may be sufficient unfolding for cross-linking to take place.[16] Figure 5 shows that cross-linking of β-lactoglobulin at the O–W interface leads to considerably higher moduli than at the A–W interface, which may be explained by the greater accessibility of the protein to the enzyme, due to

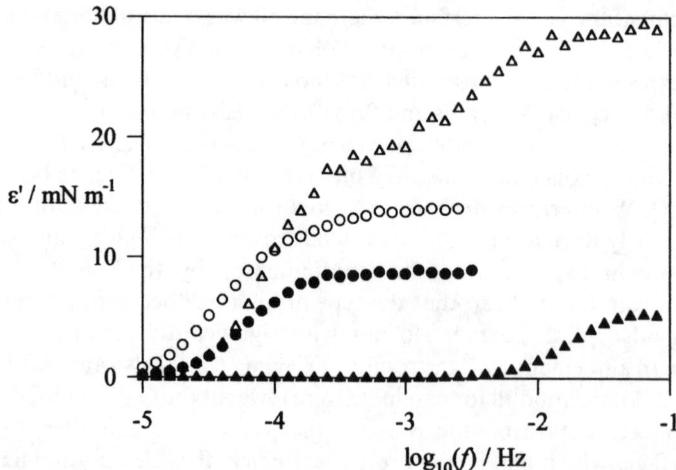

Figure 5 *Dynamic dilatational storage moduli, ε', after cross-linking the same films as in Figure 4 with transglutaminase for 2 h: sodium caseinate (●) and β-lactoglobulin (○) at the A–W interface; sodium caseinate (▲) and β-lactoglobulin (△) at the O–W interface*

its greater degree of unfolding. Other work on spread monolayers has also shown[23] a corresponding shift in the π–A isotherm to higher A on cross-linking. Caseinate films show more limited increases in moduli at the O–W interface. Possibly this is another consequence of the greater extent of unfolding at the O–W interface, such that the protein molecules are so strongly adsorbed that cross-linking by the enzyme is sterically prohibited. However, cross-linking may also occur in the bulk for caseinate, and so that the interpretation of the results is not so straightforward.

Sources of Discrepancy between Measurements: Impurities and Non-linearity

The presence of low-molecular-weight surface-active impurities, which may occur in either the protein sample or the oil phase, could have a significant effect on both the π–A isotherms and the surface rheological measurements. Certainly the charcoal treatment to which the β-lactoglobulin was subjected to in this study does seem to have a significant effect on the dynamic properties of such proteins.[24] Treatment with activated alumina or silicates is a standard method for removing surface-active impurities from oil.[25] In all the measurements made at the O–W interface described so far, the oil (n-tetradecane) was not subjected to further purification after purchase. This was because it was found that, whether the oil was subjected to treatment with activated alumina or not, after contacting the oil with pure water in the Langmuir trough for 1–2 h and compressing the interface to 25% of the original area, a small (4–5 mN m^{-1}) increase in π always appeared. This indicated the persistent presence of a low concentration of surface-active contaminant in the hydrocarbon which was difficult to remove. On the other hand, the values of the static air–oil surface tension and the static O–W interfacial tension were not reliable indicators of the presence of this contamination, since they were always close to typical literature values for 'pure' oils (see Table 1).

Table 1 *Interfacial tensions and dilatational elasticities,*[a] ε_0, *at various oil–water interfaces*

Oil	$\gamma_{A\text{-}O}$ *(24 h)*[b]/ *mN m*$^{-1}$	$\gamma_{O\text{-}W}$ *(24 h)*[c]/ *mN m*$^{-1}$	ε_0 *(24 h)*/ *mN m*$^{-1}$	ε_0 *(48 h)*/ *mN m*$^{-1}$
n-Tetradecane (unpurified)	26.5	50.01	29.0	37.5
Sunflower (unpurified)	33.1	10.8	25.0	35.2
Sunflower ($\frac{1}{2}$ × purified)[d]	33.0	16.7	11.0	15.0
Sunflower (1 × purified)	–	28.5	–	–
Sunflower (2 × purified)	33.1	31.0	22.1	25.0

[a] Measured at the interface between 10^{-3} wt% β-lactoglobulin and the oil at f = 0.05 Hz, ΔA = 8%.
[b] Surface tension between oil and air after 24 h.
[c] Interfacial tension between oil and pure water after 24 h.
[d] Equal volume mixture of unpurified and 1 × purified.

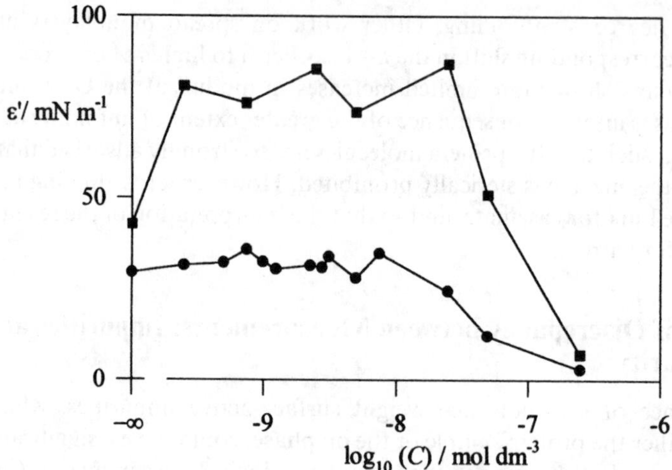

Figure 6 *Dynamic dilatational storage moduli, ε', for adsorbed films of β-lactoglobulin at the A–W interface (●) and O–W interface (■) as a function of the concentration,* C, *of added $C_{12}E_8$. Measurements made via the step area technique:* f = *6.31×10^{-3} Hz, ΔA = 10%, 30 °C, bulk protein concentration 10^{-3} wt%, with oil phase* n*-tetradecane*

Only relatively small amounts of surface-active impurities are necessary to affect the dilatational and shear rheology of protein films.[25] For example, Figure 6 shows the effect of deliberately adding the water-soluble surfactant $C_{12}E_8$ to systems containing adsorbed films of β-lactoglobulin. It can be seen that a marked drop in modulus occurs at a low concentration of surfactant, the values of moduli for proteins being typically higher than those for surfactants.

Table 1 shows the dilatational elasticity, ε_0, for β-lactoglobulin adsorbed at the *n*-tetradecane–water interface and at the sunflower O–W interface, using sunflower oil that was subjected to increasing cycles of purification. For untreated sunflower oil ε_0 was lower than with the untreated tetradecane. Even with the most highly purified ('2 × purified') sunflower oil, the value of ε_0 was still lower than for the unpurified tetradecane. The O–W interfacial tension measurements, $\gamma_{O\text{-}W}$, indicated that the '2 × purified' sunflower oil was essentially surfactant free, because $\gamma_{O\text{-}W}$ was constant at 33.1 mN m^{-1} for 24 h. Although the *n*-tetradecane was 'unpurified', the value of $\gamma_{O\text{-}W}$ for this oil was almost steady, falling only 3 mN m^{-1} to 50.1 mN m^{-1} over a period of 24 h. Thus the difference in ε_0 for these two fairly pure oils must be ascribed to differences in structure of the adsorbed protein at these two interfaces, rather than different impurities in the oils.

For the sunflower oil subjected to different degrees of purification, ε_0 changed in a complicated way. The unpurified oil had the highest value of ε_0, the '$\frac{1}{2}$ purified' oil the lowest value of ε_0, and the '2 × purified' oil an intermediate value of ε_0. The surface-active impurities will be of lower molecular weight: fatty acids, monoglycerides, phospholipids, *etc.* At intermediate concentrations (*i.e.*, in the '½ purified' oil) a mixed protein–surfactant film will exist, with the protein structure and mobility possibly being modified through surfactant binding. This apparently lowers ε_0, the dominant effect possibly being the more rapid exchange of the impurities between the bulk phase and the interface, in order to establish the equilibrium tension as the interface expands and contracts. At high concentrations of these impurities (*i.e.*, in the unpurified oil) there will be a much higher surface concentration of low-molecular-weight surfactants (possibly completely displacing protein) or more protein–surfactant complexes. Either way this appears to give some enhancement of ε_0. Thus, the chemical nature of the oil, and/or the level of impurities in the oil, may affect the dilatational rheology at the O–W interface in a complicated way.

It is known[1] that the interfacial shear rheology of adsorbed protein films is highly non-linear. This can be expected in view of the complex, network-like structure of such films. Figure 7 shows some recent dilatational measurements, which also indicate non-linearity, obtained via both the oscillatory and step area change methods.[23] The rheology appears to be approximately linear (*i.e.*, modulus independent of the strain) above a certain strain. Below this strain, the rheology appears to be highly non-linear, though extension to even higher strains might show further changes in ε_0. It is therefore important to ensure that comparisons of measurements made on different instruments are made at the same value of the strain, ΔA. The technological significance of this behaviour is that strains and rates of strain imposed during emulsion and foam processing will far exceed those employed here, and so the moduli under these practical conditions may be considerably different from those determined in laboratory film studies.

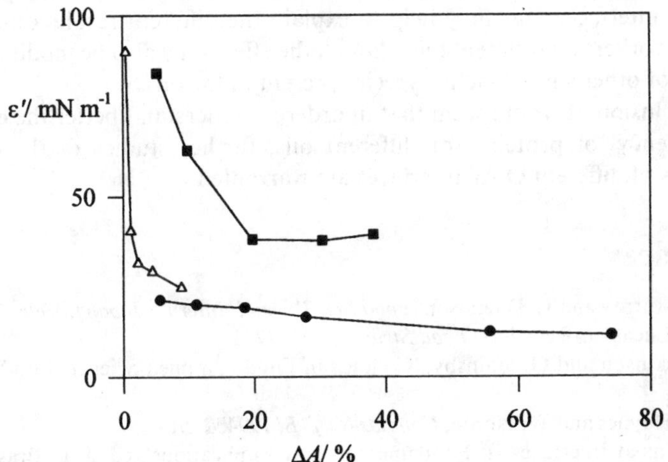

Figure 7 *Dynamic dilatational storage moduli, ε', of adsorbed β-lactoglobulin films as function of % strain, ΔA:* n-*tetradecane–water interface at 30 °C via step area change (■); pure sunflower oil interface at 25 °C via oscillating area (●); air–water interface at 25 °C via oscillating area (△). Bulk protein concentration* 10^{-3} *wt% in each case, at frequency 0.05 Hz*

Figure 8 *Proposed conformations of a typical globular protein on initial adsorption at air–water and oil–water interfaces*

4 Concluding Remarks

The experiments described here clearly indicate that the dynamic behaviour of films of both globular and random coil proteins have important differences at A–W and O–W interfaces. Overall, proteins appear to be more unfolded and possibly more flexible at the O–W interface, and this results in higher dilatational moduli for globular proteins. This effect is most likely due to the better solvency of the oil for the side-chains of the hydrophobic amino acid residues of the proteins. This is understood to produce the sort of structural changes as indicated schematically in Figure 8, where it is important to note that none of the polypeptide is appreciably solubilized *in* the oil phase. The polarity of the peptide bond makes this extremely unlikely: even polyphenylalanine has very poor solubility in oil. This picture is different from the one presented by Graham and Phillips,[10] who suggested that lower areas per molecule for β-casein at the O–W interface were due to looping of hydrophobic segments *into* the oil. Either way, it is to be expected that the chemical nature of the oil and its affinity for the hydrophobic segments will affect the unfolding of the protein at the O–W interface. This may help to explain the different results obtained by different workers on different oils, though the effect may also be modified by the presence of other surface-active species present in the oils.

In conclusion, it would seem that in order to understand better the emulsifying 'efficiency' of proteins for different oils, further studies of the dynamic properties of different O–W interfaces are warranted.

References

1. B. S. Murray and E. Dickinson, *Food Sci. Technol. Intern. (Japan)*, 1996, **2**, 131.
2. E. H. Lucassen-Reynders, *Food Struct.*, 1993, **12**, 1.
3. E. Dickinson and G. Stainsby, 'Colloids in Food', Applied Science, London, 1982, p. 285.
4. C. A. Haynes and W. Norde, *Colloids Surf. B*, 1994, **2**, 517.
5. 'Proteins at Interfaces II: Fundamentals and Applications', ed. J. L. Brash and T. Horbett, American Chemical Society, Washington DC, 1995.
6. J. L. Brash, *Curr. Opin. Colloid Interface Sci.*, 1996, **1**, 682.
7. J. Benjamins and F. van Voorst Vader, *Colloids Surf.*, 1992, **65**, 161.
8. B. S. Murray, in 'Proteins at Liquid Interfaces', ed. R. Miller and D. Möbius, Elsevier, Amsterdam, 1998, p. 179.

9. J. Castle, E. Dickinson, B. S. Murray, and G. Stainsby, *ACS Symp. Ser.*, 1987, **343**, 118.
10. D. E. Graham and M. C. Phillips, *J. Colloid Interface Sci.*, 1979, **70**, 403, 415, 427.
11. D. E. Graham and M. C. Phillips, *J. Colloid Interface Sci.*, 1980, **76**, 227, 240.
12. D. J. Adams, M. T. A. Evans, J. R. Mitchell, M. C. Phillips, and P. M. Rees, *J. Polym. Sci. Part C*, 1971, **34**, 167.
13. B. S. Murray, *Langmuir*, 1997, **13**, 1850.
14. A. Williams and A. Prins, *Colloids Surf. A*, 1996, **114**, 267.
15. J. Benjamins, A. Cagna, and E. H. Lucassen-Reynders, *Colloids Surf. A*, 1996, **114**, 245.
16. M. Færgemand, B. S. Murray, and E. Dickinson, *J. Agric. Food Chem.*, 1997, **45**, 2514.
17. B. S. Murray and P. V. Nelson, *Langmuir*, 1996, **12**, 5953.
18. G. Loglio, U. Tesei, and R. Cini, *J. Colloid Interface Sci.*, 1984, **100**, 393.
19. J. R. Mitchell, L. Irons, and G. J. Palmer, *Biochim. Biophys. Acta*, 1970, **200**, 138.
20. J. R. Mitchell, in 'Developments in Food Proteins', ed. B. J. F. Hudson, Elsevier, London, 1986, vol. 4, p. 291.
21. G. Garofalakis and B. S. Murray, *Colloids Surf. B*, in press.
22. B. S. Murray, C. Lallemant, and A. Ventura, *Colloids Surf. A*, in press.
23. M. Færgemand and B. S. Murray, *J. Agric. Food Chem.*, 1998, **46**, 885.
24. D. K. Sarker, P. J. Wilde, and D. C. Clark, *Colloids Surf. A*, 1996, **114**, 227.
25. A. G. Gaonkar, *J. Am. Oil Chem. Soc.*, 1989, **66**, 1090.

Surface Activity at the Air–Water Interface in Relation to Surface Composition of Spray-Dried Milk Protein-Stabilized Emulsions

By Anna Millqvist-Fureby, Norman Burns, Karin Landström, Pia Fäldt,* and Björn Bergenståhl†

INSTITUTE FOR SURFACE CHEMISTRY, P.O. BOX 5607, SE-114 86 STOCKHOLM, SWEDEN

1 Introduction

Many food powders (*e.g.*, milk powders) are produced by spray-drying of emulsions made up of protein, fat and lactose. Several important technical properties depend on particle–particle interactions (*e.g.*, flowability)[1] or on particle–liquid interactions (*e.g.*, wettability, dispersability).[1,2] It can be assumed that these properties are determined to a large extent by the chemical composition of the outermost surface of the particles. Spray-dried powders of biological origin (*e.g.*, milk powder) have been studied by methods such as scanning electron microscopy, which provide information on the physical structure at the particle surfaces.[3–8] Free fat extraction has been employed to analyse the extractability of the fat in the powder particles.[9,10] Useful information on the powders can be obtained from such analyses, but they do not provide any quantitative information on the chemical composition of the powder surface. An expansion of our understanding has been made possible by the introduction of ESCA (electron spectroscopy for chemical analysis) for the chemical analysis of surface composition.[11,12] This ESCA technique allows for the characterization of the surface in terms of the principal chemical composition.

From theoretical and experimental arguments, a theory has been derived[12] on how the surface of the drying droplet is formed. According to this theory a saturated solution of dissolved solids is formed close to the droplet surface, and a solid crust is generated at the droplet surface by precipitation of the least soluble substance from this saturated solution.[13] The crust formation is less

*Present address: Aromatic AB, P.O. Box 440 40, SE-100 73 Stockholm, Sweden
†Present address: Department of Food Technology, Lund University, P.O. Box 124, SE-221-00 Lund, Sweden

distinct in the case of mixed biological powders containing carbohydrate and/or protein. According to the previously mentioned theory, the surface of the particles should consist mainly of lactose in the case of a milk powder or a similar spray-dried emulsion rich in lactose. However, the analysis of spray-dried dairy-based emulsions by ESCA has revealed that the powder surfaces are generally dominated by protein, whilst fat is largely encapsulated inside the particles.[12,14,15] These observations are corroborated by an interesting earlier study of Müller,[16] who investigated milk powders by transmission electron spectroscopy (TEM) after treatment with osmium tetroxide (which binds to the protein and fat and thus improves contrast in the TEM). The surface of spray-dried skim milk and whole milk powders was observed to consist of a thin layer ($\approx$10 nm) of casein aggregates and other proteins.

Studies of the surface tension of the drying liquid in relation to the surface composition have suggested an alternative to the 'crust theory' for the surface formation in spray-drying. This is based on the assumption that the composition of the air–water interface of the sprayed droplet is reflected in the composition of the dried particle surface.[17] Accordingly, the air–water interface of the sprayed droplets is dominated by the most surface-active component in the solution, in this case protein, which therefore becomes over-represented at the powder surface. The decrease in surface tension, as compared to pure water, depends on the type of protein. Caseins are more surface-active than whey proteins, in the sense that they give a lower surface tension at the air–water interface.[15,17] Hence, it could be assumed that the caseinate samples would give a higher surface coverage than whey proteins. It is well known that caseinates are more efficient emulsifiers than are whey proteins.[18,19] We note that the fat is not accessible for adsorption at the air–water interface since it is present as emulsified fat droplets and has a very poor solubility in the water phase.

The pH of the solution affects the solution properties of the protein, and hence the surface tension and the adsorption kinetics. In this investigation we have combined measurements of surface tension and surface analysis by ESCA to elucidate the relationship between the surface tension at the air–water interface and the surface composition of the dried emulsions, for the cases of three different dairy-based proteins at various pH values.

The ESCA Technique

Electron spectroscopy is a well-established technique for elemental analysis of solid surfaces, and its application to organic powders has been described elsewhere.[11] The underlying principle of the technique is the photoelectric effect. A sample in high vacuum is irradiated with monochromatic X-ray photons. On collision with the sample, a photon's energy is completely transferred to an orbital electron. If the binding energy of this electron is less than the photon energy, the electron will be emitted from the atom with a kinetic energy equal to the difference between the photon energy and the binding energy of the emitted electron. An analysis of the kinetic energy of these emitted electrons provides information on which elements are present in the sample, since the binding energy is characteristic for each element and orbital. The emitted electrons have to travel through the material to reach the surface, and in this process inelastic scattering of the energy occurs. Thus only the electrons emitted close to the surface of the material can escape into vacuum. In general, the surface layer that is analysed is 2 to 10 nm deep, but the actual depth of analysis depends on the type of material and the geometry of the surface. For organic materials and spherical particles, the penetration depth is estimated to be approximately 3 to 6 nm.[11]

Assuming that the elemental composition of the surface is a linear combination of the elemental compositions of the different molecular species in the sample, the data on the elemental composition can be used to estimate the molecular composition of the surface layer by solving a matrix equation,

$$\mathbf{C}\gamma = \mathbf{c}, \tag{1}$$

where **C** is the matrix containing the elemental compositions of the molecular species (atom%), **c** is the vector containing the elemental surface composition (atom%), and γ is the surface coverage of the different molecular species (%). To allow for the use of over-determined systems, we apply the least square method when solving equation (1), *i.e.*

$$\gamma = (\mathbf{C}^{\mathrm{T}}\mathbf{C})^{-1}\mathbf{C}^{\mathrm{T}}\mathbf{c}. \tag{2}$$

Full details of the calculation method have been described elsewhere.[11]

2 Materials and Methods

Materials

Sodium caseinate (spray-dried, 95% protein, 1.1% fat, 0.2% lactose, 3.6% ash, Miprodan 31) and calcium caseinate (spray-dried, >93.5% protein, 1.2% fat, 0.2% lactose, 4.0% ash, Miprodan 40) were obtained from MD Foods (Denmark), and whey protein concentrate (76% protein, 9.5% fat, 5.0% carbohydrate, 3.0% ash, WPC80) was obtained from Norske Meierier (Oslo, Norway). Rapeseed oil was obtained from Karlshamm AB (Karlshamm, Sweden). Lactose and sodium chloride were purchased from Merck (Darmstadt, Germany). Lactic acid was obtained from KEBO (Stockholm, Sweden). Doubly distilled water was used throughout.

Preparation of Emulsions

The dry solid content of all emulsions were, by weight, 40% lactose, 30% rapeseed oil and 30% protein (sodium caseinate, calcium caseinate, or whey protein concentrate). Solutions of lactose and protein were made to a concentration of 10% and adjusted to pH 7 with 1 M NaOH. The lactose and protein solutions were then mixed in amounts corresponding to lactose/protein proportions (by weight) of 4/3. Rapeseed oil was then added to a total solids content of 13.5%.

The mixtures were pre-homogenized in a high speed colloid mill, Ultraturrax (IKA Labortechnik, Staufen, Germany), for 5 minutes. The resulting emulsions were then further homogenized in a high-pressure homogenizer, Microfluidizer TM 110 (Microfluidic Inc., Newton, MA, USA), at 800 bar. Each emulsion was recycled 8 times. The homogenization device was placed in a water bath (~20 °C) and the liquid was passed through a cooling coil at 20 °C after passing the homogenization valve and before returning to the reservoir; thus the temperature of the homogenized liquid did not rise above ~45 °C.

The pH was adjusted by slow (dropwise) addition of 10 mM lactic acid to the cooled emulsion (6–10 °C) for pH values between 7 and 5. For the lower pH values, adjustment was made by fast addition of concentrated lactic acid, followed by pH adjustment with 1 M NaOH. After addition of lactic acid, water was added to adjust the total emulsion solids content to 10%.

Spray-Drying of the Emulsions

Emulsion samples at various pH values were spray-dried in a laboratory spray-drier built at the Institute for Surface Chemistry. The dimensions of the drying chamber are 0.5 × 0.15 m. The spray-drier operates co-currently and has a spray-nozzle with an orifice 1 mm in diameter. Inlet gas temperature was 180 °C. Outlet gas temperature was kept in the range 70–75 °C by adjusting the gas flow. Liquid feed to the drier was 11 ml min^{-1}. The flow of drying air was 0.8 m^3 min^{-1}. Powder was collected in a cyclone (60 mm in diameter) at the outlet, yielding 3–5 g of powder per 100 ml of solution. The powder samples were stored in closed containers over dry silica gel in a desiccator.

Electron Spectroscopy for Chemical Analysis (ESCA)

The ESCA measurements were performed with an AXIS HS photoelectron spectrometer (Kratos Analytical, UK). The instrument uses a monochromatic Al Kα X-ray source. The pressure in the vacuum chamber during analysis was less than 10^{-7} torr. In the present investigation, a take-off angle of the photoelectrons perpendicular to the sample holder was used throughout. The powders were loosely packed in 40 μl aluminium pans, and the surface was levelled. The area analysed consisted of a circular region of approximate diameter 1.3 mm, measurements being made at three different points for each sample. The standard deviation of the analysis was 0.2% for C, 0.05% for O, and 0.2% for N.

Measurement of Surface Tension

The surface tension at the air–water interface of the protein solutions at various pH values was determined according to the du Noüy ring method using a Sigma

70 tensiometer (KSV Instruments Ltd, Finland). The equilibrium time required for steady state surface tension measurements was 12–15 minutes.

3 Results and Discussion

Analysis of Surface Composition of Dried Emulsions

We have studied the effect of three different milk protein products on the surface composition of powders obtained by spray-drying emulsions consisting of lactose, rapeseed oil and milk protein. Furthermore, the effect of emulsion pH on the surface composition has been studied. The surface coverages of protein, fat and lactose at pH values between 3 and 7 are presented in Figure 1.

Sodium caseinate is efficient at encapsulating the rapeseed oil, which is present to a level of 35% or less at the surface (Figure 1b). The protein (Figure 1a) and lactose (Figure 1c) coverages are somewhat increased at high pH, whilst the fat coverage is significantly reduced at high pH. The caseins are soluble at low and high pH, and show some aggregation into structures with a molecular weight of approximately 2.5×10^5 D at high pH[20] which are supposed to be similar to the sub-micelles of casein micelles. The caseinate is precipitated at pH 4 to 5.5, which is around the casein isoelectric point (pH 4.6).

The protein coverage of powders prepared with calcium caseinate as the stabilizing agent ranges from a high level ($\approx$80%) at high pH (pH $\geqslant$ 5.5) to a low level ($\approx$50%) at low pH (pH $\leqslant$ 4.0), the shift between levels occurring close to the agglomeration pH of 5.5 (Figure 1a). The fat coverage (Figure 1b) shows the opposite dependence on pH, whilst the lactose coverage (Figure 1c) is very low throughout the whole pH range, except at pH 5.0. The calcium caseinate forms larger aggregates at high pH, whilst it is well solubilized at low pH, possibly into sub-micelle-like aggregates. This was clearly observed by eye, the protein solution appearing milky at pH $\geqslant$ 5.5, but being slightly opalescent at pH 3.0 when the large aggregates were dissolved. At intermediate pH (4–5.5), the calcium caseinate was found to agglomerate and precipitate. The original casein micelles of milk are not restored in reconstituted calcium caseinate, since colloidal calcium phosphate is not reformed.[21] From these results it seems that the larger casein aggregated structures are superior to the dissolved caseins for the purpose of encapsulating the fat phase.

The protein and fat coverages of the whey protein-containing powders were found to be very similar in magnitude throughout the pH range studies, and there are only small variations in the levels of all the components (see Figure 1). The whey proteins are soluble over the entire pH range studied, with a slight decrease in solubility at around pH 4,[22] and they do not form supermolecular aggregates.

Sodium caseinate and calcium caseinate (the latter at high pH) both show a more pronounced over-representation on the powder surface than does the whey protein. The caseins are known to be more surface-active than the whey proteins in spray-drying,[14,15,17] and to have better emulsifying properties,[18,19] and hence it could be expected that they would be superior in encapsulating the

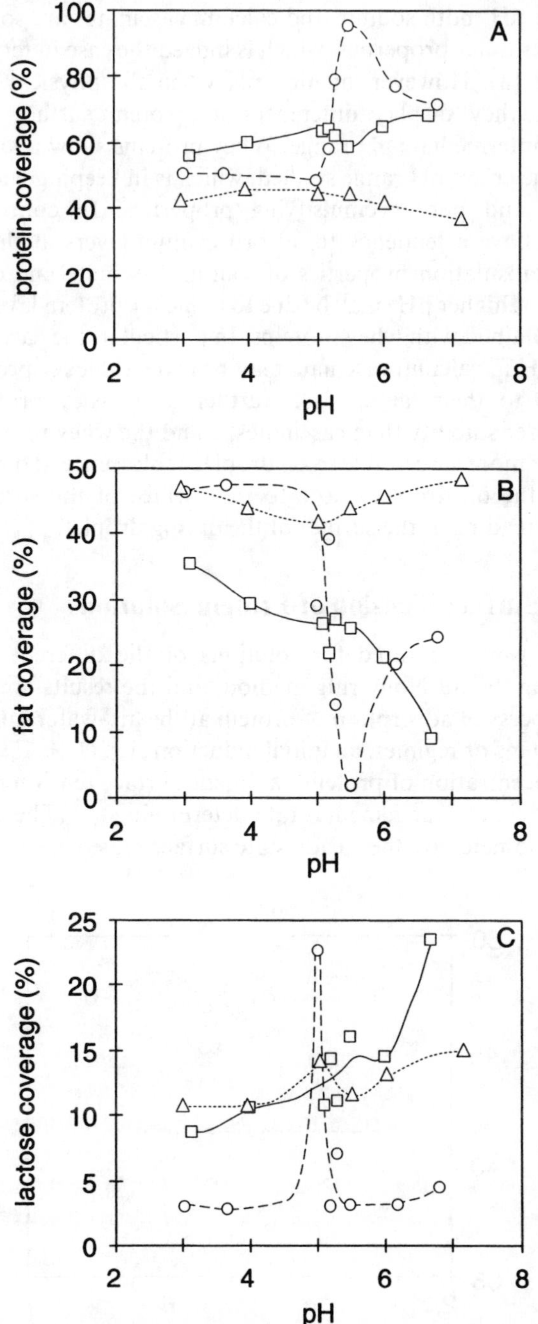

Figure 1 *Surface coverage of components on the spray-dried emulsions. (A) protein coverage, (B) fat coverage, and (C) lactose coverage. (□) Na-caseinate stabilized emulsions, (△) whey protein stabilized emulsions, and (○) Ca-caseinate stabilized emulsions*

fat phase. At low pH, both sodium and calcium caseinates are soluble and are expected to show similar properties, which is indeed the case in terms of protein coverage (Figure 1a). However, at high pH when the physical forms of the caseinates differ, they display differences in protein surface coverage, in particular in the intermediate pH range. Whey proteins show a lower coverage of protein over the entire pH range studied, which is in keeping with their lower surface activity, and poorer emulsifying properties. In contrast to whey proteins, caseins have a tendency to adsorb in multilayers at the interface.[23] The superior encapsulation properties of sodium caseinate and calcium caseinate (albeit only at higher pH) may be due to a thicker protein layer close to the surface than is obtained with whey proteins. In particular, the casein aggregates present at high pH in calcium caseinate may provide a thicker protein layer on the surface, due to their larger size. Furthermore, whey protein leads to emulsions of poorer stability than caseinates,[24] and the whey protein-stabilized emulsions may be more susceptible to oil droplet coalescence and rupture of the protein layer of the oil droplets close to the surface of the sprayed droplet, resulting in fat spreading at the surface of the drying droplet.

pH Effects on Surface Tension of Protein Solutions

Surface tensions were measured for solutions of the different milk protein preparations using the du Noüy ring method, and the results are presented in Figure 2. The process of adsorption of protein at the air–water interface can be divided in three steps or regimes: an initial induction step ($\Gamma < \Gamma_{max}/2$, where Γ is the surface concentration of protein), a kinetic surface tension regime, and a regime where protein rearrangement is rate-determining.[23,25] The data obtained with the ring tensiometer are the steady state surface tensions.

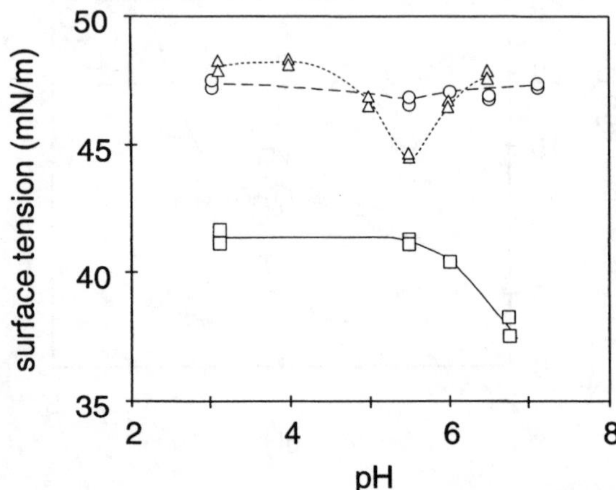

Figure 2 *Surface tension as a function of pH measured by the ring tensiometer for the individual protein solutions: (□) Na-caseinate stabilized emulsions, (△) whey protein stabilized emulsions, and (○) Ca-caseinate stabilized emulsions*

Sodium caseinate showed an increase in surface tension between pH 5.5 and 6.0 (Figure 2). This is in contrast to the findings of Wüstneck *et al.*,[26] but the protein concentration in the present study is considerably higher (3% as compared to 0.024%), which might affect the relative surface tensions at the different pH values. However, the equal surface tensions at pH 6.8 and 6.0 are in accordance with results obtained by Tornberg.[19] It was not possible to measure the surface tension at pH values from 4 to 5, due to severe precipitation of the protein (and thus a significantly lowered protein concentration and possibly altered composition of the remaining soluble protein fraction).

Calcium caseinate showed small variations in surface tension when pH was varied, with values always higher than those of sodium caseinate. Previously casein micelles and sodium caseinate have been observed[19] to give similar surface tensions at pH 7.0, albeit at a lower protein concentration of 0.1%, and so the size of the protein aggregates does not appear to have any appreciable effect on the rate of adsorption at the air–water interface.

Whey protein, on the other hand, showed a higher surface tension than sodium caseinate, with a decrease in surface tension at pH 5.5. Wüstneck *et al.*[26] found no dependence on pH for β-lactoglobulin. However, Waniska[27] observed a minimum in surface tension for β-lactoglobulin at pH 5.3, which is similar to our results. In a separate investigation, the surface tension was also found to be depressed on decreasing pH from 7.0 to 6.0 for whey protein.[19]

The Surface Tension in Relation to the Surface Composition

In Figure 3, the surface coverage of protein is seen to be correlated with the surface tension of the corresponding protein solutions. In the case of sodium

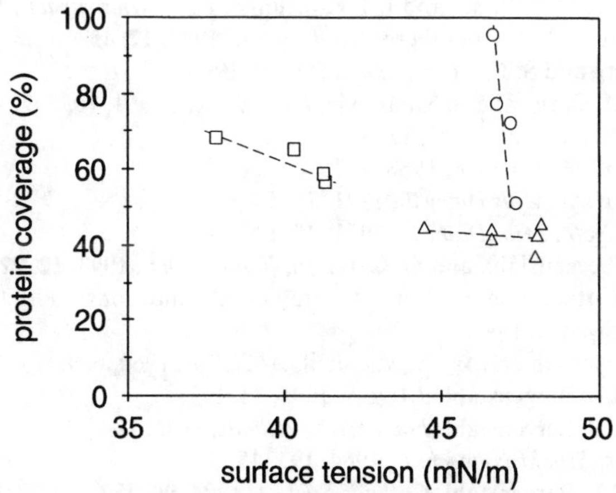

Figure 3 *Surface coverage of protein on spray-dried emulsions in relation to the surface tension of the corresponding protein solutions. Linear least-square fits are added for clarity: (□) Na-caseinate stabilized emulsions, (△) whey protein stabilized emulsions, and (○) Ca-caseinate stabilized emulsions*

caseinate and whey proteins, the expected effect of surface tension is confirmed, *i.e.*, a lower surface tension provides a higher surface coverage of protein on the dried particles. A comparison of powders obtained by spray-drying mixtures of lactose + sodium caseinate[17] and lactose + whey protein[15] shows that over-representation of protein on the powder surface is more pronounced in the former case. This may be due to the somewhat poorer adsorption characteristics of the whey protein, observed as a higher surface tension.

Calcium caseinate adsorption at the spray-dried droplet surface seems to be more weakly correlated with the surface tension, which varies over a very narrow range for this protein. This result suggests that the difference in aggregate size has a major effect on the protein coverage of the surface. If diffusion were rate-limiting in the sprayed droplets, the calcium caseinate at pH $\geqslant 5.5$ would provide the lowest surface coverages due to the diffusion rate being lower for larger particles; however the contrary behaviour is observed. This indicates that additional transport mechanisms exist, *e.g.*, by convection originating from temperature and pressure gradients. Due to the high shear-rates in the spray nozzle, it can be assumed that the liquid in the spray droplets is not quiescent, and it may even be turbulent. Under such turbulent conditions, protein particles are known to adsorb more rapidly than individual proteins at an oil–water interface.[28] Clearly, then, properties of the proteins other than their surface activity, as well as properties of the emulsions, also affect the surface composition of the powders.

References

1. N. King, *Dairy Sci. Abstr.*, 1965, **27**, 91.
2. T. J. Buma, *Neth. Milk Dairy J.*, 1971, **25**, 33.
3. M. Rosenberg, Y. Talmon, and I. J. Kopelman, *Food Microstruct.*, 1988, **7**, 15.
4. D. L. Moreau and M. Rosenberg, *Food Struct.*, 1993, **12**, 457.
5. M. Rosenberg and S. L. Young, *Food Struct.*, 1993, **12**, 31.
6. M. Kaláb, M. Caric, and S. Milanovic, *Food Struct.*, 1991, **10**, 327.
7. M. Kaláb, *Food Struct.*, 1993, **12**, 95.
8. Z. Saito, *Food Microstruct.*, 1988, **7**, 75.
9. T. J. Buma, *Neth. Milk Dairy J.*, 1971, **25**, 53.
10. T. J. Buma, *Neth. Milk Dairy J.*, 1971, **25**, 123.
11. P. Fäldt, B. Bergenståhl, and G. Carlsson, *Food Struct.*, 1993, **12**, 225.
12. P. Fäldt, 'Surface composition of spray-dried emulsions', Ph.D. Thesis, Lund University, Sweden, 1995.
13. D. H. Charlesworth and W. R. Marshall, *AIChE J.*, 1960, **6**, 9.
14. P. Fäldt and B. Bergenståhl, *JAOCS*, 1995, **73**, 171.
15. P. Fäldt and B. Bergenståhl, *Food Hydrocolloids*, 1996, **10**, 421.
16. H. R. Müller, *Milchwissenschaft*, 1964, **19**, 345.
17. P. Fäldt and B. Bergenståhl, *Colloids Surf. A*, 1994, **90**, 183.
18. J. E. Kinsella, *CRC Crit. Rev. Food Sci. Nutr.*, 1984, **21**, 197.
19. E. Tornberg, A. Olsson, and K. Persson, in 'Food Emulsions', ed. K. Larsson and S. Friberg, Marcel Dekker, New York, 1990, p. 247.
20. L. K. Creamer and G. D. Berry, *J. Dairy Res.*, 1975, **42**, 169.

21. J. A. Lucey, C. Gorry, B. O'Kennedy, M. Kalab, R. Tan-Kinita, and P. F. Fox, *Int. Dairy J.*, 1996, **6**, 257.
22. P. Walstra and R. Jenness, 'Dairy Chemistry and Physics', New York, 1984, p. 301.
23. E. Dickinson, 'An Introduction to Food Colloids', Oxford University Press, Oxford, 1992.
24. S. Y. Lee, C. V. Moor, and E. Y. W. Ha, *J. Food Sci.*, 1992, **57**, 1210.
25. B. C. Tripp, J. J. Magda, and J. D. Andrade, *J. Colloid Interface Sci.*, 1995, **173**, 16.
26. R. Wüstneck, J. Krägel, R. Miller, V. B. Fainerman, P. J. Wilde, D. K. Sarker, and D. C. Clark, *Food Hydrocolloids*, 1996, **10**, 395.
27. R. D. Waniska, 'Glycosylated β-lactoglobulin: chemical, structural, interfacial and functional properties', Ph.D. Thesis, Cornell University, Ithaca, NY, 1982.
28. P. Walstra and H. Oortwijn, *Neth. Milk Dairy J.*, 1982, **36**, 103.

Protein–Lipid Interactions at the Air–Aqueous Phase Interface

By Juan M. Rodríguez Patino and Mª. Rosario Rodríguez Niño

DEPARTAMENTO DE INGENIERÍA QUÍMICA, FACULTAD DE QUÍMICA, UNIVERSIDAD DE SEVILLA, C/PROF. GARCÍA GONZÁLEZ, S/N, 41012 SEVILLE, SPAIN

1 Introduction

Food dispersions contain many emulsifiers or surfactants, the most important being proteins and low-molecular-weight surfactants. The colloidal stability of the dispersed phase, in the form of a foam or an emulsion, is determined by the chemical and physical properties of surface-active molecules like lipids and proteins. Proteins stabilize foams and emulsions by forming intermolecular interactions between the adsorbed protein molecules which encapsulate the dispersed phase. The barrier is formed as intermolecular interactions between the adsorbed protein molecules are established. These contribute significantly to its rheological properties and immobilize proteins in the adsorbed layer. In contrast, lipids stabilize the dispersed droplet or bubble by formation of a densely packed, but much less rigid, monomolecular layer which is stabilized by dynamic processes (*i.e.* the Gibbs–Marangoni effect). The complex mechanisms involved in the formation and stabilization/destabilization of dispersions make fundamental studies on applied systems difficult. One approach has been to clarify the basic physical and chemical properties of these dispersions by the study of simple model systems. The existence of protein–surfactant interactions at the air–aqueous phase interface has been studied in detail by different methods, including tensiometry, surface dilational rheology, and drainage and diffusion in thin liquid films. Attention is here directed towards the effect of surfactants (monostearin, monoolein, and Tween 20) and some food additives (ethanol and sucrose) on the protein–lipid interactions at interface.

2 Surface Properties of Protein + Lipid Mixed Films at Equilibrium

Surface tension measurements were used to determine the equilibrium adsorption of BSA (>96% pure, Fluka) and a soluble lipid (Tween 20, Fluka) and to explore the presence of interactions at the interface between BSA and soluble or insoluble lipids (monostearin and monoolein). The surface activity is expressed by the surface tension, σ, or the surface pressure, $\pi = \sigma_o - \sigma$, where σ_o and σ are the aqueous subphase surface tension and the surface tension of the aqueous solutions of BSA or BSA + lipid mixed films, respectively. Measurements were performed with a Krüss digital tensiometer K10 based on the Wilhelmy plate method.[1]

Surface Tension of Protein + Insoluble Lipid

The effect of the protein/lipid ratio on the surface activity of mixed BSA + lipid systems on water (Mille-Q, Millipore) and 1 M aqueous solutions of ethanol and sucrose at 20 °C is shown in Figure 1. In these experiments the lipid concentration spread on a previously adsorbed BSA film was maintained constant at 11.9 and 8.9 molecule nm^{-2} for monostearin (>99% pure, Sigma) and monoolein (>99% pure, Sigma), respectively. These surface densities are higher than those corresponding to monolayer saturation by the lipid, as detected by the values of the equilibrium surface pressure[2] of monostearin, Π_e (MS), and monoolein, Π_e (MO), which are indicated in Figure 1 by means of arrows. The variation of the protein/lipid ratio is due to the BSA added to the bulk phase over the range $1\text{–}5 \times 10^{-7}$ wt%. The BSA concentration dependence on π for BSA + lipid mixed systems shows a sigmoidal behaviour. The surface activity of the BSA + lipid mixed systems depends both on the protein/lipid ratio and the aqueous phase composition.

On water (Figure 1A), the π values approach those of pure BSA films at higher relative BSA concentrations in the mixed systems, as the monolayer was saturated by the protein. At lower relative BSA concentrations ($<1 \times 10^{-4}$ wt%), π is practically the same value as the π_e of the pure lipid. Thus it can be suggested that, at the lipid surface densities spread here, the protein is removed and the interface is saturated by a collapsed monostearin or monoolein film with liquid-condensed or liquid-expanded structure,[1,2] respectively. Above this protein concentration (1×10^{-4} wt%) and up to monolayer saturation by BSA (over the range $10^{-2} - 10^{-1}$ wt%) significant further reduction in π was observed. The effect results in an inflection in the π–log(BSA) curve in the intermediate region. The general features described earlier support indirect evidence of the existence of BSA–lipid interactions at the interface.

The surface activity of the mixed systems on ethanol aqueous solutions (Figure 1B) was found to be lower than that on water (Figure 1A), a phenomenon opposite to that observed on sucrose aqueous solutions (Figure 1C). This fact could be due to the effects of solutes in the subphase on protein–lipid interactions as will be discussed later.

Surface Tension of Protein + Soluble Lipid

When a soluble lipid (Tween 20) as well as BSA are present in water and aqueous solutions of ethanol and sucrose, both emulsifiers will form adsorbed

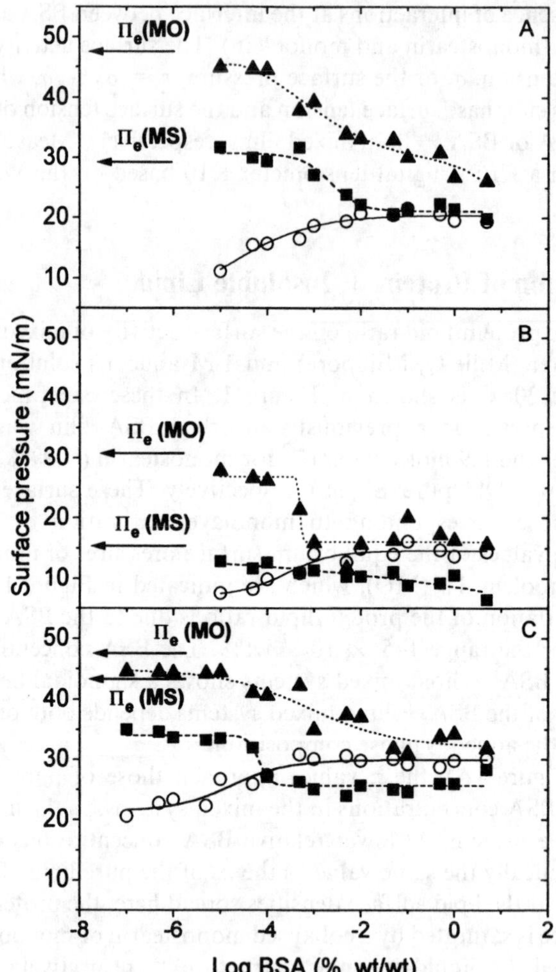

Figure 1 *The effect of the spreading of monostearin, MS (■), and monoolein, MO (△), on a film of BSA (○) previously adsorbed on (A) water and on aqueous solutions of (B) ethanol (1 M) and (C) sucrose (1 M). Temperature: 20 °C. The arrows indicate the equilibrium surface pressure for monostearin, Π_e (MS), and monoolein, Π_e (MO)*

films at the air–water interface. The BSA–Tween 20 interactions at the interface can be deduced from the surface tension *versus* Tween 20 concentration dependence (Figure 2). In our experiments,[1] the amount of BSA injected in the subphase under a film of previously adsorbed Tween 20 was maintained constant at 0.1 wt%. Therefore, the variation of the BSA/Tween 20 ratio is due to the variation of the Tween 20 concentration in the aqueous phase. It can be deduced that the BSA–Tween 20 interactions depend on the BSA/Tween 20 ratio and on the aqueous phase composition.

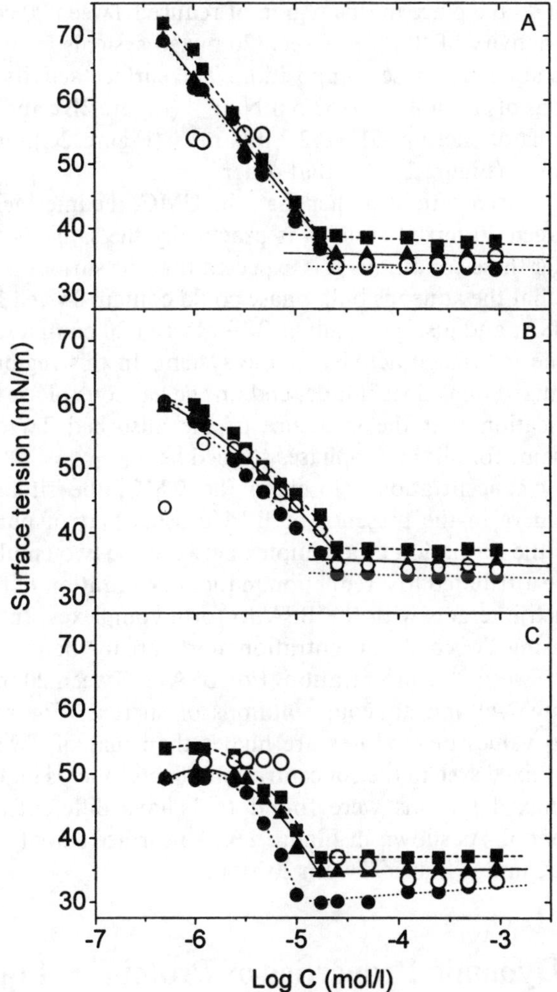

Figure 2 *Surface tension versus logarithm of the molar concentration of Tween 20 in (A) water and in aqueous solutions of (B) ethanol (1 M) and (C) sucrose (0.5 M), in the presence (○) and the absence (■, ▲, ●) of BSA at 0.1 wt%. Temperature (°C): (■) 5, (○, ▲) 20, and (●) 30*

At the lower Tween 20 concentration—well below that of the critical micelle concentration, CMC—the interfacial activity of the BSA + Tween 20 mixed systems was found to be higher than that of the pure lipid. It can be seen (Figures 2A and 2B) that the σ values for the BSA + Tween 20 mixed systems are lower than for Tween 20 alone. However, σ is higher than that of BSA alone[1] on the same aqueous solution. This is an indication that in this region the protein predominates at the interface, but with a different conformation from that of pure BSA. Moreover, it is likely that the formation of a BSA + Tween

20 complex could take place in this region of reduced Tween 20 concentration. The interfacial activity of BSA + Tween 20 mixed systems in this region also depends on the aqueous phase composition. The surface activity is higher for aqueous solutions of ethanol, 43.9–53.5 mN m^{-1} (Figure 2B), and decreases for aqueous solutions of sucrose, 51.8–52.3 mN m^{-1} (Figure 2C), and for water, 53.5–55.0 mN m^{-1} (Figure 2A), in that order.

At Tween 20 concentrations higher than the CMC, the interfacial activity of the BSA + Tween 20 mixed systems is practically the same as that for pure Tween 20. Under these conditions it is expected that the surface is covered with Tween 20, and that the aqueous bulk phase could contain Tween 20 monomers and micelles, BSA, and also probably a BSA–Tween 20 complex, but without any effect on the interfacial activity of the system. In this region the surface activity of the mixed films does not depend on the aqueous phase composition. This is an indication that the structure of the adsorbed Tween 20 film is essentially the same for all the subphases studied here.

At Tween 20 concentrations closer to the CMC, the situation is more complex. The curve in the presence of BSA crosses that of pure Tween 20, which indicates the formation of a complex between the two emulsifiers, as the crossover can be attributed to a reduction in the concentration of free Tween 20 in solution which interacts with the BSA to form complexes. In this region σ falls with increasing Tween 20 concentration more rapidly than in the preceding region at lower Tween 20 concentration. For BSA + Tween 20 mixed systems on water (Figure 2A) and aqueous solutions of sucrose (Figure 2C), it was observed that σ values of mixtures are higher than that of Tween 20 alone, although in the mixed system the concentration of emulsifier is higher. The BSA + Tween 20 mixed systems were found to behave differently in aqueous solutions of ethanol. As shown in Figure 2B, the surface activity of the mixed systems was the same as that of Tween 20 alone.

3 Surface Dynamic Properties of Protein + Lipid Mixed Films

The existence of protein–lipid interactions at the interface can have consequences for the surface dynamic properties (dynamic surface tension and surface dilational properties). That is, these properties are very sensitive to the existence of protein–lipid interactions at the interface. The surface rheological parameters (*i.e.* the surface dilational modulus E, and its elastic and viscous components) and the surface tension were measured according to the method of Kokelaar *et al.* as a function of time and radial frequency, as described elsewhere.[3,4]

In general the results show some interesting features from a rheological point of view:[3,4] (i) the surface dilational modulus E is effectively elastic (*i.e.* having a very low viscous component), and (ii) the frequency dependence of E, and the storage and loss moduli, describes viscoelastic behaviour of the surface over the experimental frequencies (0.08–3.4 rad s^{-1}) which is almost fully elastic.

However, the surface dilational properties do depend on the interfacial and bulk aqueous compositions.

Surface Dynamic Properties of BSA + Insoluble Lipids

The surface dynamic properties were determined after a solution of monostearin or monoolein had been spread on a film of BSA previously adsorbed from the subphase.

BSA + monostearin films at the air–aqueous phase interface. The effect of monostearin concentration on the surface dynamic properties, E and σ, of mixed BSA + monostearin films on aqueous ethanol solutions (as an example), as a function of time after monostearin spreading, is shown in Figures 3A and 3B, respectively. The spreading of monostearin on a film of BSA caused a sudden drop followed by a rapid increase in σ. However, E goes through a maximum at 10–15 minutes and then decreases as the time further increases. At longer time, σ decreases and E increases with the amount of monostearin spread on the interface. Moreover, at 60 min after monostearin spreading (Figure 4), E increases with the concentration of monostearin spread on the interface, but it decreases with the content of ethanol or sucrose in the bulk phase, especially at the lower monostearin surface densities where the monolayer is not saturated by the lipid (Figure 4). From a practical point of view, the consequences of this phenomenon are more important for mixed films on ethanol aqueous solutions (Figure 4A)—ethanol produces a low E value for BSA adsorbed films[5]—than for aqueous sucrose solutions (Figure 4B).

BSA + monoolein films at the air–aqueous phase interface. As with monostearin (Figure 3B), after monoolein was spread on an adsorbed BSA film, the value of σ was found to increase with time to a plateau value (data not shown[5]). However, E decreases with time, approaches a minimum value within 10–30 min, and then increases to a plateau value at longer times (Figure 3C). The results may be explained with a different hypothesis: (i) readsorption of BSA molecules after an initial displacement by the lipid, (ii) the existence of lipid–protein complexes at the interface, and/or (iii) relaxation phenomena associated with monolayer molecular loss. Mechanisms (i) and (ii) are more likely to occur in protein + lipid mixed films on aqueous ethanol solutions than on a pure water interface.[3] Figure 5 summarizes the effect of ethanol and sucrose concentrations on E after monoolein spreading on an adsorbed BSA film. In comparison with pure BSA films, E increases more for BSA + monoolein mixed films on ethanol aqueous solutions (at ethanol concentrations higher than 0.5 M (Figure 5A)), but the opposite was observed on aqueous sucrose solutions (Figure 5B). This phenomenon is different to that observed for BSA + monostearin films (Figure 4A). Thus, it can be concluded that the existence of a more condensed structure, and the presence of crystalline structures upon collapse of monostearin films, could displace the BSA molecules from the interface.[3]

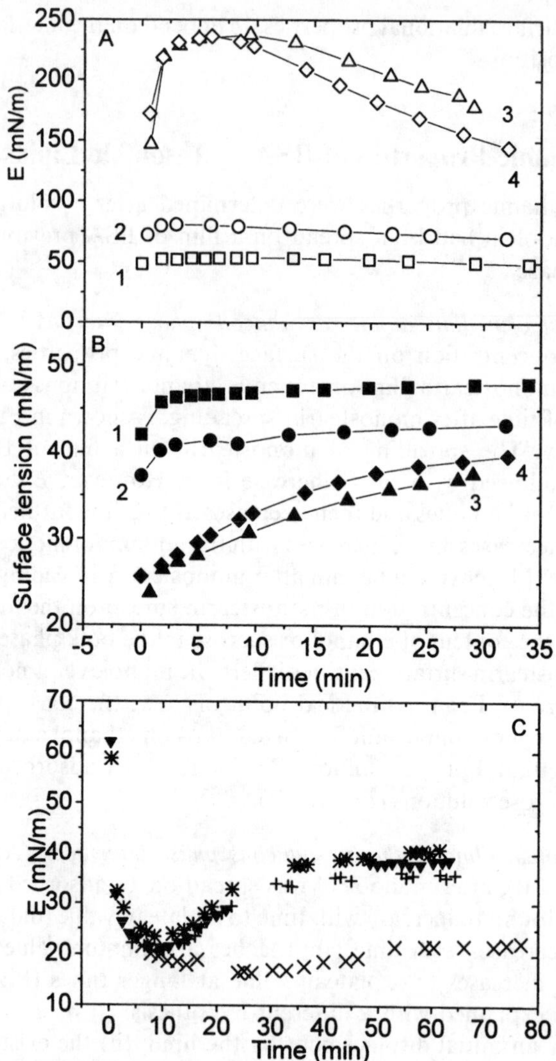

Figure 3 *Time dependence of (A) surface dilational modulus and (B) surface tension for monostearin spread on a film of BSA (0.1 wt%) adsorbed from a 1 M ethanol aqueous solution. Monostearin spread at various concentrations (molecule nm^{-2}): (1) 1.64; (2) 3.29; (3) 4.93; and (4) 6.58. (C) Time dependence of surface dilational modulus for monoolein spread (at 1.21 molecule nm^{-2}) on a film of BSA (0.1 wt%) adsorbed from aqueous ethanol solutions of various concentrations: (∗) 0.1 M, (▼) 0.5 m, (+) 1 M, and (×) 2 M. Temperature: 20 °C. Angular frequency: 0.81 rad s^{-1}*

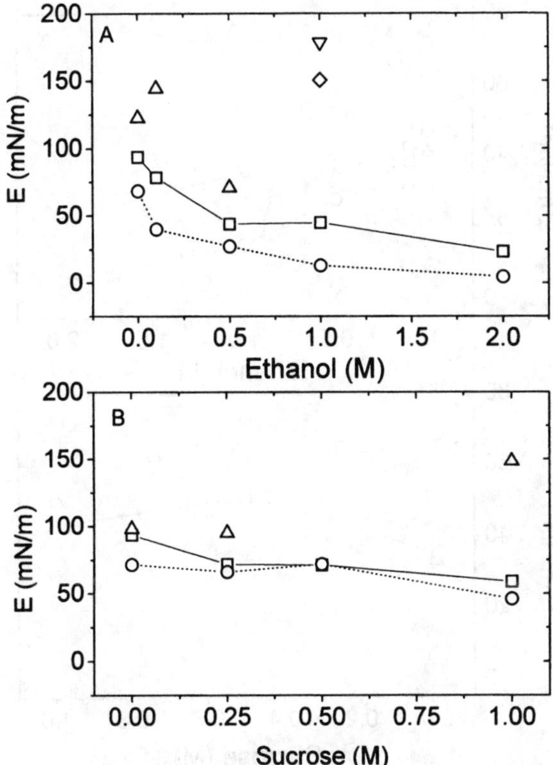

Figure 4 *Surface dilational modulus as a function of (A) ethanol and (B) sucrose solutions for monostearin spread on a film of BSA (0.1 wt%) at angular frequency of 0.81 rad s^{-1}. Temperature: 20 °C. Monostearin spread at various concentrations (molecule nm^{-2}): (□) 1.64, (△) 3.29, (◇) 4.94, and (▽) 6.68. Discontinuous line: adsorbed BSA pure film (○)*

Surface Dynamic Properties of BSA + Tween 20

In these experiments the dynamic surface properties were determined during the adsorption of a mixture of BSA + Tween 20 from the subphase.[4] During the adsorption, σ decreases, and E increases with time and then tends to a plateau. However, the rate of change of σ and E depends on the aqueous phase composition (Figure 6A). The Tween 20 concentration dependence on E is influenced by the type of solute in the subphase (Figure 6B). On water and sucrose solution (0.5 M) a sudden drop in E was found after addition of Tween 20, the value reaching a minimum at 5 μM Tween 20, and ultimately increasing to a plateau value as the Tween 20 concentration increases above 20 μM. On an aqueous ethanol solution (1 M), E increases with Tween 20 concentration and tends to a plateau value at a Tween 20 concentration of around 10 μM. That is, at higher relative BSA concentrations in mixed films the value of E is higher, which agrees with a stronger protein–protein interaction. However, lower E

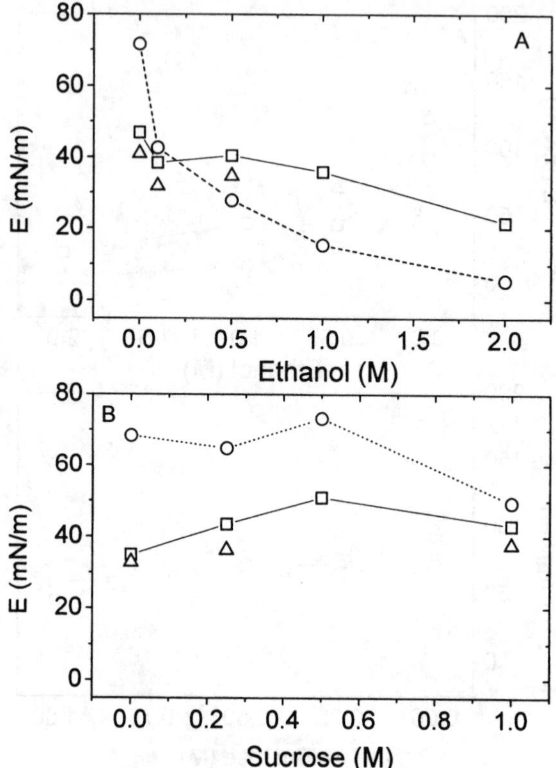

Figure 5 *Surface dilational modulus as a function of (A) ethanol and (B) sucrose solutions for monoolein spread on a film of BSA (0.1 wt%) at angular frequency of 0.81 rad s^{-1}. Temperature: 20 °C. Monoolein spread at various concentrations (molecule nm^{-2}): (□) 1.21, and (△) 2.42. Discontinuous line: BSA adsorbed pure film (○)*

values were observed at higher relative Tween 20 concentrations, which agrees with weaker surfactant–surfactant interactions. The minimum observed in mixed films at Tween 20 concentrations of 5–7.5 μM—concentrations which are of the same order of magnitude but lower than the CMC (see arrows in Figure 6B)—could be associated with a transition in the adsorbed layer structure due to protein–surfactant interactions[4] with specific properties, such as drainage and surface diffusion in thin films.

4 Drainage and Molecular Diffusion in Protein + Lipid Mixed Thin Films

The drainage characteristics and surface lateral diffusion properties of thin liquid films formed from BSA + Tween 20 were studied using optical microscopy including epi-illumination and fluorescence recovery after photo-

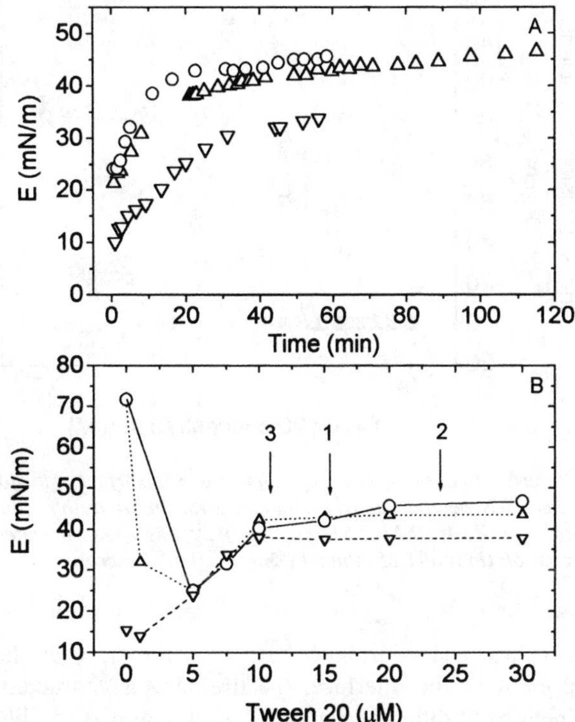

Figure 6 *Surface dilational modulus* E *as a function of (A) time and (B) Tween 20 concentration at 60 minutes of adsorption time, for mixtures of BSA (0.1%) and Tween 20 adsorbed on (○) water and aqueous solutions of (▽) ethanol (1 M) and (△) sucrose (0.5 M). Angular frequency: 0.81 rad* s^{-1}*. Temperature: 20 °C. The arrows indicate the CMC of Tween 20 on (1) water and on aqueous solutions of (2) ethanol (1 M) and (3) sucrose (0.5 M)*

bleaching (FRAP).[6] The drainage of thin films (data not shown) stabilized by Tween 20 demonstrates the typical features of a non-interacting surfactant,[6] exhibiting behaviour commonly referred to as mobile or 'chaotic' drainage. Such mobile surface behaviour allows rapid drainage of liquid from the film. This is due to the free surface diffusion of the surfactant molecules at the film surface (Figure 7). In contrast, thin films stabilized by BSA show initial drainage characteristics typical for protein-stabilized films, with many static concentric rings present, and drainage that is slow and uniform. The drainage behaviour of mixed BSA + Tween 20 films was found to be intermediate between the chaotic Tween 20 drainage and the uniform BSA drainage. The films showed distortion of the uniform concentric rings, together with interfacial aggregates or interfacial segregation as a function of BSA/Tween 20 ratio and aqueous phase composition.[6] Thin film drainage and FRAP techniques provide complementary data relating the behaviour of thin liquid films. Under conditions where protein–protein interactions are dominant, no diffusion is observed (Figure 7).

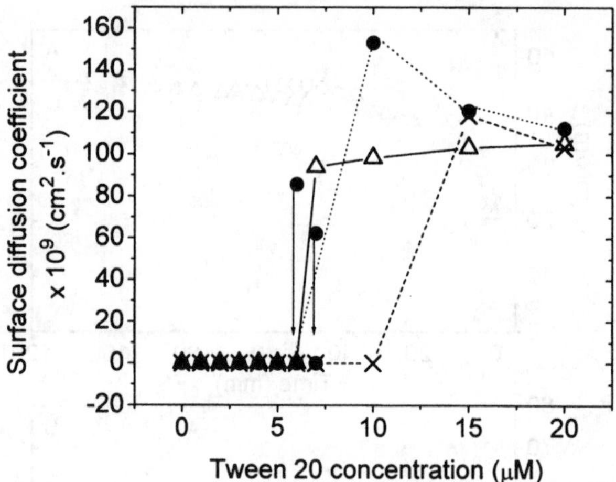

Figure 7 *The Tween 20 concentration dependence of the surface diffusion coefficient of adsorbed 5-N-(octadecanoyl) aminofluorescein in a thin film stabilized by mixtures of BSA (0.1 wt%) + Tween 20 adsorbed on water (△), and on aqueous solutions of 1 M ethanol (●) and 0.5 M sucrose (×)*

These films also show a slow drainage rate. In contrast, when the Tween 20 is significantly present at the interface, the drainage is characterized by rapid movement of regions of different thickness and the film at equilibrium shows a measurable surface diffusion coefficient (Figure 7). Again, this transition occurs at Tween 20 concentrations higher than, but of the same order of magnitude as, the CMC.

5 Concluding Remarks

The presence of water-soluble surfactant (Tween 20) or the spreading of insoluble lipid (monostearin or monoolein) on a BSA film adsorbed at the air–aqueous phase interface induces dramatic changes in the interfacial characteristics (surface activity, surface dilational properties, and interfacial mobility) of the mixed films. From the results of this investigation it may be inferred that interfacial composition is not the only parameter that can influence the properties of protein + lipid mixed films at the air–aqueous phase interface. Protein–surfactant interactions at the interface depend on the overall protein/surfactant ratio. There is experimental evidence that the interfacial characteristics of the mixed films are determined by either protein or lipid at the limits of high or low protein/lipid ratio, respectively. In the intermediate region, the existence of protein–surfactant interactions dominate the interfacial characteristics of the mixed films. Moreover, it has been demonstrated that the aqueous phase composition has a role in determining the extent of protein–surfactant interactions and, as a consequence, in determining the interfacial characteristics of mixed films. It is likely that this is due to the effect of solutes in the subphase on the competitive adsorption behaviour of protein, surfactant, and protein–surfactant complex, and on the protein structure at the interface and in the bulk phase.

Acknowledgements

This research was supported in part by DGICYT (through grants PB94-1459 and PR95-175). The authors thank Dr. D. C. Clark and Mr. P. J. Wilde (IFR, Norwich) for providing facilities and support for this work.

References

1. M^a. R. Rodríguez Niño and J. M. Rodríguez Patino, *J. Amer. Oil Chem. Soc.*, 1998, **75**, 1233, 1241.
2. J. M. Rodríguez Patino and M^a. R. Martín Martínez, *J. Colloid Interface Sci.*, 1994, **167**, 150.
3. M^a. R. Rodríguez Niño, P. J. Wilde, D. C. Clark, and J. M. Rodríguez Patino, *Langmuir*, 1998, **14**, 2160.
4. M^a. R. Rodríguez Niño, P. J. Wilde, D. C. Clark, and J. M. Rodríguez Patino, *J. Agric. Food Chem.*, 1998, **46**, 2177.
5. M^a. R. Rodríguez Niño, P. J. Wilde, D. C. Clark, and J. M. Rodríguez Patino, *J. Agric. Food Chem.*, 1997, **45**, 3010, 3016.
6. P. J. Wilde, M^a. R. Rodríguez Niño, D. C. Clark, and J. M. Rodríguez Patino, *Langmuir*, 1997, **13**, 7151.

Caseinate-Stabilized Emulsions: Influence of Ageing, pH and Oil Phase on the Behaviour of Individual Protein Components

By Jeffrey Leaver, Andrew J. R. Law, and David S. Horne

HANNAH RESEARCH INSTITUTE, AYR KA6 5HL, SCOTLAND, UK

1 Introduction

Milk is a rich source of proteins whose surface activity makes them prime ingredients in a variety of processed foods. These include emulsions and foams, where the protein molecules, frequently in combination with low-molecular-weight surfactants, stabilize the oil–water and air–water interfaces. During high pressure emulsification, protein molecules rapidly adsorb to, and stabilize, the newly formed interfaces. However, the composition and topography of the proteins at the time of adsorption may be very different from that after storage, which in commercial emulsions may be for weeks or months. Limitations in large-scale fractionation techniques mean that milk proteins are always used industrially as mixtures, either in the form of skimmed milk (frequently after drying or concentration) or after separation into the caseinate and whey protein fractions. A distinct hierarchy has been shown to exist with regard to the interfacial behaviour of caseins. β-Casein is the most surface active of the milk proteins, and when emulsions are made with a mixture of α_{S1}- and β-caseins, it preferentially adsorbs at the surface of the emulsion droplets.[1] β-Casein exchanges readily with α_{S1}-casein adsorbed to the surface of an oil droplet, whereas displacement of β-casein by α_{S1}-casein is much more limited.[1] In soya oil-in-water emulsions stabilized with sodium caseinate, which is a natural mixture of α_{S1}-, α_{S2}-, β- and κ-caseins, the β-casein gradually exchanges with the less surface-active α_{S1}-casein, particularly if there is a large excess of unadsorbed protein.[2]

In simple model emulsions hydrocarbon oils are commonly used in place of the triglycerides which are ingredients of commercial emulsions, the assumption being that the behaviour of adsorbed protein molecules is independent of the nature of the oil phase. However, a comparison of model single-protein emulsions made with hydrocarbon oils with those made with triglyceride oils has revealed differences in the behaviour and structure of the adsorbed protein molecules.[3,4]

The structure and behaviour of proteins at oil–water interfaces have been determined using a variety of physical and biochemical techniques. Of these methods, competitive displacement of adsorbed protein by the non-denaturing surfactant Tween 20 is a relatively simple way of determining changes in the binding of proteins at oil–water interfaces. In this paper we report on the influence of the oil phase on the displacement of individual protein components from caseinate-stabilized oil-in-water emulsions, and on the effects which ageing and pH have on this displacement behaviour. We show that hydrocarbon oils are not necessarily good substitutes for triglycerides when studying the behaviour of proteins in food emulsions.

2 Materials and Methods

Polyoxyethylene sorbitan monolaurate (Tween 20, Surfact-Amps grade) was purchased from Pierce, Rockford, IL, USA. Soybean oil and *n*-tetradecane were from Sigma Chemical Co. Ltd., Poole, Dorset, UK. Caseinate was prepared by acid precipitation of protein from skimmed bulk milk obtained from the Institute's herd of Friesian cattle. After neutralization with NaOH the protein was freeze-dried and stored at −15 °C.

Protein-stabilized oil-in-water emulsions were prepared by microfluidization at 450 bar using a M120E laboratory scale microfluidizer (Christison Scientific Equipment Ltd., Gateshead, Tyne-and-Wear, UK) at a protein concentration of 7 mg ml^{-1} in imidazole buffer (20 mM, pH 7.0) and an oil concentration of 20 vol%. The caseinate emulsions were diluted five-fold with imidazole buffer, divided into aliquots, and the pH adjusted by addition of 0.5 M HCl or NaOH. These emulsions were not washed. All buffers contained sodium azide (0.01%) and emulsions were stored at 4 °C in sealed containers. Emulsion droplet size was determined using a Mastersizer (Malvern Instruments, Malvern, Worcs, UK).

Appropriate volumes of Tween 20 were added to vigorously mixed samples of the emulsions at room temperature. After mixing for a minimum of 1 hour, the samples were centrifuged at 30 000 *g* in a Sorvall SS-34 rotor fitted with adaptors to enable it to take 4 ml centrifuge tubes. Sub-phases were removed using a syringe and protein composition was determined by reverse phase HPLC after diluting the samples with an equal volume of 7 M urea in 20 mM bis-tris propane, pH 7.5, containing 5 μl ml^{-1} of 2-ME. The HPLC method was essentially that described by Visser *et al.*,[5] but an Apex WP C18 column (Jones Chromatography Ltd., Hengoed, Mid-Glamorgan, UK) was used. Protein displacement was plotted against the molar ratio (R) of Tween 20 to total protein.

3 Results and Discussion

Influence of Oil Phase on Protein Structure during Ageing

Typical RP-HPLC profiles of the proteins displaced from caseinate-stabilized tetradecane emulsions stored at 4 °C for various periods are shown in Figure 1.

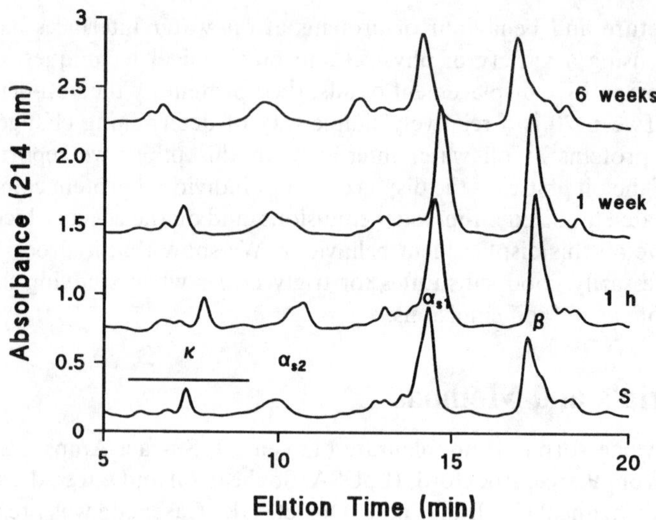

Figure 1 *RP-HPLC profiles of proteins displaced from caseinate-stabilized* n-*tetradecane emulsions by Tween 20 at* R = *88 after ageing at pH 7 for 1 hour, 1 week and 6 weeks. S = caseinate standard; the individual casein peaks are marked*

The pH is 7.0 and the *R* value is 88. The appearance of the proteins displaced from the 1-week-old emulsion is very similar in Figure 1 to that from the 1-hour-old emulsion, and even after 6 weeks storage the changes are slight. In contrast, profiles of the proteins displaced from the corresponding soya oil emulsion showed very pronounced changes (Figure 2). Even after 1 hour of storage, the α_{S1}- and β-casein peaks were broader than in the original caseinate analysis. After 1 day, shoulders eluting at lower acetonitrile concentrations were observed, and after 2 days these were resolved as separate peaks. After 4 days storage, a general smearing of the protein pattern was observed, and it was no longer possible to distinguish the individual protein components reliably. This indicates that the proteins in the triglyceride emulsions were rapidly being modified. Since these changes did not appear to occur in the *n*-tetradecane emulsion, the most obvious explanation would appear to involve the binding of free fatty acids in the soya oil to hydrophobic patches on the proteins. However, since the interfacial protein is displaced from the interface by the high concentration of detergent prior to being dissolved in 3.5 M urea and separated by RP-HPLC using an acetonitrile gradient, the combination of which would be expected to disrupt hydrophobic interactions, it was felt that the changes probably resulted from a different interaction. A second possible explanation for the changes is that they were due to proteolysis of the interfacial protein by proteolytic enzymes such as plasmin which were carried through to the freeze-dried caseinate from the original milk. SDS-polyacryamide gels of 4-week-old emulsions show that the proteins at the interface have molecular weights similar to those of the original caseinate components; also the gels failed

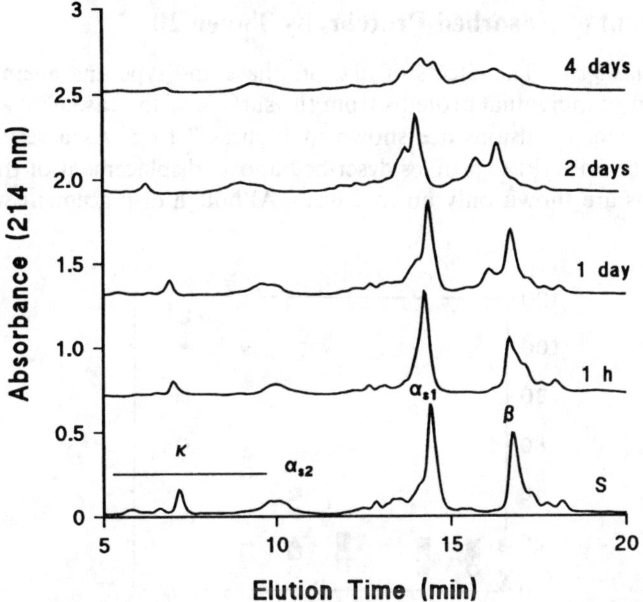

Figure 2 *RP-HPLC profiles of proteins displaced from caseinate-stabilized soya oil emulsions. Conditions as in Figure 1*

to identify any low molecular weight peptides which would have resulted from proteolysis.

Another possible explanation for the observed changes is that the proteins in the soya oil emulsion were being modified covalently. Gas chromatographic/mass spectrometric analysis of freshly prepared caseinate-stabilized soya oil emulsion revealed the presence of a number of volatile compounds which were not present in either the original stock soya oil or in *n*-tetradecane emulsions prepared under the same conditions. Several of these volatile components were identified as enals (α,β-unsaturated aldehydes), the major ones being 2-heptenal and 2,4-decadienal. As a consequence of the electrophilic centres at their β- and carbonyl carbon atoms, enals have been shown[6] to react readily with nucleophilic side chains of proteins such as lysine to form Schiff's bases and Michael adducts via condensation and addition reaction, respectively. These enal compounds are probably produced by peroxidation of polyunsaturated fatty acyl chains of the triglycerides during or immediately after microfluidization. The effect of these enal/protein interactions would be to convert a relatively hydrophilic, charged amino-acid side-chain to a hydrophobic, alkyl 'tail'. This should have a significant influence on the hydrophobic/hydrophilic balance of the protein molecules and hence on their interfacial properties. In addition, the nature of the peptides released during digestion of these modified proteins will be altered, which may be important on nutritional grounds. These modifications will be discussed in more detail in a forthcoming publication.

Displacement of Adsorbed Proteins by Tween 20

Soya oil emulsions. The effects of pH, oil phase and type and ageing on the displacement of individual proteins from the surface of the caseinate-stabilized soya oil-in-water emulsions are shown in Figures 3 to 5. As a result of the changes in the RP-HPLC profiles described above, displacement of the individual proteins are shown only up to 2 days. Although displacement was mea-

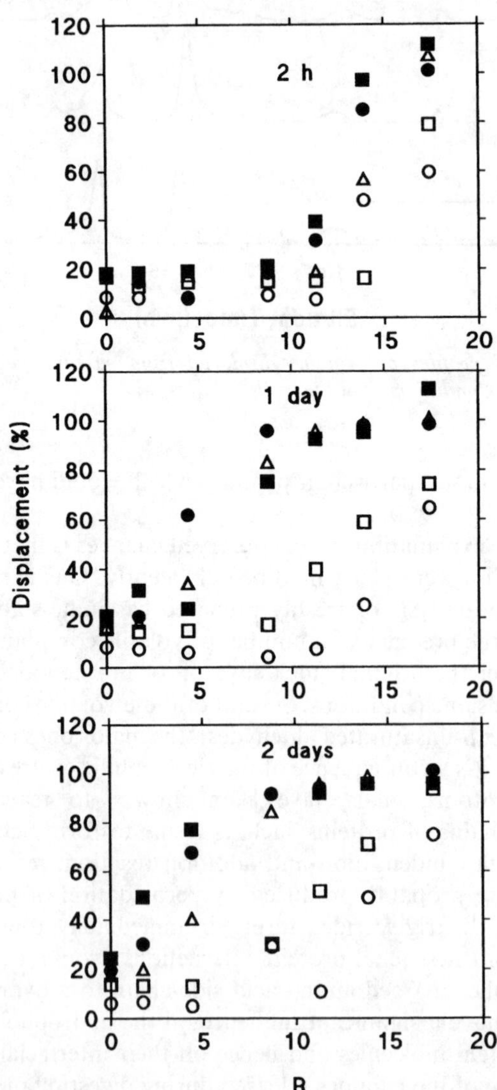

Figure 3 *Influence of ageing and pH on the displacement of* α_{S1}*-casein as a function of surfactant/protein molar ratio* R *for soya oil emulsions:* ○, *pH 6;* □, *pH 7;* △, *pH 8;* ●, *pH9;* ■, *pH 10*

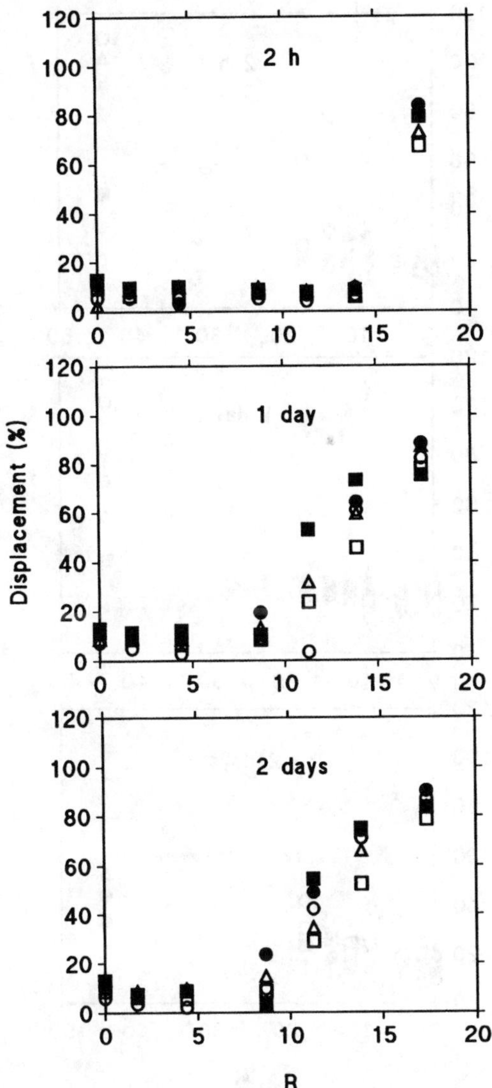

Figure 4 *Influence of ageing and pH on the displacement of β-casein from soya oil emulsions. Symbols as in Figure 3*

sured up to $R = 88$, only the values up to $R = 20$ are shown in the profiles, except for the displacement of κ-casein, where values are shown up to $R = 50$.

The emulsion droplets showed little tendency to aggregate. The initial $d_{0.5}$ value of 0.36 μm was constant for at least 13 weeks. In general, the ease of displacement of proteins from the soya oil–water interface was in the order α_{s1}- > β- > κ-casein. Compared to freshly prepared emulsions, both α_{s1}- and β-casein were more readily displaced after storage for 1 day (Figures 3 and 4). Ageing was found slightly to decrease the displacement of κ-casein at low Tween

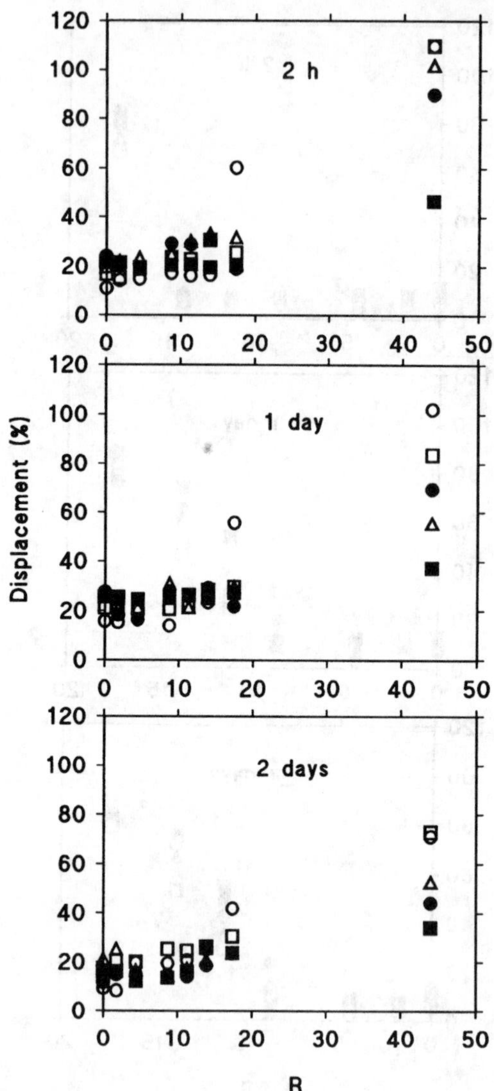

Figure 5 *Influence of ageing and pH on the displacement of κ-casein from soya oil emulsions. Symbols as in Figure 3*

levels, but significantly to decrease it at higher concentrations (Figure 5). Displacement is also influenced by pH. β-Casein displacement from the freshly prepared emulsion is relatively independent of pH. κ-Casein displacement is only slightly affected by pH at low R values, but at higher Tween 20 concentrations the protein is more readily displaced at lower pH. However, α_{S1}-casein displacement is more obviously pH-sensitive, even in the freshly prepared emulsion, being more readily displaced at higher pH as the net negative charge on the protein increases. The combined effects of pH and ageing on α_{S1}-casein

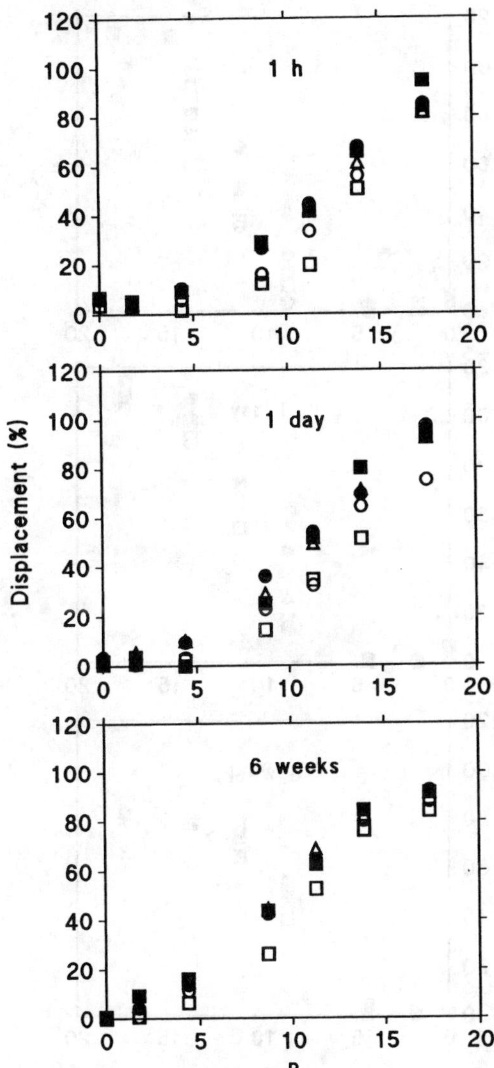

Figure 6 *Influence of ageing and pH on the displacement of* α_{S1}*-casein from* n-*tetradecane emulsions. Symbols as in Figure 3*

displacement are to convert the shape of the displacement curve from sigmoidal at all pH values in the fresh emulsion to hyperbolic at pH $\geqslant 8$ in the 1- and 2-day-old emulsion. α_{S1}-Casein is more readily displaced than β-casein at higher pH, but at pH 6 and 7 the displacement of both proteins is similar. Whether these ageing effects are due to proteins rearranging at the interface, or to the covalent modifications detailed above, or to both, is still unresolved. If it were solely due to covalent modifications, it would be expected that the presumed attachment of alkyl chains would make the proteins more hydrophobic and

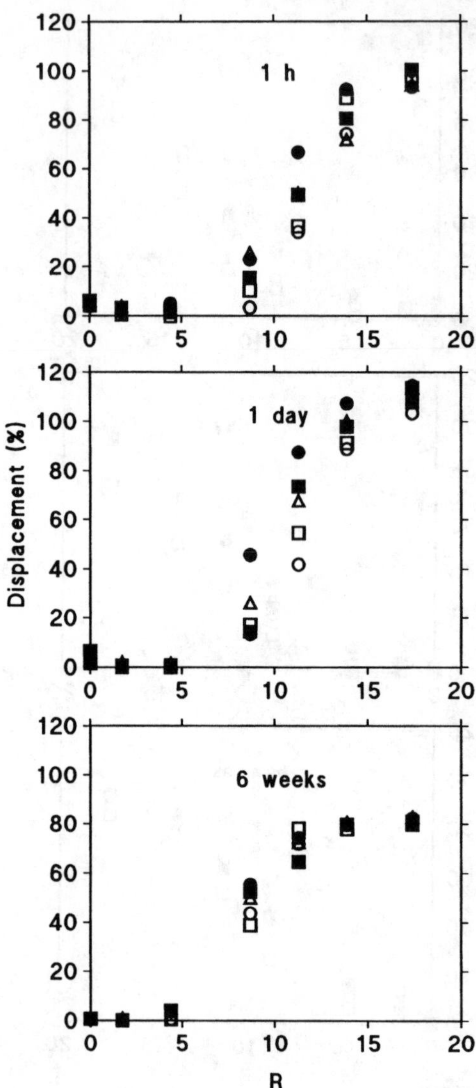

Figure 7 *Influence of ageing and pH on the displacement of β-casein from* n-*tetradecane emulsions. Symbols as in Figure 3*

hence more difficult to displace. From the profiles presented, this is obviously not the case. However, although displacement of individual proteins over longer time periods cannot be measured due to the loss of resolution, a measure of total displacement could be obtained from the total peak areas as measured by HPLC. This showed that, with the pH 7 emulsion at R = 88, the total protein displaced decreases by 50% after storage for 30 days in accordance with the predicted changes in the properties of the adsorbed protein.

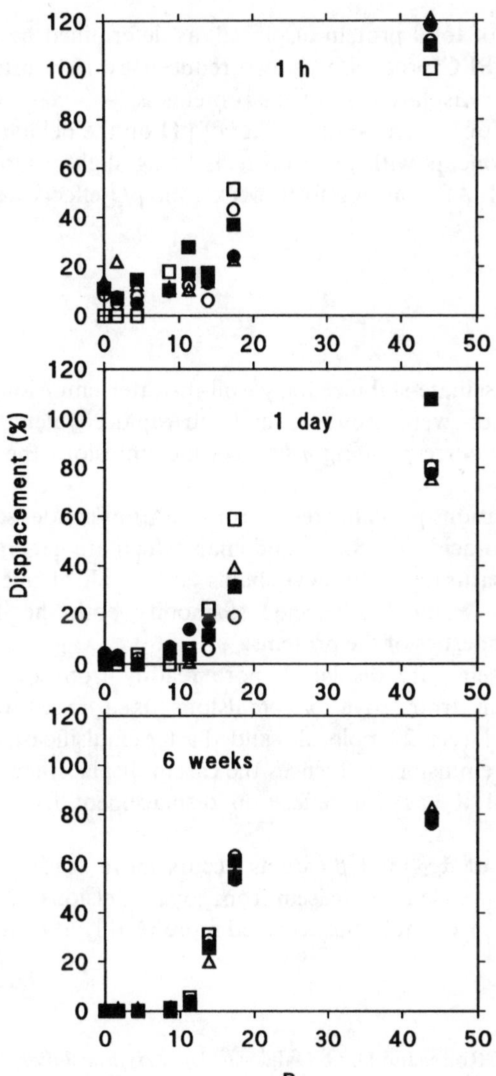

Figure 8 *Influence of ageing and pH on the displacement of κ-casein from* n-*tetradecane emulsions. Symbols as in Figure 3*

n-*Tetradecane emulsions.* *n*-Tetradecane emulsions, with initial average droplet size very similar to that of the soya oil emulsions, were also found to be stable throughout the duration of the experiment. Individual proteins, except for the α_{S1}-casein component at higher pH, were found generally to be displaced more easily from the *n*-tetradecane–water interface than from the soya oil–water interface (Figures 6 to 8). Although ageing had little effect on the shape of the displacement curves with these emulsions, after 6 weeks the maximum amounts of *κ*- and *β*-casein displaced were slightly reduced. The maximum amount of total protein displaced, as determined by summation of the areas on the HPLC profiles, was also reduced by 13% after this period. The order of ease of displacement of the proteins is β- > α_{S1}- > κ-casein. As with soya oil emulsions, there is some effect of pH on the behaviour of the α_{S1}- and β-casein components with these proteins being slightly more readily displaced at higher pH. After ageing for 6 weeks, the pH effects were less noticeable.

4 Conclusions

1. Proteins in caseinate-stabilized soya oil-in-water emulsions prepared by microfluidization were shown to undergo rapid covalent modifications, whereas in the corresponding *n*-tetradecane emulsions these changes did not occur.
2. These modifications probably result from addition/condensation reactions between amino-acid side-chains and enals which are generated by peroxidation of unsaturated fatty acyl chains as a result of microfluidization. These changes would be expected to modify both the interfacial and nutritional properties of the proteins.
3. Individual caseins are displaced more readily from the *n*-tetradecane emulsions than from soya oil emulsions as a result of competitive adsorption of Tween 20 molecules added after emulsification.
4. Ageing of the emulsions influences the casein displacement from soya oil emulsions, but it has little effect on displacement from *n*-tetradecane emulsions.
5. Displacement of α_{S1}- and β-caseins occurs more readily at higher pH, especially in the case of α_{S1}-casein from soya emulsions. Displacement of κ-casein from soya emulsions occurred more readily at lower pH.

References

1. E. Dickinson, S. E. Rolfe, and D. G. Dalgleish, *Food Hydrocolloids*, 1988, **2**, 397.
2. E. W. Robson and D. G. Dalgleish, *J. Food Sci.*, 1987, **52**, 1694.
3. J. Leaver and D. G. Dalgleish, *J. Colloid Interface Sci.*, 1992, **149**, 49.
4. E. M. Stevenson, D. S. Horne, and J. Leaver, *Food Hydrocolloids*, 1997, **11**, 3.
5. S. Visser, C. J. Slangen, and H. S. Rollema, *J. Chromatogr.*, 1991, **548**, 361.
6. C. T. Houston, A. G. Baker, M. V. Novotny, and J. P. Reilly, *J. Mass Spectrom.*, 1997, **32**, 662.

Adsorption of Proteins at the Gas–Liquid and Oil–Water Interfaces as Studied by the Pendant Drop Method

By A. V. Makievski, R. Miller,[1] V. B. Fainerman,[1] J. Krägel,[1] and R. Wüstneck[2]

INSTITUTE OF TECHNICAL ECOLOGY, BLVD. SHEVCHENKO 25, DONETSK, 340017, UKRAINE
[1]MAX-PLANCK-INSTITUT FÜR KOLLOID- UND GRENZFLÄCHENFORSCHUNG, RUDOWER CHAUSSEE 5, D-12489 BERLIN-ADLERSHOF, GERMANY
[2]UNIVERSITÄT POTSDAM, INSTITUT FÜR FESTKÖRPERPHYSIK, POTSDAM, GERMANY

1 Introduction

The dynamic behaviour of proteins at gas–liquid and liquid–liquid interfaces is important for many colloidal systems. For example, the structure of adsorption layers affects the processes of formation and stability of foams and emulsions.[1–3] Dynamic surface tension studies of adsorbed and spread layers of proteins and also their interfacial rheological behaviour can provide information on the formation mechanisms and structure of protein adsorption layers.[4,5]

Recently, new equations of state for surface layers and adsorption isotherms of proteins at liquid–fluid interfaces have been derived. These new equations take into account the non-ideality of enthalpy and entropy of mixing in the adsorption layer resulting from the difference in the sizes of protein and solvent molecules. The effect of the electric charge of the protein molecules with its predominant contribution to the surface pressure is discussed extensively in ref. 6.

In the present work experiments of dynamic and equilibrium surface tension and surface relaxation measurements for β-casein, β-lactoglobulin and human serum albumin solutions are performed using the axisymmetric drop shape analysis (ADSA) method. The data obtained for the solution–air and solution–*n*-tetradecane interfaces are interpreted on the basis of a thermodynamic adsorption model for protein layers published recently.[6]

2 Materials and Methods

β-Casein (C6905, lot 12H9550), β-lactoglobulin (L-0130, lot 91H7005) and human serum albumin (HSA, A8301, lot 94H8270) were supplied by Sigma (Germany). Molecular weights of the substances are 24000, 18400 and 69000 daltons, respectively. β-Casein and human serum albumin were used without further purification. The samples of β-lactoglobulin were purified according to a procedure given elsewhere.[7] Phosphate buffer solutions used were prepared by mixing appropriate stock solutions of K_2HPO_4 and NaH_2PO_4 for HSA, and K_2HPO_4 and KH_2PO_4 for β-casein and β-lactoglobulin. The surface tension of the buffer solution at the air–water interface at pH 7 was 72.5 mN m^{-1}. The *n*-tetradecane used in the present study was purchased from Fluka and purified by rinsing through an alumina column several times. The interfacial tension of the purified *n*-tetradecane in contact with water was 51.5 mN m^{-1}. All experiments were performed at room temperature (22 °C).

The adsorption of proteins was studied by dynamic surface and interfacial tension measurements using the axisymmetric drop shape analysis (ADSA).[8,9] For the water–air interface the drops were formed at the tip of a PTFE capillary immersed into a cuvette filled with a water-saturated atmosphere. For the water–oil interface studies the capillary was immersed into the oil phase contained in the cuvette. The drop volumes were *ca.* 0.03 cm^3. The standard deviation of the ADSA method in these studies was ± 0.25 mN m^{-1}.

3 Results and Discussion

Equilibrium State of Adsorbed Layers

The dependencies of dynamic interface pressure Π for HSA, β-casein and β-lactoglobulin at the solution–air and solution–oil interfaces for various concentrations are shown in Figures 1–3. These dependencies indicate that the nature of the interface affects significantly the dynamics of the adsorption process and the equilibrium adsorption characteristics. In particular, for HSA and β-casein, both the rate of interfacial pressure increase and the equilibrium values of Π at the solution–oil interface are higher than the corresponding values at the solution–air interface. For β-lactoglobulin, however, a similar behaviour is observed at high bulk concentrations only, while at low bulk concentrations the observed increase in surface tension was more pronounced for the solution–air interface.

To estimate the equilibrium surface pressures, the curves $\Pi = \Pi(t)$ were extrapolated onto $t \rightarrow \infty$. The values obtained from two extrapolation procedures ($\Pi(t^{-1/2})$ and $\Pi(t^{-1})$ as $t \rightarrow \infty$) were found[10] to be close to one another (differences were smaller than ± 0.5 mN m^{-1}). In the following the average of these two extrapolation values was used. In Figures 4–6 the experimental equilibrium surface pressure isotherms for HSA, β-casein and β-lactoglobulin at pH 7 at the solution–air and solution–oil interfaces are plotted *versus* the initial protein concentration in the solution. For comparison, the

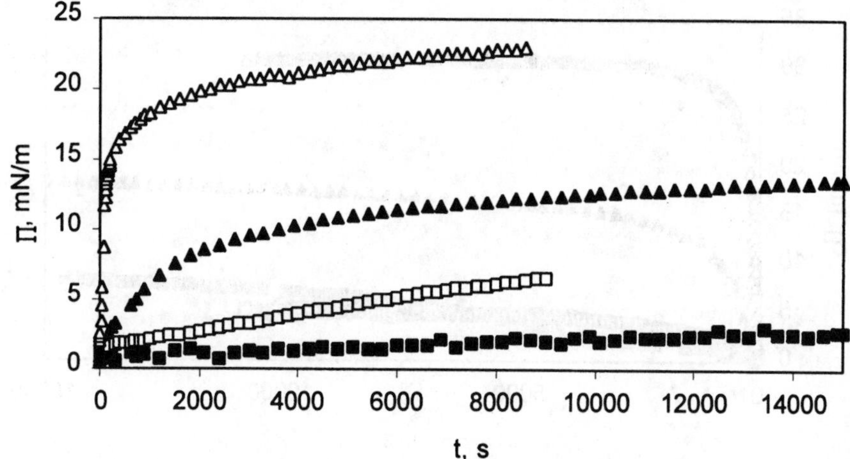

Figure 1 *Dynamic surface pressure* Π(t) *of HSA solutions for two concentrations:* 2×10^{-8} *mol* l^{-1} (■,□); 2×10^{-7} *mol* l^{-1} (▲,△). *Closed symbols correspond to the liquid–air interface; open symbols correspond to the liquid–liquid interface*

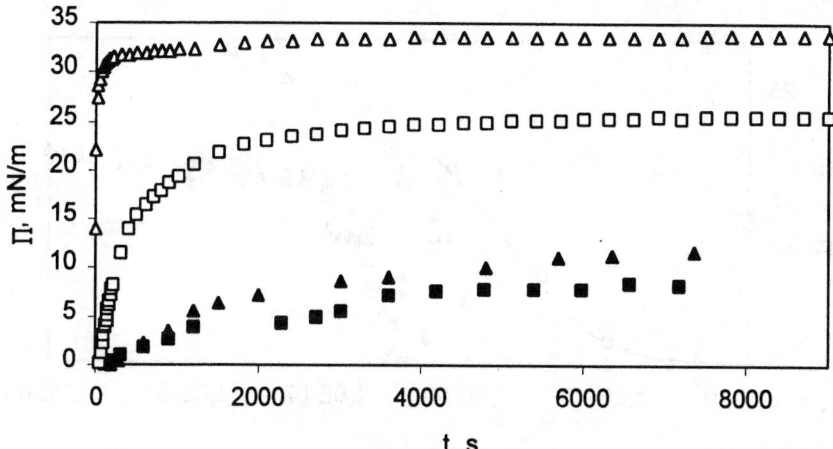

Figure 2 *Dynamic surface pressure* Π(t) *of β-casein solutions for two concentrations:* 2×10^{-7} *mol* l^{-1} (■,□); 2×10^{-6} *mol* l^{-1} (▲,△). *Closed symbols correspond to the liquid–air interface; open symbols correspond to the liquid–liquid interface*

results of Graham and Phillips[11] for BSA (whose structure is similar to that of HSA) and for β-casein at the solution–oil interface (*n*-decane and toluene) and the data of Wüstneck *et al.*[12] for β-lactoglobulin at the solution–air interface are also presented. In these studies the plate and ring methods were used, with a 2 cm solution depth. Under these conditions, the adsorption of protein only slightly depletes the solution bulk. Moreover, the results of Graham and

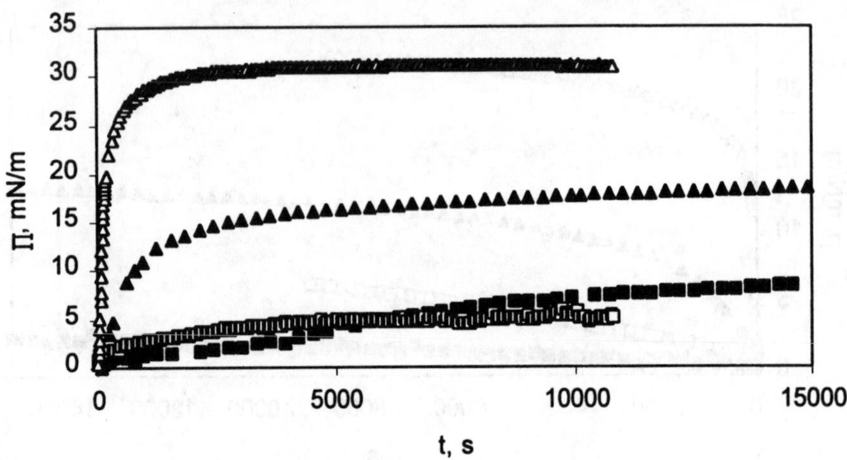

Figure 3 *Dynamic surface pressure* Π(t) *of β-lactoglobulin solutions for two concentrations:* 1×10^{-7} *mol* l^{-1} (■,□); 1×10^{-6} *mol* l^{-1} (▲,△). *Closed symbols correspond to the liquid–air interface; open symbols correspond to the liquid–liquid interface*

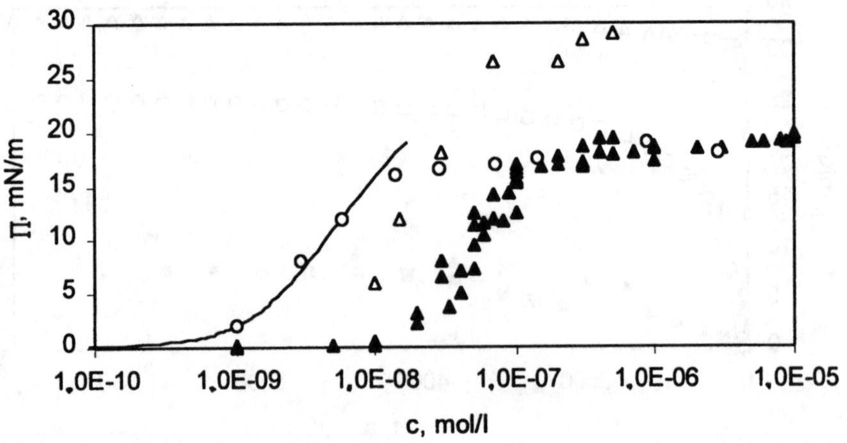

Figure 4 Π versus c *isotherm for albumins at solution–air and solution–oil interfaces:* ▲,△, *our data for the two interfaces;* ○, *data from Graham and Phillips*[11] *for solution–oil interface; the curve is calculated from the data reported by Graham and Phillips*

Phillips[11] are presented in a corrected form, that is, plotting *versus* the final concentration of the solution considering the loss due to adsorption. It is seen that the surface pressure isotherms measured by the pendant drop method are significantly 'shifted' towards higher protein concentrations when presented as a function of the initial protein concentration. This initial concentration can differ from the equilibrium concentration due to the adsorption of proteins at

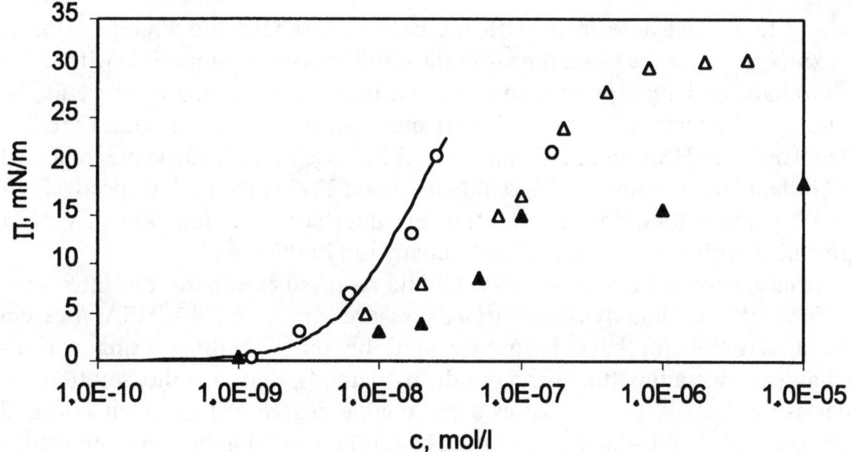

Figure 5 Π versus c *isotherm for β-casein at solution–air and solution–oil interfaces:* ▲,△, *our data for the two interfaces;* ○, *data from Graham and Phillips*[11] *for solution–oil interface; the curve is calculated from the data of Graham and Phillips*

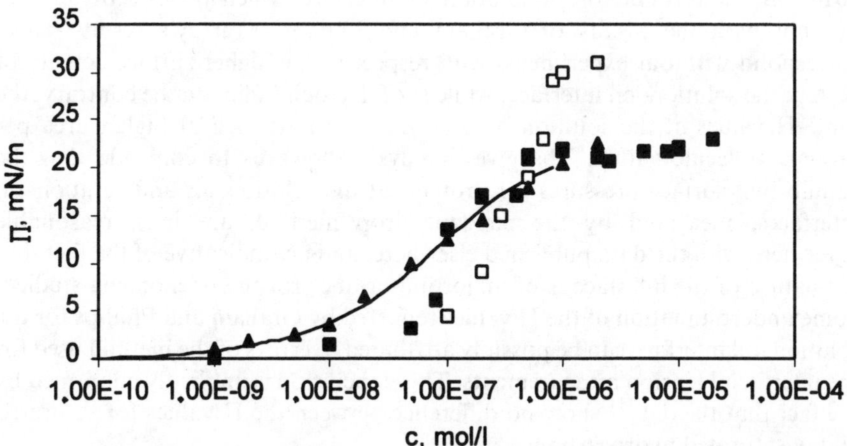

Figure 6 Π versus c *isotherm for β-lactoglobulin at solution–air and solution–oil interfaces:* ■,□, *our data for the two interfaces;* ▲, *Wüstneck's data*[12] *for solution–oil interface; the curve is calculated from Wüstneck's data*

the drop surface. The equilibrium concentration c of the protein within the drop (*i.e.*, the concentration in the adsorbed equilibrium state) is related to the initial concentration c_0 via the expression $c = c_0 - \Gamma_\Sigma\,(S/V)$, where S and V are the area and volume of the drop, respectively, and Γ_Σ is the adsorbed amount of the protein. This correction of the concentration (the values of the adsorption were taken from the Γ_Σ–Π isotherms[11]) results in a shift of the experimental isotherms shown in Figures 4 and 5 towards lower bulk concentrations, leading to perfect agreement with the data of Graham and Phillips[11] for $\Pi < 15\,\mathrm{mN\,m^{-1}}$. At the same time, our data still remain incompatible with those of Graham and Phillips in respect to the maximum surface pressure at the solution–oil interface. In particular, at high concentration our values are 27–28 $\mathrm{mN\,m^{-1}}$ for HSA and 31–32 $\mathrm{mN\,m^{-1}}$ for β-casein, while the values reported by Graham and Phillips are 19–20 $\mathrm{mN\,m^{-1}}$ and 23–24 $\mathrm{mN\,m^{-1}}$, respectively. It is to be noted that, for the solution–air interface, our data are in perfect agreement with those presented by Graham and Phillips.[11]

Let us also compare our results with data reported elsewhere. The interfacial tension isotherms and dynamic surface pressure curves for BSA, HSA, β-casein and β-lactoglobulin have been measured by many authors using various techniques (including the pendant drop method), but for the solution–air interface only.[13–23] In most cases a reasonable degree of agreement is found between these results, the data reported by Graham and Phillips, and our results corrected for the loss of protein due to adsorption. Spread monolayers of BSA and β-lactoglobulin at the solution–air and solution–oil interfaces were studied by Murray.[18] The maximum surface pressures for BSA at the solution–air interface reported therein, *ca.* 20 $\mathrm{mN\,m^{-1}}$, reasonably agree with those characteristic of protein adsorption layers. At the same time, a value of $\Pi \cong 30\,\mathrm{mN\,m^{-1}}$ was found for the solution–oil interface, which agrees with our data but not with the results of Graham and Phillips. Murray's results[18] also correspond with our experiments with respect to the higher surface activity of BSA at the solution–oil interface, while for β-lactoglobulin, on the contrary, the same Π values at the solution–air interface were attained at higher area per protein molecule values. The given analysis allows us to conclude that the equilibrium surface pressures for proteins at the solution–air and solution–oil interfaces, measured by the pendant drop method, are in a reasonable agreement with the data published elsewhere, and are indicative of the fact that the nature of the interface is of major importance for the phenomena studied. Some underestimation of the Π values reported by Graham and Phillips for the solution–oil interface can be possibly attributed to errors of the method used for the interfacial tension measurements. This conclusion is implicitly confirmed by the fact that the data[11] show no difference between the Π values for saturated and unsaturated hydrocarbons.

Comparison with a Theoretical Equation of State Model

The main concepts of the theoretical model used to describe the adsorption of proteins at a solution/fluid interface are given elsewhere.[24–26] The theory takes into account two features characteristic of polyelectrolytes: the existence of multiple states of adsorbed molecules in the surface layer with different partial molar areas, and the existence of significant unbound electric charges. This model differs significantly from other semi-empirical models given in the literature, for example, the models of Guzman *et al.*[27] and Douillard and Lefebvre,[28] which are based on the concept of two-layer adsorption of proteins, or the approach of Graham and Phillips.[29] The number of adjustable parameters involved in these theories[27,28] allow the authors to attain a best fit with the experimental results.[11] The theory proposed by Uraizee and Narsimhan[30] is thermodynamically more rigorous, since it assumes the non-ideality of the surface layer and accounts for the inter-ion interactions. This theory, however, relies on an empirical relationship between the composition of the adsorbed layer and the value of the adsorption. Other known theoretical models are discussed elsewhere.[26] The theory used here assumes that the composition of the surface layer is controlled by the surface pressure.[24–26] The relation between the composition of the adsorbed layer and the surface pressure is the key equation of this theory. This relation (in contrast to that employed by Uraizee and Narsimhan[30]) is derived essentially from thermodynamic considerations, involving the analysis of chemical potentials of the protein molecule existing either in the surface layer or in the bulk solution. The simplest form of this relation (assuming that the surface activity of the protein molecule does not depend on its adsorbed state) is:

$$\Gamma_i = \Gamma_\Sigma \frac{\exp\left[-\frac{(i-1)\Pi\omega_1}{RT}\right]}{\sum_{i=1}^{n}\exp\left[-\frac{(i-1)\Pi\omega_1}{RT}\right]}. \quad (1)$$

Here Γ_i is the adsorbed amount in the ith state, Γ_Σ is the total adsorbed amount of protein in all states, Π is the surface pressure, ω_1 is the minimum surface area per protein molecule, n is the number of adsorption states, and R and T are the gas constant and the absolute temperature. It is seen from equation (1) that the increase in interfacial pressure leads to a displacement of the states with larger molar surface area out of the adsorbed layer and to an increase of the adsorbed layer thickness. The equation of state and the adsorption isotherm were derived[25,26] assuming the non-ideality of entropy and enthalpy contributions, and including effects related to the electric double layer. To make these equations applicable to the analysis of experimental results, the equations have to be significantly simplified. In the equation of state of the interfacial layer, the contribution from the non-ideality of the enthalpy to the interfacial pressure can be neglected as compared to the contribution of inter-ion interactions, while in the adsorption isotherm equation the contributions from these two effects are of opposite sign and so tend to cancel out. The resulting simplified expressions for the equation of state of the surface layer and the protein adsorption isotherm are

$$\Pi = -\frac{RT}{\omega_\Sigma}\left[\ln(1-\Gamma_\Sigma\omega_\Sigma) - a_{\text{el}}\Gamma_\Sigma^2\omega_\Sigma^2\right], \quad (2)$$

$$b_1 c = \frac{\Gamma_1\omega_\Sigma}{(1-\Gamma_\Sigma\omega_\Sigma)^{\omega_1/\omega_\Sigma}}, \quad (3)$$

where c is the equilibrium bulk concentration, a_{el} is a parameter characterizing the electrical charge effect, and b_1 is a constant characterizing the surface activity of the protein. Here the average molar surface area of the adsorbed protein is defined as weight average over all states of the molecules within the surface layer:

$$\omega_\Sigma = \omega_1 \frac{\sum_{i=1}^{n} i \exp\left(-\frac{i\Pi\omega_1}{RT}\right)}{\sum_{i=1}^{n} \exp\left(-\frac{i\Pi\omega_1}{RT}\right)}. \tag{4}$$

Amongst the parameters that enter equations (1)–(4), only a_{el} can be varied over a rather wide range to achieve the best possible fit with the experimental data, because the other parameters, such as $\omega_{max} = \omega_{min} + (n-1)\cdot\Delta\omega$, $\omega_{min} = \omega_1$ and $\Delta\omega$, are unambiguously determined by the size and flexibility of the protein molecule. These parameters were chosen so as to obtain the best fit of the theoretical equations to the experimental surface pressure and adsorption isotherms. For all the proteins studied, the derived values of a_{el} (per one amino acid group or per molecular mass unit) were found to be quite similar to one another. For HSA, at the two interfaces studied, the values are ω_{max} = 80–100 nm^2 (per one protein molecule), ω_{min} = 40–50 nm^2, and a_{el} = 120–180; for β-casein solutions the values are ω_{max} = 100–120 nm^2, ω_{min} = 6–7 nm^2, and a_{el} = 70–80; and for β-lactoglobulin solutions the values are ω_{max} = 15–20 nm^2, ω_{min} = 7–8 nm^2, and a_{el} = 40–50. The theoretical curves calculated from equations (1)–(4) with these parameter values are presented in Figures 4–6. The values of the adsorption equilibrium constant b_1 for HSA and β-casein at the solution–oil interface are somewhat higher than at the solution–air interface, while for β-lactoglobulin the values of this constant are essentially the same at both interfaces.

It is noted that, for all the proteins studied, a so-called model of discrete adsorption states was used.[25,26] For HSA molecules the number of possible states should be two. However, for β-casein and β-lactoglobulin the number of intermediate states can be rather high. The increment of molar surface area, $\Delta\omega$, *i.e.*, the difference between the molar surface areas of neighbouring states for these proteins, is 0.8–1.5 nm^2 per molecule. These $\Delta\omega$ values agree with those obtained by MacRitchie[31,32] from kinetic experiments.

Estimation of Protein Adsorption from Mass Balance

For concentration $c_0 < 10^{-7}$ mol l^{-1} (β-casein and β-lactoglobulin), and for $c_0 < 3 \times 10^{-8}$ mol l^{-1} (HSA), the values of the final (equilibrium) concentration calculated from the relation $c = c_0 - \Gamma_\Sigma(S/V)$ were found to be 10% or less, as compared with the initial concentration c_0. For some experiments at low concentrations the values were found to be negative, which is the result of obviously too large values of Γ_Σ introduced into the theory. Therefore, it appears that, for $c_0 \gg c$, an essentially new procedure can be applied for the analysis of protein adsorption from drop experiments. When adsorption and diffusion equilibrium have been established, we can make the approximation that almost all protein is concentrated within the surface layer. The condition of

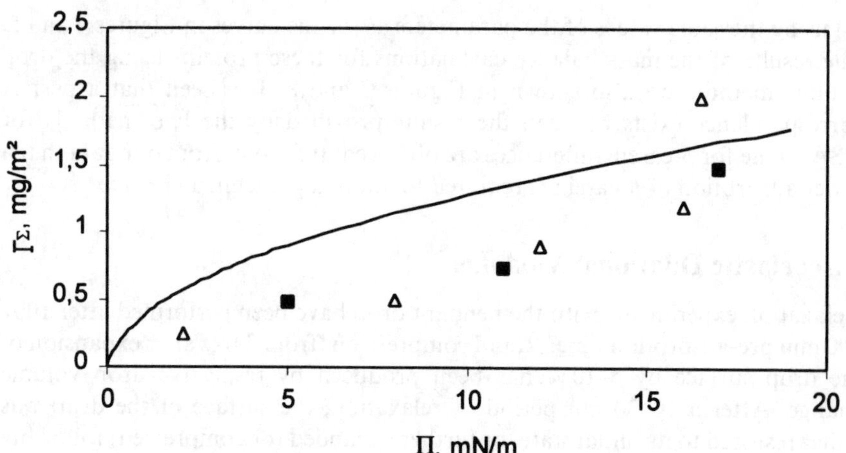

Figure 7 Π versus Γ_Σ *isotherm for two albumins at the liquid–oil interface:* ■, *our data based on mass balance (HSA);* △, *experimental data from Graham and Phillips*[11] *and calculated curve based on these data (BSA)*

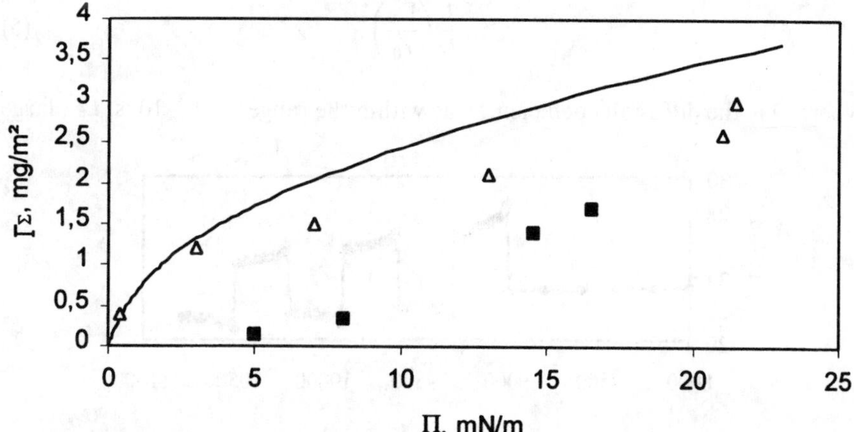

Figure 8 Π versus Γ_Σ *isotherm for β-casein at the liquid–oil interface:* ■, *our data based on mass balance;* △, *experimental data from Graham and Phillips*[11] *and calculated curve based on these data*

mass balance then yields $c_0 = \Gamma_\Sigma(S/V)$. Thus, at low protein concentration the adsorption of protein can be estimated from the initial value of protein concentration within the drop. In Figures 7 and 8 the experimental data reported by Graham and Phillips[11] for HSA and β-casein at the solution–oil interface are plotted in the form of Π as a function of Γ_Σ. In this study the adsorption measurements were performed using radiolabelling and ellipsometry methods, independent of the surface tension measurements. The theoretical dependencies presented in Figures 7 and 8 were calculated from equations (1)–

(4) using the same values of the parameters as for the curves in Figures 4 and 5. The results of the mass balance calculations for these proteins using the drop volume method are also shown in Figures 7 and 8. It is seen that a perfect correspondence exists between the results provided by the two methods for HSA, while for β-casein differences are observed. It follows from our data that a lower adsorption of β-casein is required to attain a predefined Π value.

Viscoelastic Dilational Modulus

Relaxation experiments with the pendant drop have been performed after 100–200 min pre-adsorption time. Rapid compression (from 3 to 5 s) or expansion of the drop surface by 5–10% has been produced by respective drop volume change. After a 10–30 min period of relaxation, the surface of the drop was either restored to its initial state, or further expanded (or compressed) following the initial compression (expansion). This operation was repeated several times. The results of these experiments performed at the solution–air and solution–oil interfaces are illustrated in Figures 9–11. It is noted that, for these experiments, the value of the diffusion relaxation time,

$$t_{\mathrm{d}} = \frac{1}{D}\left(\frac{\Gamma_{\Sigma}}{c_0}\right)^2, \tag{5}$$

where D is the diffusion coefficient, was within the range of 10^3–10^6 s, *i.e.*, large

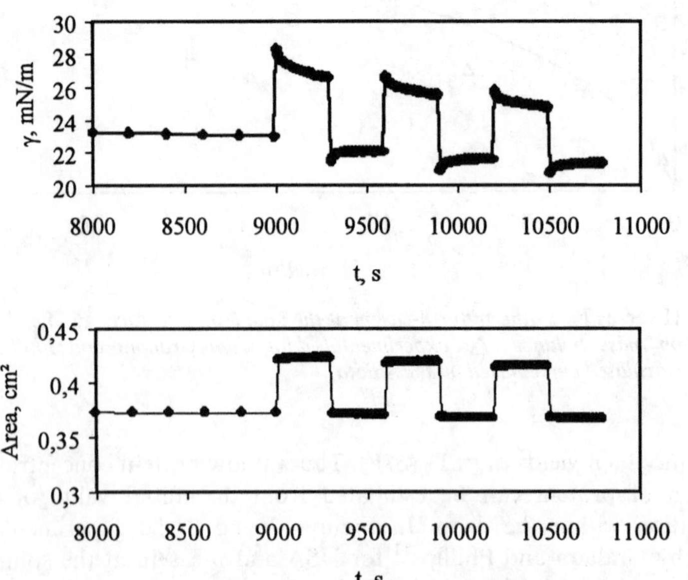

Figure 9 *Dependence of dynamic surface tension γ (top graph) and drop area (bottom graph) on time* t *for periodic trapezoidal deformations of the drop surface (5×10^{-7} mol l^{-1} HSA, solution–oil interface). Initial compression*

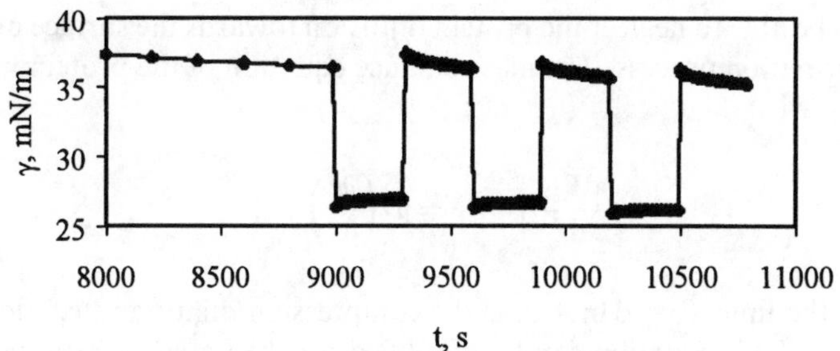

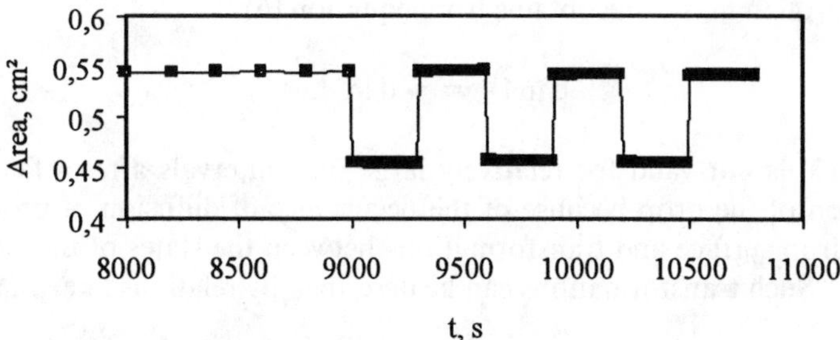

Figure 10 *Dependence of dynamic surface tension γ (top graph) and drop area (bottom graph) on time* t *for periodic trapezoidal deformations of the drop surface (3×10^{-8} mol l^{-1} HSA, solution–oil interface). Initial compression*

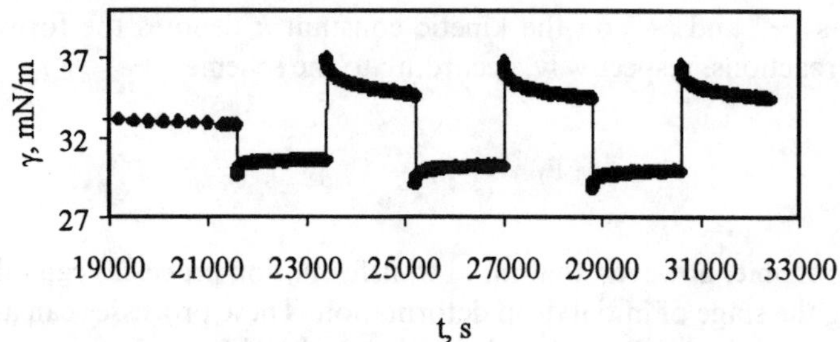

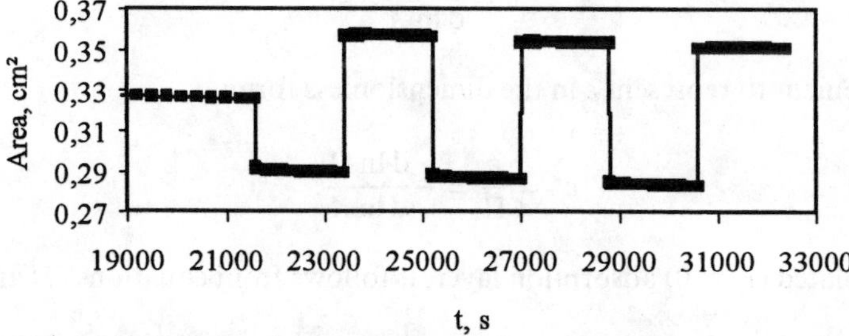

Figure 11 *Dependence of dynamic surface tension γ (top graph) and drop area (bottom graph) on time* t *for cyclic deformations of the drop surface (5×10^{-8} mol l^{-1} β-casein, solution–oil interface). Symmetric expansion–compression deformations*

enough to be able to neglect the protein diffusion towards the surface during the drop deformation process. The mass balance equation of the protein within the surface layer is

$$\frac{d\Gamma_\Sigma}{dt} + \theta\Gamma_\Sigma = D\left(\frac{\partial c}{\partial x}\right)_{x=0}, \tag{6}$$

where t is the time, $\theta = d\ln A/dt$ is the compression/dilation rate, and A is the surface area. This equation can be simplified for the initial deformation of the surface by neglecting the diffusion flux. Therefore, assuming the conservation of adsorbed protein mass, one obtains from equation (6):

$$d\ln\Gamma_\Sigma = -d\ln A. \tag{7}$$

Equation (7) is not valid for relatively large time intervals after a fast initial deformation of the drop because of the occurrence of diffusion of protein to/from the drop surface and transformations between the states of the adsorbed molecules.[6] Such transformations can be described by relations like

$$\frac{d\Gamma_i}{dt} = -\Gamma_i\left(k_i^- + k_{i+1}^+\right) + \Gamma_{i-1}k_i^+ + \Gamma_{i+1}k_{i+1}^- + I_i, \tag{8}$$

where I_i is the diffusion flow of the ith state from the bulk solution, and the superscripts '+' and '−' on the kinetic constant k denotes the forward and backward reactions, respectively, according to the scheme:

$$\Gamma_{i-1} \underset{k_i^-}{\overset{k_i^+}{\leftrightarrow}} \Gamma_i \underset{k_{i+1}^-}{\overset{k_{i+1}^+}{\leftrightarrow}} \Gamma_{i+1}. \tag{9}$$

It is quite possible, however, that rapid transformation processes can take place also during the stage of initial drop deformation. These processes can affect the viscoelastic modulus of the surface layer, ε, calculated from the relation

$$\varepsilon = \frac{d\gamma}{d\ln A}. \tag{10}$$

It is convenient to represent ε in the dimensionless form:

$$\varepsilon^* = \frac{\varepsilon}{\Pi} = -\frac{d\ln\Pi}{d\ln A}. \tag{11}$$

For an isolated ($I_i = 0$) adsorption layer, it follows from equations (7) and (11) that

$$\varepsilon^* = \frac{d\ln\Pi}{d\ln\Gamma_\Sigma}. \tag{12}$$

For some particular cases equation (12) yields rather simple expressions. For example, for an ideal gaseous monolayer with $\Pi = RT\Gamma_\Sigma$, we have $\varepsilon^* = 1$.

It has been found that, for the proteins studied, the values of a_{el} are high for the two interfaces considered. Therefore one can neglect the first (logarithmic) term as compared to the second term in the equation of state [equation (2)], provided that Π is high enough. This leads to an approximate expression

$$\Pi = \frac{RT}{\omega_\Sigma}\left[a_{el}\Gamma_\Sigma^2\omega_\Sigma^2\right]. \tag{13}$$

Introducing equation (13) into equation (12), one obtains $\varepsilon^* = 2$ assuming the formation of a protein adsorption layer without conformational changes. If the contribution of the first term in equation (2) cannot be neglected (which is the case for low and medium surface pressures), then it follows from equation (12) that we have $\varepsilon^* > 2$. On the contrary, transitions between the adsorbed states induced by the surface layer deformation, which leads to a reduction in $\Delta\Pi$ at constant adsorption, can result in a decrease of ε^* as compared with the theoretical value, derived above from the equation of state.

To compare the approximate viscoelastic modulus $\varepsilon^* = 2$ with the experimental value, we use the data of Figures 9–11. The differential quotient in equation (11) was employed in form of a finite difference, *i.e.*,

$$\varepsilon^* = -\frac{\Delta\Pi}{\Pi}\frac{A}{\Delta A}. \tag{14}$$

The experimental dependencies on Π obtained for HSA at $c_0 < 10^{-6}$ mol l^{-1} at both interfaces, and for β-casein at $c_0 < 10^{-7}$ mol l^{-1}, for the solution–oil interface, are presented in Figure 12. One can see that, for $\Pi > 5$ mN m^{-1}, the experimental ε^* values only slightly depend on the surface pressure, and lie between 1 and 3, and mainly in the interval 1.5–2.5. One can thus conclude that equation (2) agrees satisfactorily with the data obtained in the relaxation experiments. The values of ε^* for the compression process were found to exceed those obtained in expansion by *ca.* 15–20%. Moreover, the values for HSA at the solution–air interface exceed those at the solution–oil interface. The value of ε^* averaged over all measured data at the solution–air interface was 2.9, and at solution–oil interface 1.8. For β-casein at the solution–oil interface the mean value of ε^* is 1.3, which is lower than that for HSA, indicating a greater flexibility of adsorbed β-casein molecules. It is noted that, for concentrated solutions of β-casein ($c_0 > 10^{-5}$ mol l^{-1}) at $\Pi > 30$ mN m^{-1}, very low values of the viscoelastic modulus were found, *i.e.*, $\varepsilon^* = 0.2$–0.3. This is indicative of extremely high mobility of β-casein segments in dense adsorbed layers at the solution–oil interface. On the contrary, the increase in HSA concentration at $c_0 > 10^{-6}$ mol l^{-1} results in a relatively modest decrease in the viscoelastic modulus at both interfaces.

The elastic moduli for BSA interfacial layers at solution–air and solution–vegetable oil interfaces were calculated by Benjamins *et al.*[33] from results of

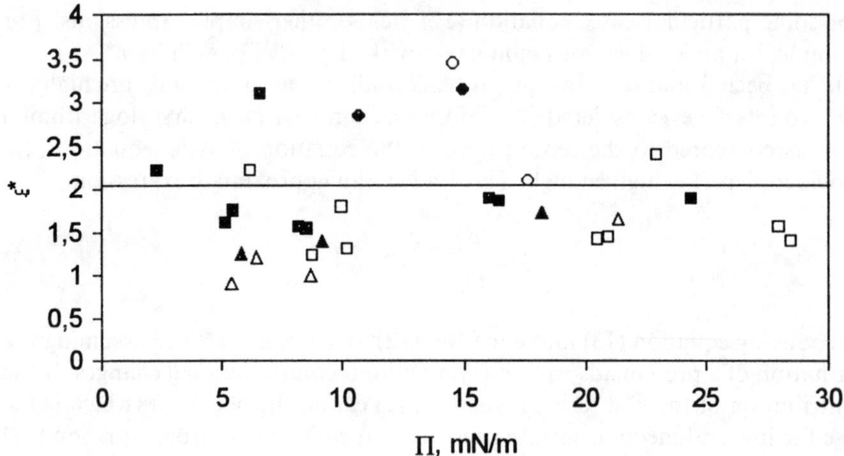

Figure 12 *Dependence of dimensionless viscoelastic modulus ε* on surface pressure Π. Open symbols = expansion; closed symbols = compression; ●,○, HSA at solution–air interface; ■,□, HSA at solution–oil interface; ▲,△, β-casein at solution–oil interface*

Langmuir trough experiments with a movable barrier and slowly oscillating drops. The oscillation frequency for the two methods was 0.1 Hz, with the amplitude of surface area variations $\leqslant 20\%$. Benjamins' data show[33] that for BSA the value of ε^* for the solution–oil interface at $\Pi < 10\,\mathrm{mN\,m^{-1}}$ is almost independent of surface pressure and is equal to 2.5–2.8. The value of ε^* decreases to 1 when Π increases up to $17\,\mathrm{mN\,m^{-1}}$. These results are in good agreement with our approximate value of $\varepsilon^* = 2$, and with the experimental data for HSA. The viscoelastic modulus obtained for the solution–air interface[33] are also somewhat higher, and quite similar to our data. For example, in the range of very low Π, the value of ε^* was 5–6. For $\Pi > 8\,\mathrm{mN\,m^{-1}}$ a decrease in the dimensionless viscoelastic modulus was observed down to $\varepsilon^* = 2.5$ at $\Pi = 20\,\mathrm{mN\,m^{-1}}$. It was shown above that, for low Π, the theoretical value of ε^* can increase, and therefore this result contradicts neither the theoretical model of equations (1)–(4) nor our experimental results.

4 Conclusions

The pendant drop method has been used to study the interfacial tension of solutions of some proteins at the solution–air and solution–oil interfaces. The comparison between our data and the results reported in other studies using plate and ring methods shows that, at low initial protein concentration in the drop and large drop surface area, the final equilibrium concentration of protein can differ substantially from the initial value. This deficiency of the pendant drop method can, however, be turned into an advantage for the case of extremely low initial concentrations, when almost all protein is adsorbed at the drop surface. In this case the adsorption can be estimated from the condition of protein mass balance. The values calculated in this way are in good agreement with the results published in the literature. Varying the size of the droplets, or using an inverse drop system (in this case the concentration of protein in solution remains almost unchanged during the adsorption process), one obtains a modified ADSA technique for the study of protein adsorption. This modification works under the condition $c_0 \gg c$ for a pendant drop of solution, or $c_0 = c$ for the inverse drop with a large enough volume of the solution around. In this way a direct and independent measurement of surface tension and adsorption becomes possible.

The corrected Π–c isotherms for HSA, β-casein and β-lactoglobulin at the two interfaces comply satisfactorily with the theoretical model, which assumes multiple states of adsorbed molecules and the existence of inter-ion interactions. The calculated model parameters agree well with known characteristics of molecules and molecular segments. The expression for the viscoelastic modulus of the surface layer derived from the equation of state is in good agreement with the results of the relaxation experiments after expansion or compression of the drop surface. While it is important that these deformations are homogeneous, such studies can be quite easily performed for both solution–air and solution–oil interfaces. This makes it possible to control the amplitude, the direction of the deformation, and the subsequent time delay of the deformed surface, and to vary these parameters over a rather wide range.

Comparing the two studied interfaces, significant differences are found. For HSA and β-casein both the rate of interfacial tension increase and the equilibrium values of Π at the solution–oil interface are higher than the corresponding values at the solution–air interface. For β-lactoglobulin, a similar kind of behaviour is observed at high bulk concentrations, but at low bulk concentrations the observed increase in surface tension is more pronounced for the solution–air interface. The relative elasticities measured by transient pendant drop experiments do not show a significant difference for the water–air and water–oil interfaces.

Acknowledgements

The work was financially supported by projects of the European Community (INCO ERB-IC15-CT96-0809) and the DFG (Mi418/9-1). A research grant of the Alexander von Humboldt Stiftung (A.V.M., BB-1034737) is also gratefully acknowledged.

References

1. A. Prins, M. A. Bos, F. J. G. Boerboom, and H. K. A. I. van Kalsbeek, in 'Studies of Interface Science', ed. D. Möbius and R. Miller, Elsevier, Amsterdam, 1998, vol. 7, p. 221.
2. B. S. Murray and E. Dickinson, *Food Sci. Technol. Int.* (Japan), 1996, **2**, 131.

3. V. N. Izmailowa and G. P. Yampolskaya, in 'Studies of Interface Science', ed. D. Möbius and R. Miller, Elsevier, Amsterdam, 1998, vol. 7, p. 103.
4. J. Benjamins and E. H. Lucassen-Reynders, in 'Studies of Interface Science', ed. D. Möbius and R. Miller, Elsevier, Amsterdam, 1998, vol. 7, p. 341.
5. R. Miller, V. B. Fainerman, J. Krägel, and G. Loglio, *Curr. Opin. Colloid Interface Sci.*, 1997, **2**, 578.
6. V. B. Fainerman and R. Miller, in 'Studies of Interface Science', ed. D. Möbius and R. Miller, Elsevier, Amsterdam, 1998, vol. 7, p. 51.
7. D. C. Clark, F. Husband, P. J. Wilde, M. Cornec, R. Miller, J. Krägel, and R. Wüstneck, *J. Chem. Soc. Faraday Trans.*, 1995, **91**, 1991.
8. Y. Rotenberg, L. Boruvka, and A. W. Neumann, *J. Colloid Interface Sci.*, 1983, **37**, 1699.
9. P. Cheng, D. Li, L. Boruvka, Y. Rotenberg, and A. W. Neumann, *Colloids Surf. A*, 1990, **43**, 151.
10. V. B. Fainerman, A. V. Makievski, and R. Miller, *Colloids Surf. A*, 1994, **87**, 61.
11. D. E. Graham and M. C. Phillips, *J. Colloid Interface Sci.*, 1979, **70**, 415.
12. R. Wüstneck, J. Krägel, R. Miller, V. B. Fainerman, P. J. Wilde, D. K. Sarker, and D. C. Clark, *Food Hydrocolloids*, 1996, **10**, 395.
13. G. Gonzalez and F. MacRitchie, *J. Colloid Interface Sci.*, 1970, **32**, 55.
14. A. J. I. Ward and L. H. Regan, *J. Colloid Interface Sci.*, 1980, **78**, 389.
15. E. Tornberg and G. Lundh, *J. Colloid Interface Sci.*, 1981, **79**, 76.
16. J. A. Feijter, J. Benjamins, and F. A. Veer, *Biopolymers*, 1978, **17**, 1760.
17. B. S. Murray and P. V. Nelson, *Langmuir*, 1996, **12**, 5973.
18. B. S. Murray, *Colloids Surf. A*, 1997, **125**, 73.
19. B. C. Tripp, J. J. Magda, and J. D. Andrade, *J. Colloid Interface Sci.*, 1995, **173**, 16.
20. P. Suttiprasit, V. Krisdhasima, and J. Mcguire, *J. Colloid Interface Sci.*, 1992, **154**, 316.
21. F. K. Hansen and R. Myrfold, *J. Colloid Interface Sci.*, 1995, **176**, 408.
22. J. M. Rodriguez Patino and M. R. Rodriguez Niño, *Colloids Surf. A*, 1995, **103**, 91.
23. V. Krisdhasima, J. Mcguire, and R. Sproula, *J. Colloid Interface Sci.*, 1992, **154**, 337.
24. V. B. Fainerman, R. Miller, and R. Wüstneck, *J. Colloid Interface Sci.*, 1996, **183**, 26.
25. A. V. Makievski, V. B. Fainerman, M. Bree, R. Wüstneck, J. Krägel, and R. Miller, *J. Phys. Chem.*, 1998, **102**, 417.
26. A. V. Makievski, R. Wüstneck, D. O. Grigoriev, J. Krägel, and D. V. Trukhin, *Colloids Surf. A*, in press.
27. R. Z. Guzman, R. G. Carbonel, and P. K. Kilpatrick, *J. Colloid Interface Sci.*, 1986, **114**, 536.
28. R. Douillard and J. Lefebvre, *J. Colloid Interface Sci.*, 1990, **139**, 488.
29. D. E. Graham and M. C. Phillips, *J. Colloid Interface Sci.*, 1979, **70**, 427.
30. F. Uraizee and G. Narsimhan, *J. Colloid Interface Sci.*, 1991, **146**, 169.
31. F. MacRitchie, *Colloids Surf.*, 1989, **41**, 25.
32. F. MacRitchie, *Anal. Chim. Acta*, 1991, **249**, 241.
33. J. Benjamins, A. Cagna, and E. H. Lucassen-Reynders, *Colloids Surf. A*, 1996, **114**, 245.

Conformational Changes of Globular Proteins in Solution and Adsorbed at Interfaces Investigated by FTIR Spectroscopy

By Rebecca J. Green, Ian Hopkinson, and Richard A. L. Jones

UNIVERSITY OF CAMBRIDGE, CAVENDISH LABORATORY, MADINGLEY ROAD, CAMBRIDGE CB3 0HE, UK

1 Introduction

The physical and chemical properties of an adsorbed protein layer must depend sensitively on the conformation adopted by the protein molecules in that layer. For example, immobilization of enzymes to solid surfaces can result in loss of catalytic activity if significant changes in conformation occur during adsorption, while conversely the stabilizing action of proteins adsorbed at an air–liquid interface may be enhanced if the molecules form a strongly interacting two-dimensional network. The conformational state of an adsorbed protein will frequently differ from that of a bulk protein due to properties of the interface such as hydrophobicity and charge, which may lead to surface denaturation of the protein. The purpose of this paper is to explore the conformational transitions undergone by proteins at surfaces in response to environmental changes, and to compare these transitions with the corresponding transitions in the bulk state.

The investigation of the conformational changes of proteins at interfaces is of significant interest particularly within the food and pharmaceutical industries. Knowledge of the composition and conformation of adsorbed protein layers is necessary for the design of biomedical devices, since it is this initial adsorbed layer that determines the long term fate of the biomaterial in the body.[1] Within the food industry, interfacial interactions become significant when considering the fouling of food processing equipment,[2] or during the production of foams and emulsions, where proteins are used as stabilizers.[3]

Thermal Denaturation of Proteins in Solution

The physical changes that occur when a solution of globular proteins is heated have been investigated extensively; one can identify two distinct transitions.[3] Initially, the protein molecules undergo a change in conformation or partial denaturation, during which the viscosity of the solution gradually increases due to the increased molecular dimensions of the partially unfolded protein molecules. This is followed by the solution undergoing liquid–liquid unmixing and separating into two phases, one of high protein concentration and the other low.[4] This leads to intermolecular association or aggregation of the individual unfolded protein molecules, and is accompanied by an exponential increase in viscosity and the onset of gelation.[5] Clark *et al.*[6] have shown that, unlike other globular proteins, hen egg white lysozyme forms a thermoplastic gel (work carried out at pD 2 with 100 mM NaCl added), producing a clear, rubbery gel at 70 °C, which melts upon further heating, and reforms during cooling. Conversely, bovine serum albumin (BSA) forms a thermoset gel, which sets irreversibly during heating, behaviour typical of most globular proteins.

The conformational changes that accompany protein aggregation and gelation have been widely investigated by the spectroscopic techniques of circular dichroism (CD), Raman and infrared spectroscopy.[4,6–9] These studies have shown that the major change in structure observed upon heating a protein solution is the formation of antiparallel β-sheet, which has been associated with the intermolecular association or gelation of the unfolded protein molecules. A mechanism put forward for the heat-induced gelation of ovalbumin[7] suggests that the aggregation is formed from partially unfolded protein molecules, through the crosslinking of intermolecular β-sheet structures, due to the exposure of hydrophobic residues.

Changes in protein conformation have often been investigated by Fourier transform infrared (FTIR) spectroscopy.[10,11] The main infrared frequencies of interest are the amide I, II and III regions at around 1650, 1550 and 1350–1200 cm^{-1}, respectively. The amide I peak in particular is used to determine structural information, such as the extent of α-helix or β-sheet character within the protein conformation. This is a composite peak due primarily to C=O bond stretching coupled with N–H bending and C–H stretching modes.[12] Clark *et al.*[6,13,14] used gross changes in the structure of the amide I peak to investigate the heat-set gelation of a number of globular proteins, and correlated their FTIR data with data from small-angle X-ray scattering and electron microscopy studies. They found that the appearance of shoulders at 1621 and 1684 cm^{-1} coincided with the onset of protein gelation and could be attributed to the formation of the antiparallel β-sheet. Such changes have been shown[9] not to occur in proteins that do not form gels on heating.

Proteins at Interfaces

The behaviour of a protein at an interface is likely to differ considerably from its behaviour in the bulk. Because of the different local environment at the interface, the protein may have the opportunity of adopting a more disordered state without incurring the energy penalty entailed by, for example, exposing its hydrophobic core to water.[15] The conformation of an adsorbed protein may well, therefore, differ from its native state, and the degree of change in conformation will depend upon the protein and its local environment. Therefore, one would expect that the type of interface would have a considerable effect on the thermal denaturation of an adsorbed protein layer. Protein adsorption and its adsorbed conformation can be investigated by the surface sensitive technique of attenuated total reflection (ATR) FTIR spectroscopy.[4,16]

Recently, there has been some interest in interfacial studies of gelation. A thin layer of gelled material may form at an interface, even when the bulk concentration is too low, as long as the local surface concentration exceeds the threshold required for gelation. Surface gelation was initially observed by Kim *et al.*[17] in polystyrene sulfonate solutions, and since then this possibility has also been considered for protein systems.[4,18]

In previous studies, ATR-FTIR spectroscopy has been used to compare the conformational changes of lysozyme upon heating in bulk solution and at an interface.[4] This work showed that the temperature at which protein molecules begin to form the intermolecular associations that are identified in the bulk with aggregation and gelation is considerably lower when the protein is adsorbed at an interface.

In this paper we report a substantial extension of this previous study, in which we contrast both the unfolding behaviour of an individual protein and the onset of the intermolecular associations which lead to gelation in the bulk and at the surface. We study both the temperature dependence and the kinetics of surface and bulk transitions for two proteins, hen egg-white lysozyme and BSA.

2 Experimental Methods

Sample Preparation

A deuterated phosphate buffer (pH 7, 0.1 M) was used as the adsorbate solution. D_2O (99.9% deuterium, Aldrich) was chosen rather than H_2O because water absorbs strongly in the amide region of the spectrum. Another advantage of using the deuterated solvent is that it allows the unfolding of the protein to be monitored. H–D exchange occurs rapidly for hydrogens near the surface of the protein and hence accessible to the solvent, but very slowly for hydrogens within the core of the protein and inaccessible to the solvent. This exchange process affects the absorption frequencies of the amide II peak by shifting it to 1450 cm^{-1}, but it has very little effect on the position of the amide I peak. The protein solution was made up in deuterated buffer the day before analysis in order to allow time for exchange of all accessible hydrogens at the surface of the protein in its native conformation. Upon adsorption and during heating, any unfolding of the protein brings some hydrogens originally inaccessible into contact with the solvent, and this manifests itself in the spectrum as a disappearance of the residual amide II peak (due mostly to N–H bending) at around 1540 cm^{-1}.

Thermal denaturation of adsorbed protein layers were performed at a 5 mg ml^{-1} protein concentration (BSA and lysozyme obtained from Sigma, UK) and zinc selenide (ZnSe) ATR crystals (Crytran, Dorset, UK). The thermal denaturation of protein in bulk solution was studied using a transmission cell with CaF_2 windows (Graseby Specac Ltd., UK) and a protein concentration of 0.1 g ml^{-1}, a concentration that exceeds the bulk critical gelation concentration of both proteins.

Infrared Spectroscopy

A Mattson Galaxy 4020 FTIR spectrometer with a deuterated triglycerine sulfate (DTGS) detector was used to collect the infrared spectra. The spectrometer was continually purged with dry air during each experiment to remove water vapour from the chamber. Data were collected as interferograms at 2 cm^{-1} resolution, of which typically 120 interferograms were collected and co-added, except during the initial stages of adsorption where 28 interferograms were co-added. Heating was carried out using a Eurotherm temperature controller producing stable temperatures to within $\pm 0.5\,°C$, and calibrated using a thermocouple.

ATR spectra were collected using a 10-internal reflection, 45° fixed angle of incidence ZnSe crystal. The ATR crystal was sealed in a Squarecol ATR cell (Graseby Specac Ltd., Kent, UK) which had been modified to allow liquid handling from outside of the infrared chamber. Spectra of the ATR cell containing deuterated buffer were used as the background against which all subsequent protein spectra were determined. Protein solution was added, allowing 2 hours adsorption time before replacing the aqueous phase with buffer-only solution and heating the adsorbed protein layer. A control experiment was performed on each surface prior to lysozyme adsorption, where the substrate was immersed in buffer-only solution and heated to 90 °C and cooled as described above.

Spectra of the bulk protein solution were determined using a Specac liquid transmission cell that had been modified to allow the cell to be heated. The cell was fitted with 4 mm CaF_2 windows and a tin spacer to give a pathlength of 6 μm. Heating was performed in the same manner as for the ATR experiments, and all protein spectra were referenced against a blank buffer solution.

3 Results and Discussion

Thermal Denaturation of Proteins in Bulk Solution

The spectra for the thermal denaturation of lysozyme in bulk solution can be seen in Figure 1. During heating of the protein two significant changes to the amide peaks are observed. Initially, the disappearance of the N–H residual amide II peak at 1540 cm^{-1} is seen which indicates unfolding of the protein. At higher temperatures, shoulders begin to appear in the amide I peak, one at 1621 cm^{-1} and a less pronounced shoulder at 1684 cm^{-1}. These shoulders reveal the formation of antiparallel β-sheet structure, and indicate the onset of

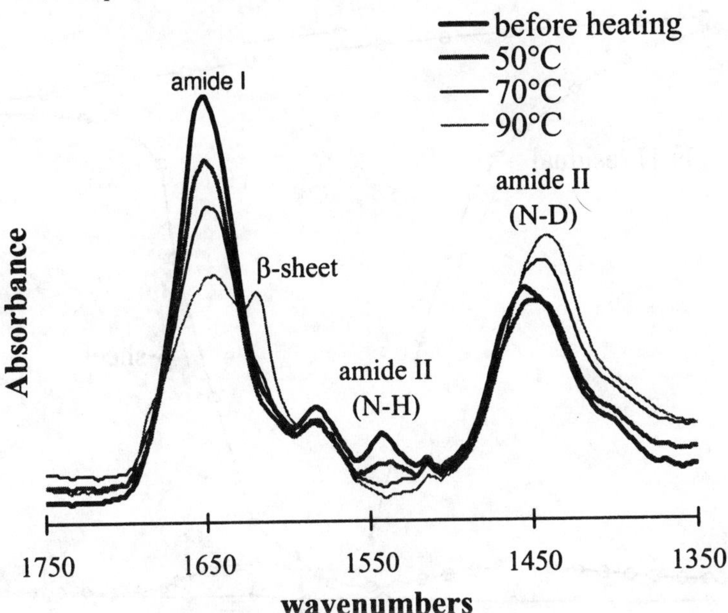

Figure 1 *The amide I and II regions of the infrared spectra (1400–1700 cm^{-1}) following the thermal denaturation of lysozyme in bulk solution*

protein aggregation. These β-sheet peaks remain prominent within the spectra during cooling of the protein solution.

Analysis of the integrals of the amide peaks reveals the rates and the temperatures at which these two transitions occur. Figure 2 shows peak area plotted against temperature for the N–H residual amide II peak (1540 cm^{-1}) and the 1620 cm^{-1} shoulder of the amide I peak (corresponding to β-sheet formation). The integrated data reveal that the protein unfolds at around 50 °C, observed by a rapid decrease in the peak area of the N–H residual amide II peak.[1] At approximately 70 °C, after completion of the unfolding transition, a second sharp transition is observed, as the appearance of the 1620 cm^{-1} shoulder of the amide I peak, which marks the initiation of protein aggregation. At around 60 °C the protein has unfolded, but does not immediately begin to aggregate, existing instead as a stable intermediate. Indeed, in separate experiments where the heating cycle has been held at 60 °C for 8 hours, complete unfolding occurred as marked by complete H/D exchange, but no significant change in secondary structure, was observed. This behaviour suggests that the protein may exist in an intermediate state; unfolded, yet not able to form strong intermolecular associations. It is tempting to associate this state with the 'molten globule' state found in other globular proteins such as α-lactalbumin. However, other studies have not revealed such an equilibrium state for hen egg-white lysozyme, though studies have suggested the existence of an analogous kinetic intermediate.[19,20]

Similar experiments for BSA (data not shown) have revealed that, on heating

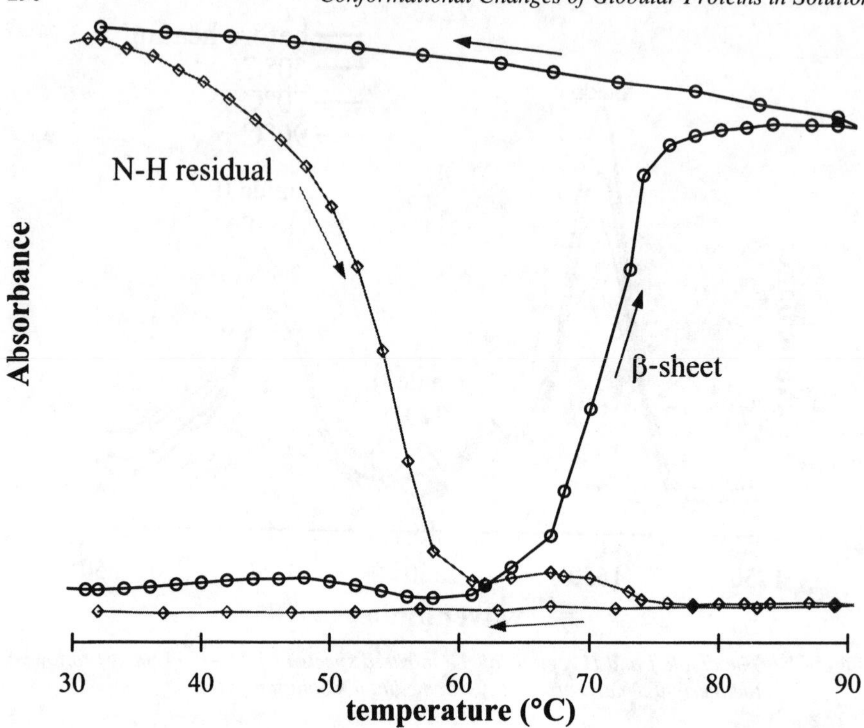

Figure 2 *Peak integrals of the β-sheet shoulder of the amide I peak (at 1621 cm^{-1}) and the N–H residual amide II peak plotted against temperature during the thermal denaturation of lysozyme in solution*

to 90 °C, the protein unfolds and aggregates in a manner similar to lysozyme. However, the rates of unfolding and β-sheet formation were found to be slower for BSA and to result in overlapping of the two transitions at 60 °C. Both lysozyme and BSA appear to form quite stable intermediates at 60 °C (data not shown) that do not significantly aggregate even during prolonged incubation unless further heated to higher temperatures. This result suggests that elevated temperatures (>60 °C) are required before intermolecular association can occur, and that even complete unfolding is not a sufficient condition for intermolecular association to take place.

Thermal Denaturation of Proteins Adsorbed to ZnSe

To allow similar investigations of possible interfacial effects upon protein unfolding and aggregation processes, lysozyme was adsorbed to a ZnSe ATR crystal. Adsorption took place over a two hour period, after which time the system was washed with buffer-only solution and the adsorbed lysozyme layer was heated. Spectra were recorded regularly during adsorption, so that the rate of adsorption and conformational change could be assessed. Figure 3a shows

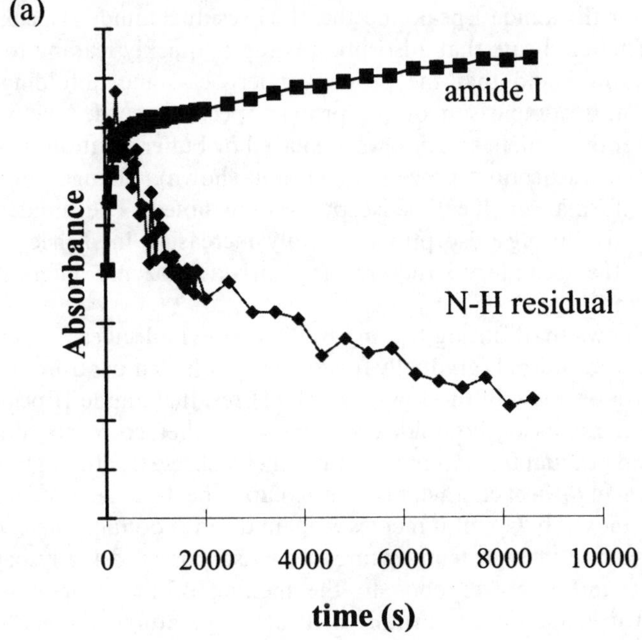

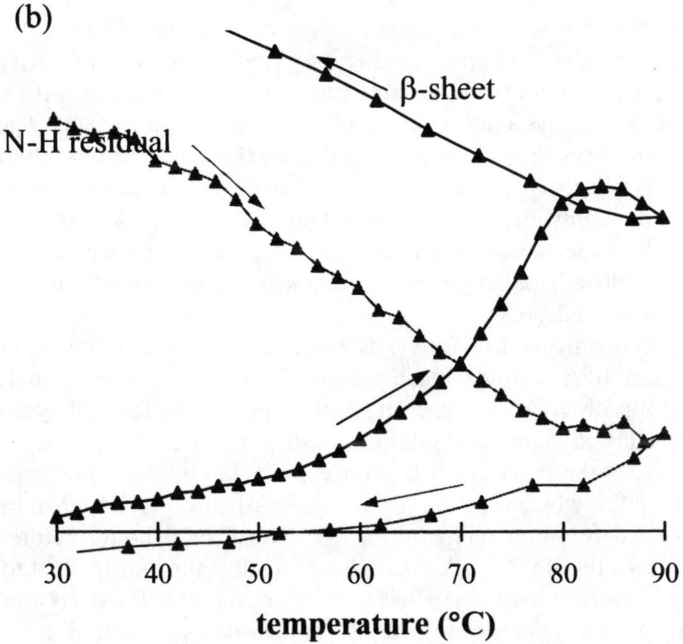

Figure 3 *Amide peak integrals during the adsorption and thermal denaturation of lysozyme at a ZnSe interface. (a) Lysozyme adsorption monitored by following the amide I and N–H residual amide II peaks as a function of time. (b) Thermal denaturation of adsorbed lysozyme observed as changes in the N–H residual amide II peak (unfolding) and the β-sheet (1621 cm^{-1}) shoulder of the amide I peak (intermolecular association)*

the integrals for the amide I peak and the N–H residual amide II peak during adsorption. These indicate that adsorption is rapid, quickly leading to saturation of the surface and that the protein undergoes some unfolding at the interface. From a comparison of the protein spectra recorded before (t = 1 min) and after the protein solution was replaced by buffer solution ($t > 2$ h) at the end of the adsorption process (data not shown), changes in protein secondary conformation due to adsorption were noted. The amide I peak broadens slightly during adsorption, possibly increasing in β-sheet content, revealing that the secondary structure of adsorbed lysozyme differs from its native structure.

Figure 3b shows that, during heating to 90 °C, the molecules in the protein layer continue to unfold gradually over the whole temperature range as manifested by a decrease in the 1540 cm^{-1} N–H residual amide II peak. Also, upon heating, the shape of the amide I peak begins to alter, corresponding to the continuous and gradual formation of antiparallel β-sheet structure. However, a slight reduction in β-sheet character is detected upon heating the protein layer to temperatures above 80 °C, but it increases again during cooling. This reduction in β-sheet structure at high temperatures was also observed for lysozyme by Clark *et al.*[13] and it corresponds to the melting of the protein gel. The aggregated protein molecules partially melt at temperatures above 80 °C but they reform on cooling, which is evidence that lysozyme forms a thermoplastic gel.

The spectra observed for lysozyme adsorbed at an interface differ considerably from those observed upon heating lysozyme in bulk solution. An adsorbed protein layer does not exhibit discrete transitions upon heating, with unfolding and intermolecular association taking place as two separate events. The thermal denaturation of lysozyme in solution indicates that the process is thermodynamically controlled, where aggregation occurs as a sharp transition at around 70 °C, and then only after the protein has completely unfolded. However, when adsorbed at an interface, lysozyme begins to aggregate before protein unfolding has been completed, and at temperatures and a rate far lower than observed for aggregation in solution.

Similar experiments for BSA adsorbed to ZnSe (data not shown) have revealed that BSA unfolds more readily during adsorption than lysozyme. Following this observation, experiments were performed for both lysozyme and BSA that allowed significantly longer adsorption times before beginning the heating stage of the experiment. In these experiments adsorption was monitored for 4 hours, after which time the surface was washed with a buffer-only solution and incubated at room temperature for a further 20 hours before heating. Figure 4 shows the amide peak areas plotted against time for the first four hours of protein adsorption to ZnSe for both lysozyme and BSA. As observed in Figure 3a protein adsorption is rapid, and during adsorption some protein unfolding occurs. However, during the four-hour period BSA shows signs of complete unfolding, whereas lysozyme only partially unfolds. For lysozyme a small increase in the width of the amide I peak is noted which suggests a shift towards β-sheet formation, and this is not observed for BSA. Incubating the

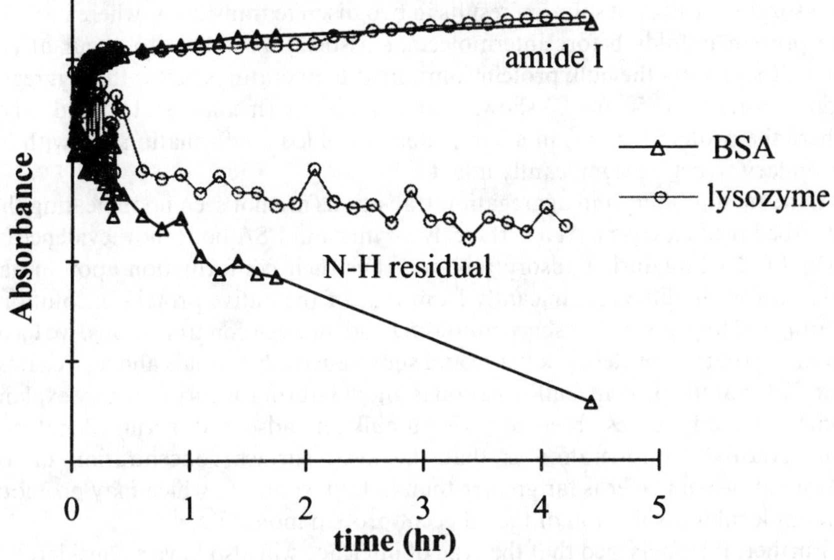

Figure 4 *The peak integrals of the N–H residual amide II peak showing the unfolding of lysozyme and BSA during adsorption to a ZnSe interface*

adsorbed protein layers at room temperature for a further 20 hours has no noticeable effect on their spectra. Upon subsequently heating the adsorbed layers to 90 °C (data not shown), unfolding and β-sheet formation occur for lysozyme at similar rates to those described previously for Figure 3b. For BSA, the already unfolded protein begins slowly and continuously to aggregate, increasing in β-sheet character, as seen by the formation of shoulders in the amide I peak. Unlike lysozyme there is no evidence of melting at higher temperatures; thus BSA irreversibly aggregates during heating.

It appears that the effect of adsorption at an interface is similar for both BSA and lysozyme, with both proteins beginning to aggregate at lower temperatures and at slower rates than when present in bulk solution. However, BSA is a less rigid molecule than lysozyme and is not as stable; it unfolds almost completely upon adsorption. Therefore, the unfolding transitions differ between the two molecules, occurring mostly during heating for lysozyme and during adsorption for BSA. Another conformational difference between BSA and lysozyme is observed when considering the rate of β-sheet formation. Lysozyme loses β-sheet character at temperatures above 80 °C, which reforms on cooling, a trend not observed during the aggregation of BSA.

4 Conclusions

Our work has highlighted significant differences between the way a protein behaves when adsorbed at an interface to its behaviour in the bulk. The heating of lysozyme or BSA in solution results in two discrete transitions, where initially the protein unfolds before intermolecular association is able to occur at *ca.* 70 °C. Incubating the bulk protein solution at temperatures below this aggregation transition (at 50–60 °C) shows that a stable intermediate state could exist, where the protein remains in a completely unfolded conformation but with its secondary structure significantly intact.

Discrete unfolding and aggregation transitions are not seen upon heating the adsorbed protein layers. At a surface, lysozyme and BSA both show evidence of partial unfolding during adsorption, and thus their conformation upon initiation of heating differs significantly from that of the native protein in solution. During heating, instead of sharp unfolding and aggregation transitions, we have observed that the protein gradually and simultaneously unfolds and aggregates. The fact that the protein conformation is altered during adsorption may explain some of the differences observed between bulk and adsorbed thermal denaturation. Another consideration is that the local surface concentration of an adsorbed protein layer is far greater than in bulk solution, which may promote intermolecular association of the adjacent protein molecules.

Further, it is envisaged that the type of interface will also have a considerable effect on the conformational changes that occur during lysozyme adsorption and thermal denaturation. Surface properties such as hydrophobicity, charge and surface hardness will affect the strength of protein binding and the extent of protein reorientation that occurs at an interface. Indeed, in continuation of the work described in this paper, we are investigating the effects of such surface characteristics on protein unfolding and intermolecular association, and this work will be described in a subsequent publication. Different polymeric interfaces have been used during this study to investigate the effect of surface hydrophobicity and also to study the effect of surface hardness, using hydrophobic solid and liquid polymers.

What this work clearly establishes is that the differences in the conformational transitions of globular proteins at the surface and in the bulk are both qualitative and quantitative in nature. Sharp unfolding transitions in the bulk become much broader transitions at the surface, and where in the bulk the onset of intermolecular associations forms a quite distinct and separate process from unfolding, for adsorbed proteins the two processes occur simultaneously.

References

1. J. L. Brash and T. A. Horbett, *ACS Symp Ser.*, 1995, **602**, 1.
2. K. L. Fuller and S. G. Roscoe, in 'Protein Structure–Function Relationships in Food', ed. R. Y. Yada, R. L. Jackman, and J. L. Smith, Blackie, Glasgow, 1994, p. 143.
3. L. G. Phillips, D. M. Whitehead, and J. Kinsella, 'Structure–Function Properties of Food Proteins', Academic Press, London, 1994.
4. A. Ball and R. A. L. Jones, *Langmuir*, 1995, **11**, 3542.
5. J. E. Kinsella, D. J. Rector, and L. G. Phillips, in 'Protein Structure–Function Relationships in Food', ed. R. Y. Yada, R. L. Jackman, and J. L. Smith, Blackie, Glasgow, 1994, p. 1.

6. A. H. Clark, D. H. P. Sanderson, and A. Suggett, *Pept. Protein Res.*, 1981, **17**, 353.
7. A. Kato and T. Takagi, *J. Agric. Food Chem.*, 1988, **36**, 1156.
8. E. Li-Chan, S. Nakai, and M. Hirotsuka, in 'Protein Structure–Function Relationships in Food', ed. R. Y. Yada, R. L. Jackman, and J. L. Smith, Blackie, Glasgow, 1994, p. 163.
9. D. M. Byler and J. M. Purcell, *Proc. SPIE-Int. Soc. Opt. Eng.*, 1989, **1145**, 415.
10. I. H. M. van Stokham, H. Linsdell, J. M. Hadden, P. I. Haris, D. Chapman, and M. Bloemendal, *Biochemistry*, 1995, **34**, 10508.
11. J. I. Boye, A. S. Ismail, and I. Alli, *J. Dairy Res.*, 1996, **63**, 97.
12. A. Dong, S. J. Prestrelski, S. Allison, and J. F. Carpenter, *J. Pharm. Sci.*, 1995, **84**, 415.
13. A. H. Clark and C. D. Tuffnell, *Int. J. Pept. Protein Res.*, 1986, **16**, 339.
14. A. H. Clark, F. J. Judge, J. B. Richards, J. M. Stubbs, and A. Suggett, *Int. Pept. Protein Res.*, 1981, **17**, 380.
15. H. J. Chen, S. L. Hsu, D. A. Tirrell, and H. D. Stidham, *Langmuir*, 1997, **13**, 4775.
16. J. S. Jeon, S. Raghavan, and R. P. Sperline, *Colloids Surf. A*, 1994, **92**, 255.
17. M. W. Kim, D. G. Peiffer, and P. Pincus, *J. Physique Lett.*, 1984, **45**, L-953.
18. E. Dickinson, 'An Introduction to Food Colloids', Oxford University Press, Oxford, 1992.
19. M. A. Williams, J. M. Thornton, and J. M. Goodfellow, *Protein Eng.*, 1997, **10**, 895.
20. C. M. Dobson, P. A. Evans, and S. E. Radford, *TIBS*, 1994, **19**, 31.

Structure, Interfacial Properties, and Functional Qualities in Foams and Emulsions of Surfactin, a Lipopeptide from *Bacillus subtilis*

By Magali Deleu, Michel Paquot, Hary Razafindralambo, Yves Popineau,[1] Herbert Budziekiewicz,[2] Philippe Jacques, and Philippe Thonart[3]

UNITÉ DE CHIMIE BIOLOGIQUE INDUSTRIELLE, FACULTÉ UNIVERSITAIRE DES SCIENCES AGRONOMIQUES DE GEMBLOUX, PASSAGE DES DÉPORTÉS 2, B-5030 GEMBLOUX, BELGIUM
[1]LABORATOIRE DE BIOCHIMIE ET DE TECHNOLOGIE DES PROTÉINES, INSTITUT NATIONAL DE LA RECHERCHE AGRONOMIQUE, 44026 NANTES CEDEX 03, FRANCE
[2]UNIVERSITY OF KÖLN, GREINSTR. 4, D-50939 KÖLN, GERMANY
[3]CENTRE WALLON DE BIOLOGIE INDUSTRIELLE, UNIVERSITÉ DE LIÈGE, 4000 LIÈGE, BELGIUM

1 Introduction

Surface-active agents are required for forming and stabilizing disperse systems such as foams and emulsions. They find applications[1–5] in an extremely wide variety of areas involving product formulation in food, cosmetics, road, pesticide, detergent, paper, and pharmaceutical industries, as well as enhanced oil recovery, transportation of heavy crude oil, and bioremediation.

In molecular terms, surface-active agents are amphiphilic compounds containing both hydrophilic and lipophilic parts.[6] Their efficiency in foaming and emulsifying is dependent on their amphiphilic structure. In general, two main types of surface-active agents can be distinguished: small surfactant molecules and amphiphilic macromolecules. Because of their small size and their simple amphiphilic structure composed of a polar head and a hydrophobic tail, small surfactant molecules diffuse and orient rapidly at fluid–fluid interfaces. They reduce efficiently the interfacial tension and promote disperse system formation.

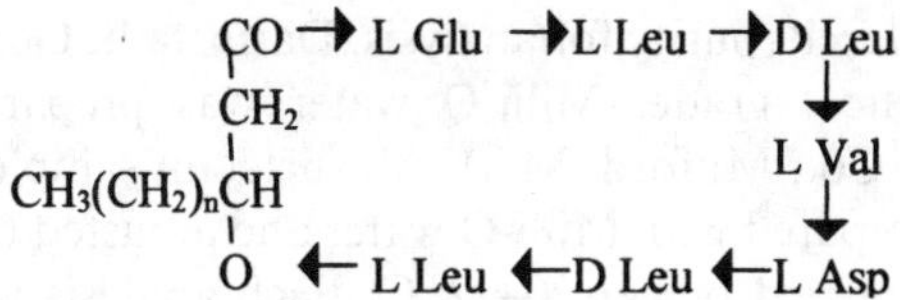

Figure 1 *Example of surfactin primary structure (n = 9, 10, or 11)*

On the other hand, amphiphilic macromolecules like proteins have a high molecular weight and a more complex multi-amphiphilic structure. They migrate less quickly to the interface, but they form a cohesive viscoelastic film via intermolecular interactions for greater long-term stability.[7,8] A perfect surface-active agent should combine the favourable features of proteins with those of the most effective small surfactant molecules.[6]

Bacillus subtilis surfactin is a typical compound which could satisfy this condition owing to its hybrid structure and intermediate size in comparison with small surfactant molecules and proteins (see Figure 1). Its structure is composed of a heptapeptide cycle closed by a β-hydroxyfatty acid that forms a lactone ring system.[9] In addition, as biosurfactant, surfactin exhibits several advantages such as biodegradability, low toxicity, and various possible structures, relative to chemically synthetized surfactants.[3,10,11]

The present paper reports a study on the effect of the hydrophobic character of the surfactin lipidic chain on the interfacial properties and functional qualities of this type of molecule in foam and emulsion systems. A comparison between foaming and emulsifying performances of surfactin variants is discussed. To acquire such information, three surfactin homologues (SuC13, SuC14 and SuC15) produced by the *Bacillus subtilis* S499 strain have been purified. Their interfacial properties at both air–water and oil–water interfaces have then been studied. Investigations include the characterization of the adsorption under dynamic conditions, and the evaluation of the foaming and emulsifying properties by studying the short-term destabilization rate of foams and emulsions.

2 Materials and Methods

Surfactins were produced by fermentation of the *B. subtilis* strain S499[12] and purified as previously described.[13] Primary structure and purity of the surfactin homologous series (>99%) were ascertained by analytical RP-HPLC (chromsphere-5 μm C18 column, 1 × 25 cm, Chrompack, Middelburg, Netherlands), amino acid analysis,[14] and electrospray mass spectrometry measurements using a Finnigan MAT 900 ST. Three surfactin homologues containing the β-hydroxyfatty acids of 13 (SuC13, MW: 1007), 14 (SuC14, MW: 1021) and 15 (SuC15, MW: 1035) carbon atoms were obtained. SuC13 and SuC15 have an iso-branched β-hydroxyfatty acid, while the SuC14 fatty acid is linear.

n-Dodecane was purchased from Sigma (purity >99%, St. Louis, MO) and *n*-hexadecane from Merck (purity for analysis, Darmstadt, Germany). All other reagents were analytical grade. Milli-Q water was prepared by Millipore apparatus (Millipore Co., Milford, MA). All surfactin samples were dissolved in 5 mM tris buffer prepared with Milli-Q water and adjusted to pH 8.0–8.5. All measurements were carried out at 20–22 °C. Each analysis was performed at least twice.

Adsorption kinetics at the air–water (A–W) and oil–water (O–W) interfaces were monitored continuously by following the decrease in surface or interfacial tension. The measurements were carried out with a drop volume tensiometer (TVT1, Lauda) in the dynamic mode.

Dynamic surface or interfacial tension *versus* time plots, $\gamma = f(t)$, were described by the relaxation equation used by several authors:[15,16]

$$\gamma_t = \gamma_m + (\gamma_0 - \gamma_m)/[1 + (t/t^*)^n]. \tag{1}$$

Here γ_0 is the surface or interfacial tension of the pure solvent, γ_t is the surface or interfacial tension at time t, γ_m is the surface or interfacial tension at meso-equilibrium, t^* is the half-time for reaching γ_m, and n is a dimensionless constant. Parameters n, t^*, and γ_m were estimated by computer fitting of the measured dynamic surface or interfacial tension data using Sigma-plot software (Jandel, Germany). By differentiating equation (1) with respect to t and substituting t for t^*, the maximum reduction rate v_{max} of γ is obtained to be

$$v_{max} = \frac{n(\gamma_0 - \gamma_m)}{4t^*} = -(d\gamma_t/dt)_{max}. \tag{2}$$

Creaming and flocculation kinetics were determined using a conductimetric method developed by Guéguen *et al.*[17] Surfactin solutions at 0.1 mg ml^{-1} (7.5 ml) and *n*-hexadecane (4.5 ml) were poured into a cylinder containing two electrodes at the base. The conductivity of the aqueous phase was measured before emulsification, and then the emulsion was formed using ultrasonic treatment (15 s, 35 W, 23 kHz). The volume of the aqueous phase was continuously monitored by conductivity measurement. The rate of destabilization of the emulsion (k_1) was measured from ϕ to $\phi + 0.1$ from ϕ *versus* time plots, where ϕ is the volume fraction of dispersed oil in the emulsion, calculated from

$$\phi = 1 - [7.5/4.5 \times (1 - C_{tc}/C_{sol})], \tag{3}$$

where C_{sol} is the conductivity of the aqueous solution before emulsification and C_{tc} is the conductivity of the emulsion at time t.

3 Results and Discussion

Adsorption at Air–Water and *n*-Dodecane–Water Interfaces

Figure 2 shows dynamic surface tension curves of SuC13, SuC14 and SuC15 at

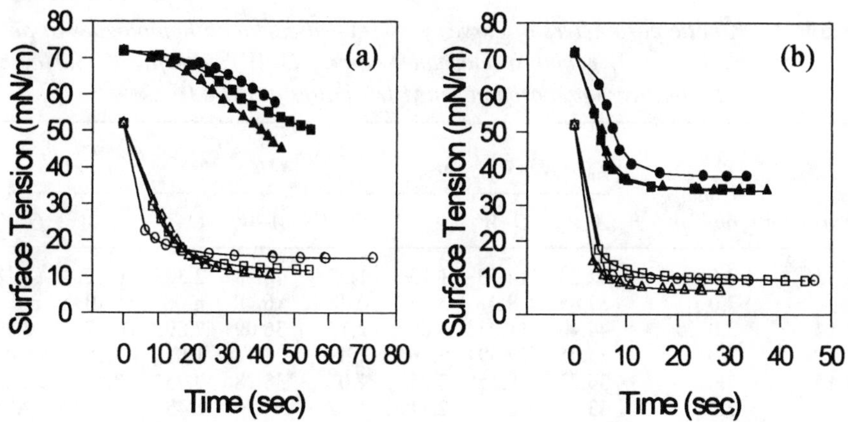

Figure 2 *Curves $\gamma = f(t)$ of surfactins: (a) at 2.0 × 10^{-8} mole cm^{-3}, (b) at 7.0 × 10^{-8} mol cm^{-3}, at A–W interface (filled symbols) and O–W interface (open symbols) for homologues C13 (●,○), C14 (■,□) and C15 (▲,△)*

air–water (A–W) and *n*-dodecane–water (O–W) interfaces at two concentrations, a lower one of 2.0 × 10^{-8} mol cm^{-3} and a higher one at 7.0 × 10^{-8} mol cm^{-3}. Generally speaking, the surfactins reduce the dynamic surface tension. These results reveal that surfactins adsorb at the A–W or O–W interfaces and present surface-active properties at these two interfaces. This general property is related to the amphiphilic character of surfactins due to the presence of a hydrophobic part consisting of a long-chain fatty acid and some lipophilic amino acids, and a hydrophilic part composed of several amino acid residues.

At lower concentrations, the effect of lipid chain hydrophobic character on the surface tension is different at the A–W and O–W interfaces. At the A–W interface, the higher the lipid chain hydrophobicity the faster the reduction of γ, while at the O–W interface, the SuC13 variant is more efficient in kinetic terms than SuC14 and SuC15. The difference between the homologous molecules could be attributed to a difference in the area occupied per molecule, which has a consequence on the surface tension. Indeed, at the A–W interface, it has been shown[18] that the increasing hydrophobicity of a tail group enhances the molecular area occupied.

At higher concentrations, differences between the three homologous surfactins are negligible at A–W as well as at O–W interfaces. This suggests that the homologous molecules occupy the same interfacial area, with perpendicular orientation of the lipid chain at either A–W or O–W interfaces at high concentration. At high and low concentrations, the decrease of surface tension is greater at the O–W interface than at the A–W interface.

From the curves at Figure 2, the kinetic parameters γ_m, n, t^* and v_{max} have been calculated. The values are listed in Table 1. As already noted from curves in Figure 2, the effect of lipid chain hydrophobicity on the surfactin kinetic behaviour is different for the two interfaces except for the value of the parameter γ_m. At low and high concentrations, γ_m decreases with strengthening of lipid

Table 1 *Kinetic parameters of curves* $\gamma = f(t)$ *of surfactin homologues at air–water (A–W) and n-dodecane–water (O–W) interfaces. Surfactin solutions were prepared in 5 mM tris buffer at pH 8.0*

Surfactin	*Conc./ mol cm*$^{-3}$	γ_m*/mN m*$^{-1}$ *A–W*	γ_m*/mN m*$^{-1}$ *O–W*	n *A–W*	n *O–W*	t*/s *A–W*	t*/s *O–W*	v_{max}*/ mN m*$^{-1}$ *s*$^{-1}$ *A–W*	v_{max}*/ mN m*$^{-1}$ *s*$^{-1}$ *O–W*
C13	2.0	47.24	14.50	1.73	1.27	48.58	2.30	0.22	5.18
	7.0	37.85	8.47	3.88	0.48	6.48	n.d.[a]	5.10	n.d.[a]
C14	2.0	44.40	11.03	3.00	2.15	36.09	7.89	0.57	2.79
	7.0	34.45	7.59	3.16	1.49	4.12	1.87	7.18	8.85
C15	2.0	36.33	8.15	3.01	2.07	38.28	9.25	0.80	2.45
	7.0	33.84	6.13	2.71	1.62	4.40	1.48	5.86	12.55

[a]n.d.: not determined. The tension for SuC13 decreases so quickly at the O–W interface that curve fitting cannot be realized.

chain hydrophobic character for both interfaces. Increased hydrophobicity increases the degree of interaction between the chains which strengthens the surfactin effect on the dynamic surface free energy.

At high concentration, the *n* value dependence on lipidic tail hydrophobicity is of opposite sign at A–W and O–W interfaces. The parameter *n* decreases with increasing chain hydrophobic character at the A–W interface whereas it increases at the O–W interface. According to Gao and Rosen,[19] the parameter *n* is related to the difference between the adsorption rate and the desorption rate. The closer the *n* value is to zero, the more the adsorption is near to equilibrium, *i.e.*, the adsorption rate is approaching the desorption rate. Based on this interpretation, increasing chain hydrophobicity at high concentration allows the equilibrium state to be reached more quickly at the A–W interface, while it slows down approach to the equilibrium state at the O–W interface. At low concentration, SuC13 has a lower *n* value than SuC14 and SuC15 at A–W and O–W interfaces. Thus, weaker hydrophobicity at low concentration allows the equilibrium state to be reached more rapidly whatever the interface type.

According to v_{max} values at low concentration, which represents the maximum rate of reducing γ at t^*, the increasing chain hydrophobicity leads to a decrease in the adsorption of surfactin at the O–W interface and to an increase at the A–W interface. The ideal chain hydrophobicity for fast reduction of γ at low concentration depends on the type of interface. SuC15 is more efficient in kinetic terms at the A–W interface while SuC13 is better at the O–W interface.

At high concentration, the more efficient in kinetic terms at the A–W interface is now SuC14 according to the values of t^* and v_{max}. At the O–W interface, it seems that SuC13 decreases γ more quickly than SuC14 and SuC15: the final γ value is reached after 5 seconds with SuC13, while it is obtained after 10–15 seconds with SuC14 and SuC15 (Figure 2).

Overall, the kinetic parameters show that the surfactant adsorption is more efficient at the O–W interface than at the A–W interface. This means that surfactins reduce the tension faster and lower, and reach more quickly the equilibrium adsorption state, when the apolar phase is liquid and not gaseous.

The global difference between the interfaces could be explained by a different lipid chain orientation and/or peptidic cycle disposition at the interface. Indeed, an apolar liquid phase is a better apolar solvent than air. According to Miller,[20] at a liquid–liquid interface there is protruding of protein parts into both adjacent liquid phases, while at an A–W interface the protruding occurs only into the water phase. Thus, the dodecane phase can exert an attraction on carbon chains of the lipid tail and on the hydrophobic amino acids residues of the peptidic ring. Consequently, the lipid chain of surfactin would be more vertically organized at the O–W interface than at the A–W interface, where the lipid tails are rather more inclined to interact laterally,[21] and the peptidic saddle[22] would be more likely to enter into the apolar phase. This could contribute to a more organized and compressed structure at the O–W interface, and so could explain the greater effect of surfactins on surface tension and on kinetic parameters at this interface.

The discrepancy between A–W and O–W interfaces with regard to the effect of increasing chain hydrophobicity could be explained on the basis of the above considerations. The variation of lipid chain hydrophobicity could influence the molecular disposition between polar and apolar phases, whatever the phase type. At the O–W interface the greater attraction of the apolar phase would tend to exert an additional effect on molecular disposition at the interface. These two effects could contribute to a more structured molecular organization at the O–W interface with increasing hydrophobicity of the lipid chain at the two concentrations investigated, which could explain that a molecule needs a higher energy to compress the film in order to create a space large enough to adsorb[23] and consequently the existence of an adsorption energy barrier for high hydrophobicity at this interface. At the A–W interface, a higher concentration is needed to overcome the adsorption energy barrier with a surfactin of high chain hydrophobicity. This could be due to the fact that the molecular organization is looser at this interface, and that, consequently, a higher concentration is required to allow the highly hydrophobic chains to interact with each other and to interfere with the adsorption of other molecules.

Foaming and Emulsifying Performances

Destabilization mechanisms encountered in foams and emulsions are rather similar. The physical changes that may occur in emulsions are creaming, flocculation and coalescence. In foams, the process of drainage of continuous phase from thin films between the bubbles effectively takes the place of creaming and flocculation in emulsions. An additional instability process in foams is Ostwald ripening.[24,25] Creaming, flocculation and drainage are destabilization phenomena that usually occur over a short time in comparison to coalescence.

Table 2 presents the effect of increasing lipid chain hydrophobicity on short-term foaming and emulsifying stability. The effect of increasing chain hydrophobicity on the destabilization rate is different for foams and emulsions. In

Table 2 *Destabilization rate of surfactin foam and emulsion. LP1 is the initial drainage rate in foams and* k_1 *is the creaming/flocculation rate in emulsions*

Surfactin	*Foams*[a] *LP1/ml min*$^{-1}$	*Emulsions* $k_1/10^{-3}$ *min*$^{-1}$
C13	2.70 ± 0.24	<0.1
C14	2.26 ± 0.03	0.675 ± 0.066
C15	2.50 ± 0.00	0.781 ± 0.020

[a]Results from Razafindralambo *et al.*[13]

foams, SuC14 exhibits the highest stability with respect to liquid drainage according to the initial drainage rate (LP1). In emulsions, SuC13 develops the best resistance to creaming/flocculation phenomena according to the creaming/ flocculation rate (k_1). Thus, the required molecular structure under the investigated conditions to stabilize foams is different from that for emulsions. In foams, there exists an optimum lipid chain hydrophobicity providing surfactin with the maximum short-term stability, while, in emulsions, a low hydrophobicity of the lipid tail, in the C13–C15 surfactin series investigated, favours short-term stability.

To prevent liquid drainage in foams, the foaming agent must form an efficient viscoelastic film. It has already been demonstrated[26] that too short a lipid chain produces insufficient cohesiveness, whereas too great a length produces too much rigidity for good film elasticity. This is in agreement with results from Razafindralambo *et al.*[13] In emulsions, the extent of flocculation mainly depends on the repulsive forces between droplets. Our results can be explained by the fact that an increase of chain hydrophobicity would modify the peptidic ring disposition at the interface, which would in turn influence the steric and/or electrostatic repulsion between droplets. A conformational analysis at the O–W interface of the three homologues could be used to assess this assumption.

4 Conclusions

Under dynamic conditions, for example in the processes of foam or emulsion formation, the dynamics of adsorption play an important role in the stabilization mechanism.[20] In the first part of this paper, a kinetic study of three homologues of surfactin at the A–W and O–W interfaces has been undertaken. From the results, all of the surfactins appear efficient for reducing the interfacial tension rapidly at the two interface types, and so they can find applications in foams as well as in emulsion systems. It would even seem that surfactins are more efficient for promoting emulsion formation than foam formation.

The kinetic behaviour of surfactin homologues is different at A–W and O–W interfaces. In general, SuC13 is more efficient, in kinetic terms, in emulsion systems, while a higher homologue is more adapted for foam systems. However, SuC15 allows a lower γ value to be reached at the two interfaces.

In the second part of this report, a discussion about the chain hydrophobicity effect on short-term stability of foams and emulsions is presented. It has been found that SuC13 develops the best short-term stability in emulsions, while SuC14 is more fitted to stabilize foams. From these results, it appears that there exists a correlation between the dynamic behaviour at high concentration and the short-term stability as regard to the chain length effect of surfactin. Conformational analysis at the interface of the three homologous surfactins is under investigation by our group in order to assess the role of molecular disposition at the interface on the surface functional properties.

It has been shown from theory[27,28] that the mechanical properties of the interfacial layer is also an important parameter in the stabilization mechanism. Experiments on surface rheological properties of surfactin monolayers would be thus of great interest.

Acknowledgements

This work received support from the European Union (BIO4-CT950176). We thank Florence Pineau for excellent assistance in the emulsifying property analysis and Regina Fuchs for the MS analysis. M.D. is FNRS assistant.

References

1. J. Aubert, A. Kraynik, and P. Rand, *Pour la science*, 1986, *juillet*, 62.
2. A. M. Kraynik, *Ann. Rev. Fluid Mech.*, 1988, **20**, 325.
3. G. Georgiou, S.-C. Lin, and M. M. Sharma, *Biotechnology*, 1992, **10**, 60.
4. J. Bibette, *L'Act. Chim.*, 1996, **4**, 23.
5. J. D. Desai and I. M. Banat, *Microbio. Mol. Bio. Rev.*, 1997, **61**, 47.
6. E. Dickinson, *Trends Food Sci. Technol.*, 1993, **4**, 330.
7. J. B. German, T. E. O'Neill, and J. E. Kinsella, *J. Am. Oil Chem. Soc.*, 1985, **62**, 1358.
8. S. Damodaran, in 'Protein Functionality in Food Systems', ed. N. S. Hettiarachchy and G. R. Ziegler, Marcel Dekker, New York, 1994, p. 1.
9. A. Kakinuma, A. Ouchida, T. Shima, H. Sugino, M. Isono, G. Tamura, and K. Arima, *Agric. Biol. Chem.*, 1969, **33**, 1669.
10. J. Vater, *Progr. Colloid Polymer. Sci.*, 1986, **72**, 12.
11. Y. Ishigami, *Inform*, 1993, **4**, 1156.
12. P. Jacques, C. Hbid, F. Vanhentenryck, J. Destain, G. Baré, H. Razafindralambo, M. Paquot, and P. Thonart, *Proceedings of the 6th European Congress on Biotechnology*, 1994, **9**, 1067.
13. H. Razafindralambo, Y. Popineau, M. Deleu, C. Hbid, P. Jacques, P. Thonart, and M. Paquot, *J. Agric. Food Chem.*, in press.
14. H. Razafindralambo, M. Paquot, C. Hbid, P. Jacques, J. Destain, and P. Thonart, *J. Chromatog.*, 1993, **639**, 81.
15. X. Y. Hua and M. Rosen, *J. Colloid Interface Sci.*, 1988, **124**, 652.
16. L. K. Filippov, *J. Colloid Interface Sci.*, 1994, **167**, 320.

17. J. Guéguen, Y. Popineau, I. N. Anisimova, R. J. Fido, P. R. Shewry, and A. S. Tatham, *J. Agric. Food Chem.*, 1996, **44**, 1184.
18. H. Razafindralambo, Ph.D. Dissertation, Faculté Universitaire des Sciences Agronomiques de Gembloux, Belgium, 1996, p. 103.
19. T. Gao and M. Rosen, *J. Colloid Interface Sci.*, 1995, **172**, 242.
20. R. Miller, *Proceedings of 2nd World Congress on Emulsion*, Bordeaux, 23–26 September 1997, vol. 4, p. 153.
21. F. Tadros and B. Vincent, in 'Encyclopedia of Emulsion Technology', ed. P. Becher, Marcel Dekker, New York, 1983, vol. 1, p. 1.
22. J.-M. Bonmatin, M. Genest, H. Labbé, and M. Ptak, *Biopolymers*, 1994, **34**, 975.
23. F. MacRitchie, 'Chemistry at Interfaces', Aacademic Press, San Diego, 1990, p. 156.
24. E. Dickinson and G. Stainsby, *Food Technol.*, 1987, **41**(9), 74.
25. P. Walstra, in 'Food Emulsions and Foams', ed. E. Dickinson, Royal Society of Chemistry, London, 1987, p. 242.
26. M. Rosen, 'Surfactants and Interfacial Phenomena', Wiley-Interscience, New York, 1989, p. 277.
27. H. A. Barnes, *Colloids Surf. A*, 1994, **91**, 89.
28. Th. F. Tadros, *Colloids Surf. A*, 1994, **91**, 39.

Rheology of Food Colloids

Factors Determining Small-Deformation Behaviour of Gels

By T. van Vliet

WAGENINGEN CENTRE FOR FOOD SCIENCES, DEPARTMENT OF FOOD TECHNOLOGY AND NUTRITIONAL SCIENCES, WAGENINGEN AGRICULTURAL UNIVERSITY, PO BOX 8129, 6700 EV WAGENINGEN, THE NETHERLANDS

1 Introduction

A large variety of intermediate products and foods are gels. Their mechanical properties as determined at small deformations vary widely from very soft and deformable to rather stiff, as can be easily experienced by hand. The small-deformation properties are frequently studied although their practical importance is limited. The most important reason for studying them is that they can account for certain aspects of the undisturbed structure of the gel, if the experiments are done well. In this context structure is defined as the spatial distribution of the relevant structural elements (building blocks) of the network and the interaction forces between them. We note that, according to this definition, a structural element may also itself have a structure. Determination of mechanical properties is especially suited to investigate the structure of materials, because they can account for both the spatial distribution of the structural elements and the interaction forces between them in contrast to most other methods. However, this also makes their interpretation more complicated.

'Small deformation' is defined as such a small relative deformation (strain) that applying it does not affect the structure of the material studied. When this is the case, the ratio of the stress involved over the accompanying strain (relative deformation) is independent of the strain (so-called linear behaviour).

A characteristic of gels is that they consist of a continuous solid-like network in a continuous liquid phase over the timescale considered. The latter aspect implies that certain products can be considered as a gel over short times but as a liquid over long times. At intermediate timescales, their reaction to an applied stress will be partly elastic and partly viscous. That is, they behave visco-elastically. A clear example is the inner part of a well-ripened Camembert or various other soft cheeses. So the dependence on time is an important characteristic of small-deformation properties of gels.

The structural elements forming the continuous network in a gel may vary from hard particles such as fat crystals to very flexible macromolecules. Small-deformation properties of gels depend of course on the characteristics of these structural elements, but it is easy to make gels from hard fat crystals in oil or from flexible gelatin molecules in water with similar resistance against a small deformation. On the other hand, for identical concentrations and conditions, as pH and temperature, the resistance against small deformations of gels containing the same structural elements may differ strongly depending on the history. Still, in spite of the large diversity of structural and mechanical properties, general rules can be given for the factors determining the latter properties.

This paper describes those aspects of the undisturbed structure that determine the small-deformation properties of gels, such as the modulus and its dependence on time.

2 Characteristic Parameters of Small-Deformation Behaviour

Consider a network upon which an external force f is applied in direction x causing a deformation.[1,2] Through a cross-section A perpendicular to x there are N strands/chains per unit area bearing the stress σ, *i.e.*, each exerting a reaction force $-(\mathrm{d}f_s/\mathrm{d}x)\Delta x$, where Δx is the distance over which the relevant structural elements of the network have moved with respect to each other. This gives:

$$\sigma = \frac{f}{A} = -N\frac{\mathrm{d}f_s}{\mathrm{d}x}\Delta x. \tag{1}$$

There is no restriction on the nature of the strands, on the elements building up the strands, or on the nature of the interaction forces involved. If the measurements are done in the so-called linear region, which is normally the case for small deformation experiments, $\mathrm{d}f_s/\mathrm{d}x$ is constant. Generally f_s can be expressed as $-\mathrm{d}F/\mathrm{d}x$, where $\mathrm{d}F$ is the change in Gibbs energy when the elements are moved apart over a distance $\mathrm{d}x$. The local deformation Δx can be related to the macroscopic (shear) strain γ by a characteristic length C determined by the geometric structure of the network. The quantity C can only be estimated easily for simplified models of the network structure. In fact it has a tensor character. As the shear modulus G is given by σ/γ, we arrive at:

$$G = NC\frac{\mathrm{d}^2F}{\mathrm{d}x^2}. \tag{2}$$

At constant temperature we have $\mathrm{d}F = \mathrm{d}H - T\mathrm{d}S$, where H is enthalpy and S is entropy, which results in the expression

$$G = NC\frac{\mathrm{d}(\mathrm{d}H - T\mathrm{d}S)}{\mathrm{d}x^2}. \tag{3}$$

From equation (3) it is clear that both enthalpic and entropic factors may contribute to the modulus.

Viscoelasticity

A general characteristic of gels is that they show predominantly solid-like (elastic) behaviour over the timescale considered. In a purely elastic material all the energy supplied during deformation is stored in the material (d^2F/dx^2 term). This energy provides the stresses required to let the material regain its original shape after the applied stress is taken away. So no energy dissipation may occur *i.e.*, no cross-links/bonds may relax due to thermal movement. Intuitively, one expects the rate of the relaxation process to depend on a frequency factor multiplied by $\exp(-F^+/kT)$ where F^+ is the Gibbs activation energy for breaking a bond. The highest possible frequency factor is the natural frequency kT/h ($\approx 6 \times 10^{12}$ s^{-1}), where k and h represent the constants of Boltzmann and Planck, respectively, and T is the absolute temperature. This gives as a first approximation for the relaxation time τ of a bond between two macromolecules:[3]

$$\tau = \frac{h}{kT} e^{F^+/kT}. \tag{4}$$

The type and strength of the bonds, and thus the values of F^+ and τ, will vary greatly between types of gels and between different types of bond in a gel. Moreover, conditions such as pH, temperature, and ionic strength, and the types of ions present, may play a role.

The effect of a relaxation process on small-deformation mechanical properties can be visualized by considering a so-called Maxwell element consisting of

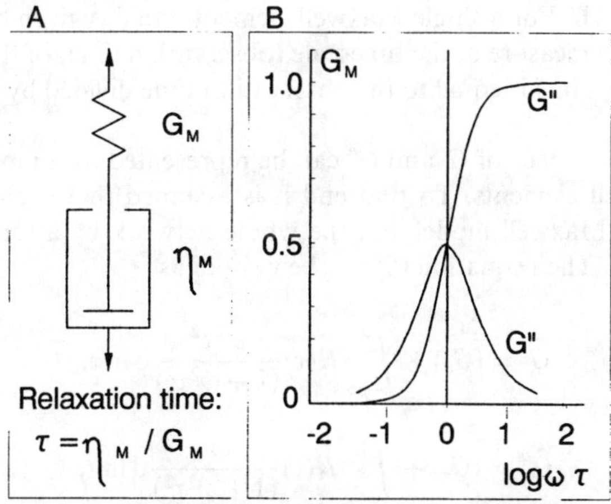

Figure 1 *Model of viscoelasticity. Maxwell element (A) and its mechanical behaviour in terms of G′ and G″ as a function of angular frequency ω on deformation in a sinusoidal way (B).* G_M *is modulus of spring element;* η_M *is dashpot viscosity*

an ideally elastic spring and a viscous dashpot (Figure 1A). If such an element is stressed over a short timescale, the resulting deformation will nearly exclusively be due to elongation of the elastic spring, while during the application of a long-lasting stress the resulting deformation is due to displacement of the piston in the dashpot. For intermediate timescales the ratio of both contributions depends on the modulus G_M of the spring and the effective resistance (which can be modelled as a viscosity η_M) experienced by the piston. The relaxation time τ_M is equal to η_M/G_M.[4,5]

The viscoelastic properties of gels can be conveniently studied by so-called dynamic tests, in which the test piece is subjected to a sinusoidal varying stress or strain, and the resulting strain or stress is measured.[4,5] For an elastic material the response will be proportional to the strain (stress and strain in-phase) and for a viscous one to the strain-rate (stress and strain 90°, or $(1/2)\pi$ rad, out-of-phase). By measuring the stress and strain amplitude and the phase difference the storage and loss moduli can be determined. The timescale of the measurement can be adjusted by varying the oscillating frequency ω. The storage modulus G' is a measure of the energy stored during a periodic application of a stress, and the loss modulus G'' a measure of the energy dissipated in the material. The tangent of the phase angle between stress and strain is $\tan\delta = G''/G'$. For a single Maxwell element, explicit expressions for G' and G'' can be derived:[4,5]

$$G'(\omega) = G_M \frac{\omega^2\tau_M^2}{(1+\omega^2\tau_M^2)}, \tag{5a}$$

$$G''(\omega) = G_M \frac{\omega\tau_M}{(1+\omega^2\tau_M^2)}. \tag{5b}$$

A graphical representation of the shape of plots of $G'(\omega)$ and $G''(\omega)$ *versus* ω is given in Figure 1B. For a single Maxwell element, $\tan\delta$ is given by $1/\omega\tau_M$. The quantity $1/\omega$ is a measure of the timescale (observation time) of the experiment; so, for this case, $\tan\delta$ is equal to the observation time divided by the relaxation time.

The time dependence of G' and G'' can be represented by an infinite series of parallel Maxwell elements. To that end it is assumed that each bond can be described by a Maxwell model and the whole network by a series of parallel linked elements. Then equation (5) can be written as

$$G' = (G_e) + \int_{-\infty}^{\infty} H(\tau)\frac{\omega^2\tau^2}{(1+\omega^2\tau^2)}\mathrm{d}\ln\tau, \tag{6a}$$

$$G'' = (G_e) + \int_{-\infty}^{\infty} H(\tau)\frac{\omega\tau}{(1+\omega^2\tau^2)}\mathrm{d}\ln\tau, \tag{6b}$$

where $H(\tau)$ is the so-called relaxation spectrum. The term $H(\tau)\mathrm{d}\ln\tau$ represents the distribution function of shear moduli associated with relaxation times τ, whose logarithms lie in the range between ln τ and ln τ + d ln τ. As a first approximation, we have

$$H(\tau) \approx \frac{1}{\pi}[G'(\omega) \sin \delta]_{\omega=1/\tau}. \tag{7}$$

Comparison of equation (7) with the definition of $H(\tau)$ shows that sin δ is a measure of the proportion of bonds with a relaxation time τ of about $1/\omega$, if δ is independent of ω.[3] If this is not the case, sin δ and, for not too high a value of δ ($<\pi/4$), tan δ may still be used as a qualitative measure of the proportion of bonds with a relaxation time τ. Because the relaxation behaviour of a bond is directly related to the Gibbs activation energy for breaking a bond, a change in tan δ at a certain timescale is a more direct indication of a change in the type and/or strength of the bonds in a gel network due to changing conditions than is a change in the modulus. The latter can be entirely due to a change in the geometrical structure of the network [equations (2) and (3)].

3 Effects of Nature of Structural Elements

Regarding the nature of the structural elements, several types of gels can be distinguished. Such a classification is irrespective of the energy content of the bonds or of their relaxation times.

Gels Formed of (Flexible) Macromolecules

By taking only the conformational entropy of the macromolecular chains between cross-links into account, Flory derived[6] the classical equation for the shear modulus

$$G = vkT, \tag{8}$$

where v is the number of elastically effective chains per unit volume. A chain is defined as the part of a macromolecule extending from one cross-link to the next one along the primary molecule. The cross-links represent the fixed points of the structure in the sense that the chain ends meeting there have to move together irrespective of the motion of the cross-link. In terms of equation (2), the quantity kT stems from the second derivative of the Gibbs energy and v is from NC. The enthalpic contribution may be neglected as the contour length L of the chain between cross-links is much longer than the root-mean-square end-to-end distance $\langle r^2\rangle^{1/2}$ of a free chain with length L.[7] Equation (8) has been shown to hold for many gels composed of synthetic polymers.[6,7] For food-grade macromolecules it holds for gelatin gels under conditions as in food and for heat-set ovalbumin gels in 6 M urea.[8,9]

Even if equation (8) holds, however, the relation between the shear modulus and the concentration of macromolecules is less straightforward than it suggests. Equation (8) can be rewritten as

$$G = \nu \frac{M_c N_{av}}{N_{av} M_c} kT = \frac{c}{M_c} RT, \tag{9}$$

where N_{av} is the number of Avogadro, M_c is the average molecular weight of the chain between two cross-links, and R is the gas constant. For instance, for a 1.95 wt% gelatin gel G' was found to be 600 N m^{-2} after ageing for 150 h at 10.4 °C.[10] The weight average molecular weight M_w was 70 kg mol^{-1}. Equation (9) gives M_c = 77 kg mol^{-1}, which is an unlikely number in view of M_w. It implies that part of the molecules is not involved in gel formation and/or that a part of the various chains (cross-links) is elastically ineffective. The latter will be the case with, for example, dangling ends and intramolecular cross-links. The part that is elastically ineffective will depend on history (cooling regime) and concentration. The storage modulus of gelatin gels has been found[11] to be proportional to the concentration squared for concentrations above 2%.

In the case of polysaccharide gels the macromolecular chains are rather stiff.[12] Especially for highly cross-linked gels, this means that the requirement $L > \langle r^2 \rangle^{1/2}$ does not hold. This has been shown to be the case for 0.94 wt% calcium alginate gels from swelling experiments and from determinations of the strain at which departure from linear behaviour occurs.[13] The value of $L/\langle r^2 \rangle^{1/2}$ was found to be only *ca.* 1.4. Often a first indication can be obtained, if the theory for flexible macromolecules holds, by using equation (9) to calculate M_c. For the calcium alginate gel G' was 2×10^3 N m^{-2}, resulting in M_c = 12 kg mol^{-1}. This implies 74 sugar units. A cross-link will involve about 20 sugar units while the length l of the statistical chain element of alginate is 25 sugar units. So between two cross-links only two statistical chain elements are present. Using $L = 2l$ and $\langle r^2 \rangle^{1/2} = l\sqrt{2}$ gives for the ratio 1.4.

Gels Formed of Hard Particles

The understanding of the relation between the structure of gels of hard particles and their small deformation properties has been greatly enhanced by the introduction of the concept that clusters are formed with a fractal structure during aggregation.[14,15] The number of primary particles N_p in a cluster with a fractal structure scales with the radius R:

$$\frac{N_p}{N_0} = \left(\frac{R}{a_{eff}}\right)^D. \tag{10}$$

Here, D is the fractal dimensionality ($D < 3$), a_{eff} is the radius of the effective building blocks forming the fractal cluster, and N_0 is the number of primary particles forming such a building block. The size of the clusters scales with R^3, and so the volume fraction of particles (ϕ_c) in the cluster decreases with increasing radius. At a certain radius R_g the average ϕ_c will equal the volume fraction of primary particles in the system, ϕ; the clusters will then fill the total volume and a gel will be formed with

$$\phi = \left(\frac{\langle R_g \rangle}{a_{\text{eff}}}\right)^{D-3}, \tag{11a}$$

or

$$\langle R_g \rangle = a_{\text{eff}} \phi^{1/(D-3)}, \tag{11b}$$

where $\langle R_g \rangle$ is a measure of the average cluster radius at the moment the gel is formed. In fact, this quantity gives an upper cut-off length, *i.e.*, the largest length scale at which the fractal regime exists.[16,17]

So, given the value of ϕ, for a full characterization of the fractal clusters forming the gel network one needs to know at least one additional parameter besides D—namely a_{eff}, R_{gel} or N_0. Gels built of fractal clusters with the same value of D but a different value of a_{eff} will exhibit a different structure at the same magnification, and hence a different permeability.[17]

The quantity a_{eff} may become different from the primary particle size if, after aggregation, the particles change their positions with respect to one another, *e.g.*, by rolling around each other until (most) particles have acquired bonds with two or more particles. This causes the chains to become several particles thick. Moreover it leads to an increase in the value of D.[18,19]

A fractal description of the clusters building a gel allows the derivation of expressions for the parameters N and C in equations (2) and (3). In general, a pair of aggregating fractal clusters will do so mainly *via* their longest projections. Because the structure of fractal clusters is scale invariant, the number of projections is independent of fractal size, and as a first approximation the number of contact points (bonds) between two clusters will be independent of their size.[14] The contact area will scale with R^2, so that the number of stress-carrying chains per unit cross-section will scale like:

$$N \propto R^{-2} \propto a_{\text{eff}}^{-2} \phi^{2(3-D)}. \tag{12}$$

The dependence of C (the local increase in distance between particles divided by the macroscopic strain γ) on the volume fraction of particles is related to the way the fractal clusters are linked to each other.[15,20] If the clusters are connected by straight chains, C is independent of R and the following expression is obtained:

$$G \propto R^{-2} \frac{d^2F}{dx^2} \propto a_{\text{eff}}^{-2} \phi^{2/(3-D)} \frac{d^2F}{dx^2}. \tag{13}$$

Because for several types of interaction forces (d^2F/d^2x) is proportional to a_{eff}^2, equation (13) predicts G to be independent of a_{eff}. The situation is different if the chains have a hinged or tortuous shape. Then C is proportional to R^{-1} or $R^{-(1+y)}$, respectively, where y is the chemical length exponent of the elastic

Table 1 *Main effects of* rearrangements *in particle gels on moduli; for explanation of symbols see text*

Stage	*Effect on geometric structure*	*Effect on moduli [Equation (2)]*
During aggregation	Increase of a_{eff}, hence also of R_g	Lower N; may (partly) be compensated by larger $\mathrm{d}^2F/\mathrm{d}x^2$
	Increase of D, hence also of R_g	Stronger dependence N (and C) on φ
After a gel is formed	Fusion of particles	Larger $\mathrm{d}^2F/\mathrm{d}x^2$ term or often, in fact, E of strands
	Straightening of strands	Smaller dependence of C on φ; increase of effective E of strands
	Displacement of particles to centre of fractal clusters[a]	Larger holes; fractal description will be lost
	Fracture of strands	Lower N; fractal description will be lost

[a]Importance in practice probably small

effective chains connecting the fractal clusters.[21] Working this out gives the following expressions for G:

$$G \propto R^{-3} \frac{\mathrm{d}^2F}{\mathrm{d}x^2} \propto a_{\text{eff}}^{-3}\, \phi^{3/(3-D)} \frac{\mathrm{d}^2F}{\mathrm{d}x^2}, \tag{14}$$

$$G \propto R^{-(3+y)} \frac{\mathrm{d}^2F}{\mathrm{d}x^2} \propto a_{\text{eff}}^{-(3+y)}\, \phi^{(3+y)/(3-D)} \frac{\mathrm{d}^2F}{\mathrm{d}x^2}. \tag{15}$$

Equations (14) and (15) predict that, for these structural models, G depends on the radius of the effective building blocks, behaviour observed, for instance, by Buscall *et al.*[22] and Bremer[23] for polystyrene latex gels.

After a gel is formed, rearrangements of the structure may continue due to ongoing cross-linking of the (dangling) chains in the clusters. Moreover, rearrangements may arise from occasional breaking of bonds and formation of new bonds elsewhere in the gel. The latter may cause transport of particles from sparsely occupied regions to more dense ones, where on average more bonds per particle can be formed. This means that the size of the holes in the gel will no longer be related to R_g. After severe rearrangements a fractal description of the clusters will no longer hold.

An overview of possible rearrangements in particle gels and their main effects on the moduli is given in Table 1.

Gels of Flexible Particles

For gels built up from aggregated protein particles, such as casein particles in casein gels, the particles forming the building blocks of the network cannot be considered any more as undeformable under the forces acting on the chains.

After aggregation, casein particles start to fuse and the interaction between the original casein micelles becomes just as stiff as the rest of the casein particles. In such a case it is inappropriate to speak further of an interaction energy between particles. In principle, one can replace the interaction term in equation (2) by the interaction energy required for deforming the protein molecules (particles) and the bonds in between. However, this is not feasible in practice at present, and probably also not in the near future. An alternative is to assign a certain modulus to the protein chain.

For a cylindrical chain of aggregated particles of length L and radius a, and where the stiffness of the particles is the same as that of the bonds in between, the Young's modulus E of such a chain is given by

$$E = \frac{\sigma}{\Delta L/L} = \frac{f}{\pi a^2} \frac{L}{\Delta L}. \tag{16}$$

In the linear region we have $\Delta L/L = \Delta x/x$. The parameter f is equal to $(\mathrm{d}^2F/\mathrm{d}x^2)\Delta x$. This ultimately leads to[15]

$$E = \frac{1}{\pi a} \frac{\mathrm{d}^2F}{\mathrm{d}x^2}, \tag{17}$$

and for the shear modulus of the gel to

$$G = NC\pi aE. \tag{18}$$

For straight strands the parameter C is given by $a/26$.[2,15] Combination of equations (12) and (18) gives:

$$G \propto \frac{\pi}{26} E\phi^{2/(3-D)}. \tag{19}$$

3 Concluding Remarks

The mechanical properties of gels as determined by applying small deformations are clearly related to the undisturbed structure of the systems, but in a more complicated way than is often realized. The spatial distribution of the structural elements (geometric structure of the network) and the interaction forces between the relevant structural elements both play a part. Moreover, the mechanical properties of the structural elements are important. Regarding the geometric structure, a very important factor is which part of the structural elements contributes effectively to the mechanical properties of the gel, and how does it depend on conditions, concentration, history, *etc*. The various factors are summarized in Table 2. Self-evidently the mechanical properties of the structural elements only play a part if the particles are not much stiffer than the bonds in between them.

Table 2 *Factors determining small deformation behaviour of gels*

Factor	*Gel property*	*Effect on mechanical properties of gel*
1. Kind of interaction forces between (relevant) structural elements	Viscoelasticity	Indirectly via Gibbs activation energy for breaking bonds
	Values of moduli	Direct effect among different gel systems; indirect effect via with factor 2 Indirect effect within one gel system as function of conditions/concentration
2. Geometric structure of network	Values of moduli	Direct effect among different gel systems Direct effect within one gel system as function of conditions/concentration
3. Mechanical properties of structural elements	Values of moduli	Direct effect

References

1. T. van Vliet and P. Walstra, *Neth. Milk Dairy J.*, 1985, **39**, 115.
2. L. G. B. Bremer and T. van Vliet, *Rheol. Acta*, 1991, **30**, 98.
3. T. van Vliet, H. J. M. van Dijk, P. Zoon, and P. Walstra, *Colloid Polym. Sci.*, 1991, **261**, 620.
4. J. D. Ferry, 'Viscoelastic Properties of Polymers', 3rd Edn., Wiley, New York, 1980.
5. R. W. Whorlow, 'Rheological Techniques', 2nd Edn., Ellis Horwood, Chichester, 1992.
6. P. J. Flory, 'Principles of Polymer Chemistry', Cornell University Press, Ithaca, 1953.
7. L. R. G. Treloar, 'The Physics of Rubber Elasticity', 3rd Edn., Clarendon, Oxford, 1975.
8. J. R. Mitchell, *J. Texture Stud.*, 1976, **7**, 313.
9. F. van Kleef, J. Boskamp, and M. van den Tempel, *Biopolymers*, 1978, **17**, 225.
10. K. te Nijenhuis, *Colloid Polym. Sci.*, 1981, **259**, 522.
11. K. te Nijenhuis, *Colloid Polym. Sci.*, 1981, **259**, 1017.
12. J. R. Mitchell, *J. Texture Stud.*, 1980, **11**, 315.
13. A. Segeren, J. Boskamp, and M. van den Tempel, *Faraday Discuss. Chem. Soc.*, 1974, **57**, 255.
14. L. G. B. Bremer, T. van Vliet, and P. Walstra, *J. Chem. Soc., Faraday Trans.*, 1989, **85**, 3359.
15. L. G. B. Bremer, B. H. Bijsterbosch, R. Schrijvers, T. van Vliet, and P. Walstra, *Colloids Surf.*, 1990, **51**, 159.
16. P. Walstra, in 'Food Rheology and Structure', ed. J. H. Windhab and B. Wolf, Vincent Verlag, Hannover, 1997, p. 107.
17. T. van Vliet, J. A. Lucey, K. Grolle, and P. Walstra, in 'Food Colloids, Proteins, Lipids and Polysaccharides', ed. E. Dickinson and B. Bergenståhl, Royal Society of Chemistry, Cambridge, 1997, p. 335.
18. P. Meakin, *Adv. Colloid Interface Sci.*, 1988, **28**, 249.
19. M. Mellema, J. H. J. van Opheusden, and T. van Vliet, this volume, p. 176.

20. T. van Vliet and W. Kloek, in 'Food Rheology and Structure', ed. J. H. Windhab and B. Wolf, Vincent Verlag, Hannover, 1997, p. 101.
21. W. D. Brown and R. C. Ball, *J. Phys. A*, 1985, **18**, L517.
22. R. Buscall, P. D. A. Mills, J. W. Goodwin, and D. W. Lawson, *J. Chem. Soc. Faraday Trans. 1*, 1988, **84**, 4249.
23. L. G. B. Bremer, Ph.D. Thesis, Wageningen Agricultural University, Netherlands, 1991.

Rheology of Highly Concentrated Protein-Stabilized Emulsions

By Y. Hemar and D. S. Horne

HANNAH RESEARCH INSTITUTE, AYR KA6 5HL, SCOTLAND, UK

1 Introduction

An emulsion is an immiscible mixture of two fluids, one of which is dispersed as fine droplets of colloidal size in the continuous phase of the other. To prevent or inhibit coalescence of the droplets formed during the preparative stage of homogenization of the mixture, a surface-active agent is added to the mixture. This adsorbs at the droplet interface to create a short-ranged interfacial repulsion between the droplets.[1] In the case of food emulsions, this surfactant is frequently but not exclusively protein, often derived from milk or eggs.

In food emulsions, the volume fraction of oil, ϕ, can range from a few percent, as in homogenized milks, to greater than 80 percent in a mayonnaise. The magnitude of the volume fraction has profound effects on the viscoelastic properties of the emulsion. Homogenized milks are freely flowing liquids with Newtonian viscosities whereas mayonnaise exhibits solid-like characteristics which allow it to maintain its shape under the relatively small stresses of gravity, but which permit it to yield and flow on the application of larger stresses when the sample is subjected to the spreading shear of a knife.

Because the flow behaviour of dilute emulsions allows them to be treated as hard-sphere dispersions, research on their stability and properties has been mainly concerned with the nature and influence of the stabilizing protein layer which controls the interdroplet interaction potential. (See Dickinson[2] for a review of recent research in this field, using milk proteins as stabilizers.) Food emulsions of high internal phase volume have not been studied to the same degree, but in recent years there has been a resurgence of interest in theoretical descriptions of the rheological behaviour of these systems. There have, however, been relatively few experimental tests of their predictions for emulsions of any type. In this paper, we consider the elastic properties of highly concentrated soya bean oil-in-water emulsions stabilized by sodium caseinate, and we compare our results with the behaviour predicted as a function of oil phase volume fraction by the theory of Princen.[3,4]

2 Theoretical Background

Highly concentrated emulsions are comprised of highly deformed droplets and exhibit a plastic-like response to shear deformation. For small deformations, these emulsions resist the shear elastically up to a yield stress τ_y and a related yield strain γ_y. When the applied stress is greater than τ_y, the emulsion starts, and continues, to flow as long as the applied stress is maintained. Conventional characterization of mayonnaise-type products seems to rely on the application of steady shear conditions and estimation of the yield point. Theoretical treatments of steady shear flow acknowledge that the strain-rate dependence of the viscous stress must reflect the complex interplay of dissipative mechanisms such as fluid flow and droplet rearrangements with storage mechanisms such as deformation,[5,6] but they have yet to be tested on emulsion systems.

On the other hand, Princen has developed[3,4] a simple geometric model for the elasticity and yield behaviour of the emulsion gel. Developed originally as a two-dimensional model system of hexagonally close-packed drops, and later extended by analogy to three dimensions, Princen[3] predicted that, for a monodisperse emulsion system with droplet radius R, the elasticity of the droplets was controlled by their internal or Laplace pressure, $\sigma/2R$, where σ is the surface tension at the oil–water interface. Further, Princen and Kiss[4] were able to relate the static shear modulus G_0 to the oil volume fraction through the relation

$$G_0 = 1.77\frac{\sigma}{R_{32}}\phi^{1/3}(\phi - \phi_0), \tag{1}$$

where R_{32} is the Sauter mean droplet radius and ϕ_0 is the maximum packing fraction of close-packed *undistorted* droplets.

3 Materials and Methods

Soya bean oil was purchased from Sigma Chemical Co. (Poole, Dorset, UK). Sodium caseinate was prepared from fresh milk from the Hannah Research Institute herd by isoelectric precipitation, followed by resuspension and neutralization. The product solution was then freeze-dried for storage.

The emulsion was prepared by mixing 100 ml of soya bean oil and 400 ml distilled water containing 10 g of sodium caseinate using a laboratory emulsifier/mixer (Silverson Machines Ltd., Chesham, Bucks, UK) for three hours at its maximum speed. A deliberately coarse emulsion was prepared because we wished to avoid producing gels which, on concentration, would be too stiff to flow under the stresses available in our rheometer. To reduce the polydispersity of our emulsion, we stood the above preparation overnight in a separating funnel, allowing a cream layer to form. The bottom and top portions of this cream layer were discarded and we retained the intermediate portion. Droplet-size distributions in the original emulsion and in this central cream layer were determined using a Malvern Mastersizer X, equipped with version 1.2b software (Malvern Instruments, Malvern, UK).

A highly concentrated parent emulsion was obtained from the retained cream layer by centrifugation on a Sorvall RC-5B centrifuge equipped with an SS34 rotor at 5000 rpm for 5 min. The oil volume fraction of this emulsion was determined by freeze-drying a portion to extract the water, and weighing the sample before and after water loss. The range of other volume fractions used in this study was obtained by diluting the parent emulsion by adding water containing 5 mg sodium caseinate per g water, this to maintain the protein coating at saturation coverage.

The rheological measurements were performed using a controlled stress Bohlin CVO rheometer equipped with stainless steel cone-and-plate geometry. The cone diameter was 40 mm and the cone angle 4°. All measurements were performed in oscillation mode at a temperature of 20 ± 0.1 °C.

4 Results

Figure 1 shows the droplet-size distributions in the original emulsion and in the cream layer used to prepare the highly concentrated parent emulsion. We can see that the creaming step has reduced the content of the smallest droplets in the distribution and consequently pushed the main peak in the normalized distribution to a larger size value.

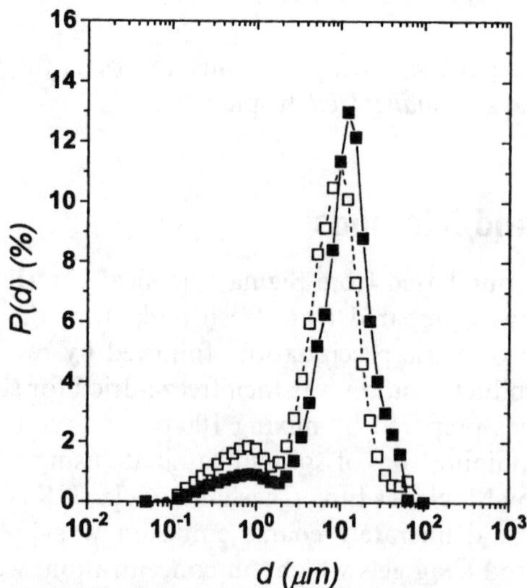

Figure 1 *Particle-size distributions,* P(d), *as a function of particle diameter* d *before creaming (open symbols) and after creaming (closed symbols). The creamed sample is from the intermediate portion of the creamed layer produced on overnight standing. This intermediate portion is then subjected to centrifugation to obtain the highly concentrated parent emulsion*

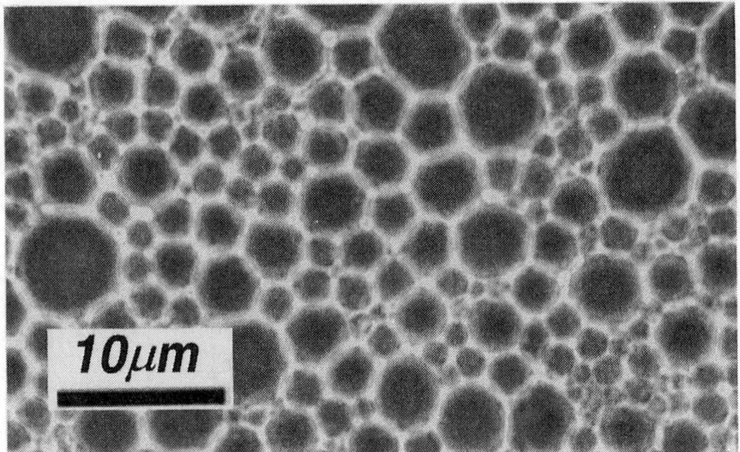

Figure 2 *Optical micrograph revealing the microstructure of the parent emulsion at* ϕ = *0.89*

Figure 2 shows an optical micrograph obtained from a sample of highly concentrated parent emulsion (measured as $\phi = 0.89$). The micrograph reveals each droplet as a separate and distinct entity but also shows the droplets to be highly polydisperse. The distortions of droplet shape brought about by packing to this high internal phase volume are particularly apparent in the mid-size range droplets where hexagonal or pentagonal structures are clearly evident.

Before measuring the storage and loss moduli (G' and G'', respectively) as a function of frequency, we first established the strain regime where the emulsion's response to applied stress was linear. We chose a fixed frequency of 1 Hz and swept from small to large strain by applying incrementally greater stresses to the emulsion.

Figure 3 shows G' and G'' as a function of the measured strain developed at a frequency of 1 Hz for three different volume fractions. At low strains, the storage modulus G' is always much higher than the loss modulus G'', confirming the dominating role of the elastic properties in these highly concentrated emulsions in this region of strain. Whilst the storage modulus shows a well defined linear region independent of developed strain, this is not the case for the loss modulus, which at the lowest stresses produces a 'noisy' result. If we determine the linear viscoelastic region solely on the basis of the storage modulus, this remains strain independent below $\gamma \approx 0.02$ for $\phi = 0.89$, below $\gamma \approx 0.006$ for $\phi = 0.75$, and below $\gamma \approx 0.002$ for $\phi = 0.70$. Clearly this yield value is also a sensitive function of volume fraction. Beyond these critical strains, there is a noticeable drop in the storage modulus, whilst at the higher volume fractions the loss modulus begins to rise perceptibly, indicating the transition to non-linear behaviour and plastic flow. At the very highest strain values plotted, the temporal strain response measured by the rheometer is no longer sinusoidal, but it becomes flattened in the peak and trough regions, another indicator of non-linear viscoelastic behaviour. Beyond the value of the

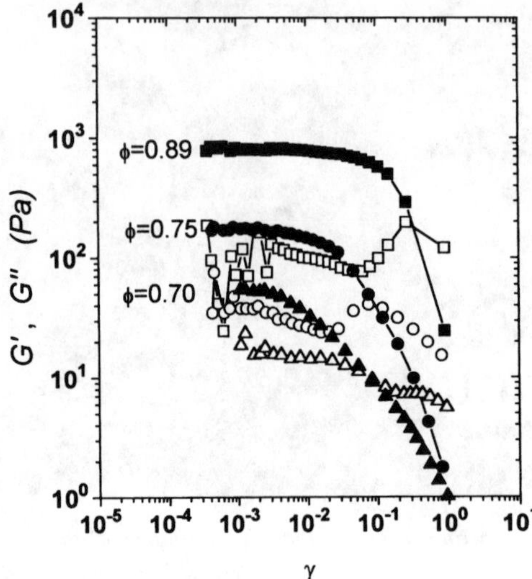

Figure 3 *Dependence on strain γ of the storage modulus G′ (solid symbols) and loss modulus G″ (open symbols) at 1 Hz for three volume fractions: 0.89 (■), 0.75 (●) and 0.70 (▲)*

last strain point plotted for each data set in Figure 3, a macroscopic slippage occurred at the sample/rheometer interface, and thereafter the measured moduli became erratic and noisy.

To characterize further the linear viscoelastic behaviour of these highly concentrated emulsions, we applied a very low stress to our samples, so that the measured strain always remained below the critical yield strain determined previously, and we measured the elastic moduli as a function of frequency. The results for the same three volume fractions used previously are shown in Figure 4, the measurements having been taken over three decades of frequency.

The two higher volume fraction samples present practically a flat response for G', essentially independent of frequency ω. At the lowest ϕ, however, $G'(\omega)$ drops off with frequency as the frequency is decreased, whilst remaining almost independent of frequency at the other end of the spectrum. This is probably due to flow of this emulsion over the long time scales to which these low frequencies are related. In all cases, the loss modulus G'' is at least a factor of 10 smaller than the storage modulus, confirming again the fact that the rheological behaviour of these emulsions is dominated by their elasticity. The loss moduli do, however, clearly increase with frequency, if somewhat noisily. This almost certainly reflects a contribution from the viscosity of the fluids of which the emulsions are comprised, a contribution which increases with frequency.

Broadly similar behaviour has been observed in other types of concentrated emulsions both in frequency sweep and stress sweep experiments. Slight differences in detail are apparent, however. Mason *et al.*[7] observed minima in

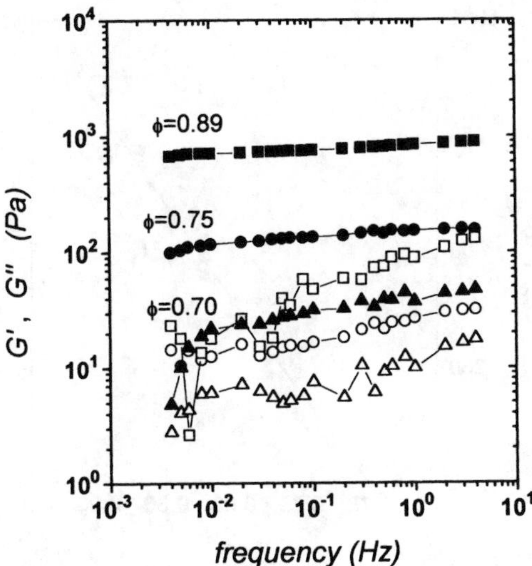

Figure 4 *Frequency dependence of the storage modulus* G′ *(solid symbols) and loss modulus* G″ *(open symbols) for three volume fractions: 0.89 (■), 0.75 (●) and 0.70 (▲). The data were obtained at a developed strain always less than the critical yield strain determined in the prior stress sweep experiments*

their plots of G'' *versus* frequency for concentrated monodisperse emulsions stabilized by surfactant, whilst Ebert *et al.*[8] found G'' independent of frequency.

In Figure 5, we demonstrate the effects of volume fraction on the elasticity of these highly concentrated emulsions. The data plotted are both the values of the storage modulus measured at a frequency of 1 Hz in a frequency sweep and the plateau values determined in the linear viscoelastic region of the strain sweep experiments. Two data sets obtained in separate dilution experiments are plotted for the 1 Hz frequency measurements as evidence of the reproducibility of the behaviour observed. Both types of rheology experiment give similar values for the storage modulus, and we take these as equivalent to the static shear modulus of the Princen formula, *i.e.* equation (1). The continuous line drawn in Figure 5 represents the fit of our data to the Princen formula. The parameters of our fit are such that $\phi_0 \approx 0.70$. This value is slightly smaller than the value of 0.712 given by Princen and Kiss,[4] but this can be explained by the differing polydispersities of the preparations (ours and theirs) or it could be a reflection of the differing stabilizing films in the highly concentrated emulsions. Our fit also gives $\sigma/R_{32} \approx 2400$ Pa for the Laplace pressure at $\phi = 1$. Assuming that the surface tension of our caseinate film is approximately 24 mN m^{-1}, similar to that of β-casein or α_{S1}-casein,[9] this gives an R_{32} value of about 10 μm. This figure is close to the maximum value in our creamed emulsion size distribution (Figure 1), but smaller than the distribution's calculated Sauter mean value of 2 μm.

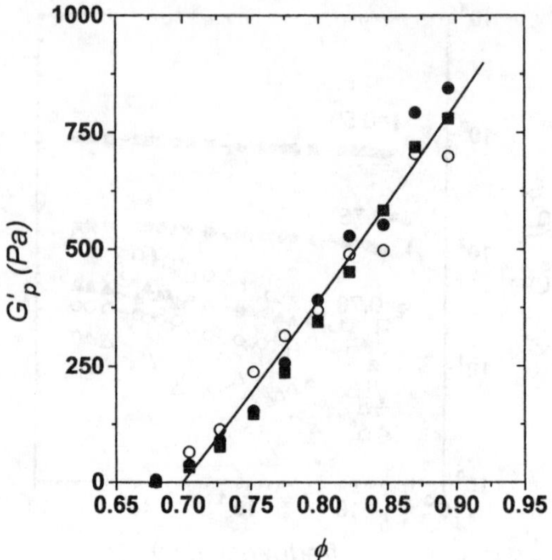

Figure 5 *Static storage modulus, G'_p, taken as the G′ value at 1 Hz in the frequency spectrum, as a function of volume fraction for two series of emulsion dilutions from the parent emulsion (●, ○). These data points were used to produce the fit to the Princen formula (solid line). The third set of data points (■) are values of G′ in the linear viscoelastic region obtained in the stress sweep experiments (also at 1 Hz)*

5 Discussion

Despite being composed of fluids of relatively low individual viscosities, these highly concentrated caseinate-stabilized soya oil-in-water emulsions possess a striking shear rigidity, characteristic of a Hookean solid. This elasticity grows by around three orders of magnitude only when the emulsion droplets are concentrated beyond the spherical close-packing limit, which we determine to be around $\phi \approx 0.70$ for our polydisperse emulsion. The rigidity is created because the droplets have been compressed, packed together, and deformed to create flat facets where neighbouring droplets touch. Shear deformations of this structure create additional droplet surface area, as modelled by Princen,[3,4] and this gives rise to the emulsion's elastic modulus. As we have shown here, with an excellent fit to Princen's formula, this elasticity is a function of the Laplace pressure, $\sigma/2R_{32}$, at contact. Importantly, no significant effect related to the stabilizing protein film itself is evident. Our emulsion droplets are of such a size that the effects of film thickness are negligible, and no effects ascribable to the nature of the film or to film interactions between opposing droplets can be discerned.

As also predicted by the Princen model, our emulsions show a volume-fraction-dependent yield stress, flowing when the applied stress exceeds a critical value. Though not considered in any detail here, this yield is also a sensitive function of the volume fraction difference above the close-packing limit. This yield stress, too, is a function of the Laplace pressure, and is not related to the breaking of any bonds between or within the particles. A more detailed treatment of the flow behaviour of these emulsions in this non-linear viscoelasticity regime will appear in a forthcoming publication.

The behaviour we observe for our model food emulsion is similar, qualitatively and quantitatively, to that observed by others for surfactant-stabilized systems, again confirming that any effects which might be related to the different types of emulsifiers are secondary. The mean droplet size calculated from the fit of our data to the Princen formula is larger than that measured by light scattering. Princen's model is based on a monodisperse system and assumes zero contact angle between the thin films separating adjacent droplets. Intuitively, since elasticity in each droplet is proportional to its Laplace pressure, introducing this as a weighting factor for each particle size in our distribution to calculate an average Laplace pressure would shift the 'mean' size calculated from the G' *versus* ϕ behaviour to a lower value, in the opposite direction to that observed. Equally, we would intuitively anticipate larger droplets to yield and flow more readily than smaller ones; yet both elasticity and yield have the same dependence on Laplace pressure in the Princen model. There is, however, also the question of contact angle θ, since any non-zero angle would reduce the effective interfacial tension by a factor $\cos \theta$, which would move the calculated size more into line with that measured. A further imponderable relates to the numerical pre-factor in the Princen equation [equation (1)]. This is an empirically determined number, and it may therefore be peculiar to the emulsions used by Princen and co-workers. It may also possess an unrealized dependence on size distribution and film thickness. The deviations between calculated and measured size are not so great, however, as to lead us to revise our conclusions as to the applicability of the model or the source of the viscoelasticity in these highly concentrated emulsions.

We have observed this viscoelastic behaviour in high volume fraction model food emulsions. Many other food products (foams, mousses and whipped toppings) also lie in the same volume fraction range, and we would anticipate that their viscoelastic characteristics would be similarly explicable using this approach.

Acknowledgements

The authors would like to thank Mrs. C. M. Davidson for technical assistance and Dr K. Hendry for the optical micrography. This research was carried out as part of the EU Framework IV Project CT96-1216 'Structure, rheology and physical stability of aggregated particle systems containing proteins and lipids' and EU financial support is gratefully acknowledged. Core funding at the Hannah Research Institute is provided by the Scottish Office Agriculture, Environment and Fisheries Department.

References

1. P. Becher, 'Emulsions: Theory and Practice', Reinhold, New York, 1965.
2. E. Dickinson, *J. Dairy Sci.*, 1997, **80**, 2607.
3. H. M. Princen, *J. Colloid Interface Sci.*, 1983, **91**, 160.
4. H. M. Princen and A. D. Kiss, *J. Colloid Interface Sci.*, 1986, **112**, 427.
5. T. G. Mason, J. Bibette, and D. A. Weitz, *J. Colloid Interface Sci.*, 1996, **179**, 439.
6. S. Radiman, C. Toprakcioglu, and T. McLeish, *Langmuir*, 1994, **10**, 61.
7. T. G. Mason, M.-D. Lacasse, G. S. Grest, D. Levine, J. Bibette, and D. A. Weitz, *Phys. Rev. E*, 1997, **56**, 3.
8. G. Ebert, G. Platz, and H. Rehage, *Ber. Bunsenges. Phys. Chem.*, 1988, **92**, 1158.
9. J. Castle, E. Dickinson, B. S. Murray, and G. Stainsby, *ACS Symp. Series*, 1987, **343**, 118.

Creaming, Flocculation, and Rheology of Casein-Stabilized Emulsions

By Eric Dickinson, Matt Golding, and Herley Casanova

PROCTER DEPARTMENT OF FOOD SCIENCE, UNIVERSITY OF LEEDS, LEEDS LS2 9JT, UK

1 Introduction

Casein, in the soluble form of sodium caseinate, is widely valued as a food ingredient having excellent emulsifying and water-holding properties. At neutral pH conditions, the adsorbed casein layer gives long-term protection against flocculation and coalescence during emulsion processing and storage.[1] The special emulsifying properties of sodium caseinate can be attributed[2] to the pronounced amphiphilicity and flexibility of the casein monomers, especially the two major components that comprise about three-quarters of total bovine milk casein—α_{s1}-casein and β-casein. Both these individual caseins are calcium-sensitive phosphoproteins containing a high proportion of proline along disordered polypeptide chains of $\sim 2 \times 10^2$ residues,[3] and both can be used alone to stabilize emulsions.[2]

A characteristic feature of casein physical chemistry is the strong tendency towards aggregation.[4] This is seen most obviously in the 'casein micelles' of milk (average radius $\sim$200 nm). Following removal of most of the calcium phosphate that holds together the internal structure of the casein micelle, the solution of (sodium) caseinate exists as a mixture of monomeric caseins and small self-assembled aggregates called 'sub-micelles' (radius $\sim$5 nm).[5,6] Individual caseins also have a strong tendency to self-associate in aqueous solution.[4] The reversible molecular association of pure β-casein through (mainly) hydrophobic interactions leads to formation of surfactant-like micelles.[7,8] In contrast, α_{s1}-casein, which is more highly charged and less distinctly amphiphilic, forms chain-like aggregates through a series of consecutive association steps.[4,9] Relatively little is known about the compositional dependence of sodium caseinate self-assembly.

Sodium caseinate emulsions are stable at pH 7 towards relatively high concentrations of 1:1 electrolytes (*i.e.* >2 M NaCl).[10] Flocculation is readily induced, however, in the presence of low concentrations of salts containing multivalent cations, *e.g.* 15–20 mM $CaCl_2$ or $BaCl_2$ (but not $MgCl_2$), or 1 mM

$Al(NO_3)_3$.[11] Model emulsions made with pure α_{s1}-casein at low protein/oil ratios are more susceptible to flocculation by salts (NaCl or $CaCl_2$) than are similar β-casein emulsions.[10,12] Thus, the presence of NaCl concentrations of >0.1 M induces extensive flocculation of α_{s1}-casein-coated droplets, whereas it has no effect on β-casein-coated droplets.[12,13] However, replacing just a small proportion of the α_{s1}-casein emulsifier by β-casein has been found[13] to lead to a substantial improvement in salt stability, so that when about half the adsorbed protein layer consists of β-casein the emulsion is stable towards >2 M NaCl in the presence or absence of Ca^{2+} (5 mM). An additional factor in these mixed systems is competitive adsorption between the individual caseins which gives a much higher concentration of β-casein at the interface than expected simply on the basis of the overall composition.[13,14] The large difference in salt stability of emulsions made with the two individual caseins can be attributed[15,16] to the higher net charge density on the adsorbed α_{s1}-casein layer and the better steric stabilizing efficiency of β-casein due to its greater amphiphilicity. Self-consistent-field theory as applied to interacting casein-coated hydrophobic surfaces predicts[17] that there is a net attractive interaction between α_{s1}-casein layers at moderate or high ionic strength, as compared with a long-range repulsion for β-casein layers.

In sodium caseinate-stabilized emulsions of high protein/oil ratio, where the protein content is considerably in excess over that required to provide monolayer saturation coverage, extensive reversible flocculation is observed to occur even in the absence of added salt.[18–20] This flocculation by non-adsorbed casein has a substantial effect on time-dependent creaming behaviour[19,20] and rheology[19,21] of these concentrated protein-stabilized emulsions. By analogy with theoretically predicted depletion flocculation of large spheres by much smaller spheres,[22,23] it can be proposed[18–20] that depletion flocculation of protein-coated emulsion droplets (diameter ~0.5 μm) is induced by the presence of much smaller caseinate sub-micelles (diameter 10–20 nm). Consistent with this interpretation is the easy reversibility towards dilution observed for this type of flocculation—in contrast to the more permanent aggregation caused by electrolyte addition or polymer bridging.

In this paper we report new results on the flocculation behaviour of emulsions containing a considerable excess of unadsorbed casein. The main factors considered here are the effects on creaming and rheological behaviour of (i) the calcium ion concentration and (ii) the composition of the casein emulsifier.

2 Sodium Caseinate Emulsions Containing Added Calcium Ions

Materials and Methods

Spray-dried sodium caseinate (5.2 wt% moisture, 0.05 wt% calcium) was obtained from DeMelkindustrie (Veghel, Netherlands). Calcium chloride 2-hydrate (AnalaR grade, 99.5%), sodium azide and buffer salts were from BDH Chemicals, and *n*-tetradecane (99%) was from Sigma Chemicals.

Oil-in-water emulsions (4 wt% sodium caseinate, 35 vol% oil) were prepared using a laboratory-scale circulatory high-pressure homogenizer. Calcium chloride (0–8 mM) was dissolved in the aqueous phase (0.05 M phosphate buffer, pH 6.8, containing 0.01 wt% sodium azide as anti-microbial agent). The average droplet diameter for each freshly prepared emulsion was $d_{32} = 0.40 \pm 0.1$ μm. Gravity creaming of emulsion samples (volume 300 ml, height 25 cm) was determined at 30 °C using the ultrasound velocity scanning technique.[24,25] Steady-state viscometry measurements were carried out at 30 °C using a Bohlin CS-50 controlled stress rheometer with a double-gap stainless-steel concentric cylindrical cell (volume 30 ml). Further details may be found elsewhere.[26]

Results

Figure 1 shows the set of time-dependent creaming profiles $\phi(h)$ for the emulsion sample with no added calcium ions. During the first few hours the oil volume fraction remains uniform [$\phi(h) \equiv 0.35$] throughout the height h of the sample. At $t = 15$ h there is an indication of ϕ increasing towards the top of the sample, and at $t = 16.5$ h there is the first evidence for a distinct serum layer at the bottom. The serum layer thickness increases steadily over the next 24 h, and then phase separation slows down again. After about one week or so, a steady state is reached with a concentrated emulsion ($\phi \approx 0.5$) occupying the upper two-thirds of the tube and a serum layer ($\phi \approx 0.03$) filling the lower part. Figure 2 shows gravity creaming for the emulsion prepared exactly as before, but with 5 mM $CaCl_2$ added prior to emulsification. Now we see that the system is

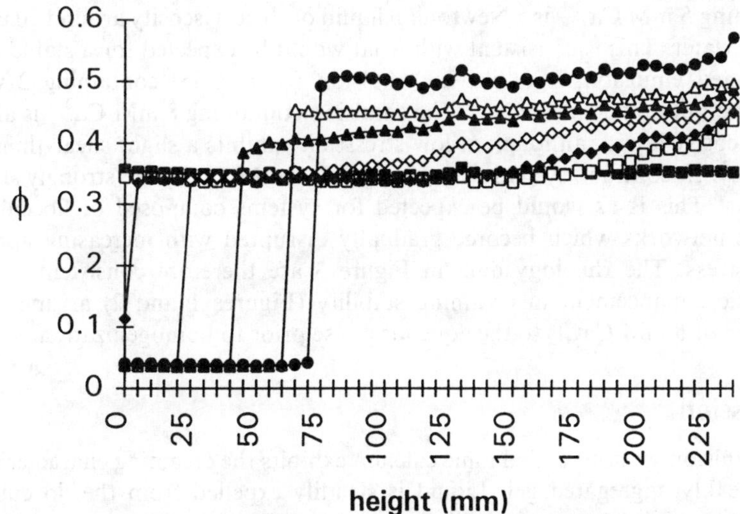

Figure 1 *Creaming profiles of sodium caseinate-stabilized emulsion (4 wt% protein, 35 vol% oil, 0.05 M phosphate, pH 6.8, 30 °C). Oil volume fraction ϕ is plotted against height* h: ■, *6 h;* □, *15 h;* ◆, *16.5 h;* ◇, *19.5 h;* ▲, *24 h;* △, *45 h;* ●, *168 h*

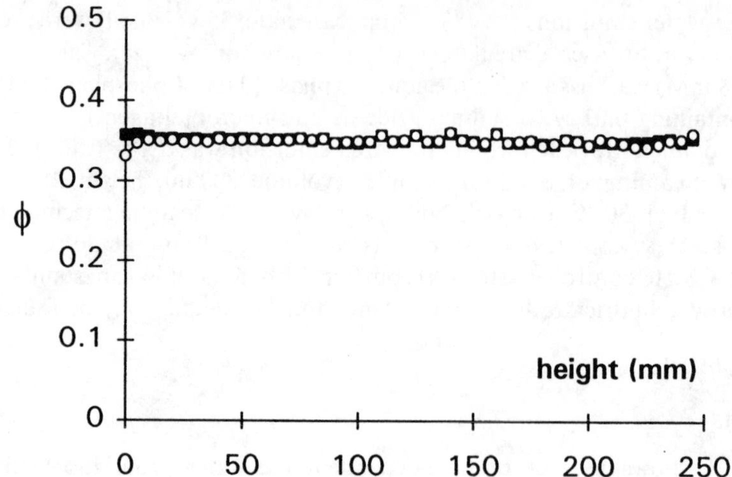

Figure 2 *Creaming profiles of sodium caseinate-stabilized emulsion (4 wt% protein, 35 vol% oil, 0.05 M phosphate, pH 6.8, 30 °C) containing 5 mM $CaCl_2$ added prior to emulsification. Oil volume fraction ϕ is plotted against height* h*: ■, 1 h; ○, 160 h*

extremely stable to creaming, with little change from the initial uniform distribution after storage for up to t = 160 h. Addition of 5 mM ionic calcium has apparently converted the flocculating emulsion (Figure 1) into one exhibiting classical non-flocculated creaming behaviour.

Figure 3 shows the rheological behaviour of the set of emulsion systems (4 wt% protein, 35 vol% oil) containing 0, 2, 5 and 8 mM $CaCl_2$. The emulsion containing 5 mM Ca^{2+} is a Newtonian liquid of shear viscosity around 10 times that of water. This is consistent with what would be expected for a stable non-flocculated emulsion, *e.g.*, like an emulsion ($\phi \approx 0.35$) containing 2 wt% caseinate but no added $CaCl_2$.[21] The emulsion containing 8 mM Ca^{2+} is also a low viscosity liquid, although at low stresses it exhibits a slight shear-thinning tendency. In contrast, systems containing 0 or 2 mM $CaCl_2$ are strongly shear-thinning. This is as would be expected for systems composed of flocculated droplet networks which become gradually disrupted with increasing applied shear stress. The rheology data in Figure 3 are therefore consistent with a dramatic enhancement in creaming stability (Figures 1 and 2) arising from addition of 5 mM $CaCl_2$ to the aqueous phase prior to homogenization.

Discussion

The emulsion without added ionic calcium exhibits the creaming characteristics of a weakly aggregated gel. Liquid is steadily expelled from the flocculated droplet network by a process akin to syneresis. Like the gentle squeezing of a water-filled sponge, the droplet network compresses and rearranges under gravity to release aqueous continuous phase into a distinct serum layer at the bottom of the cell.[27,28] Comparison of Figures 1 and 2 shows that the pre-

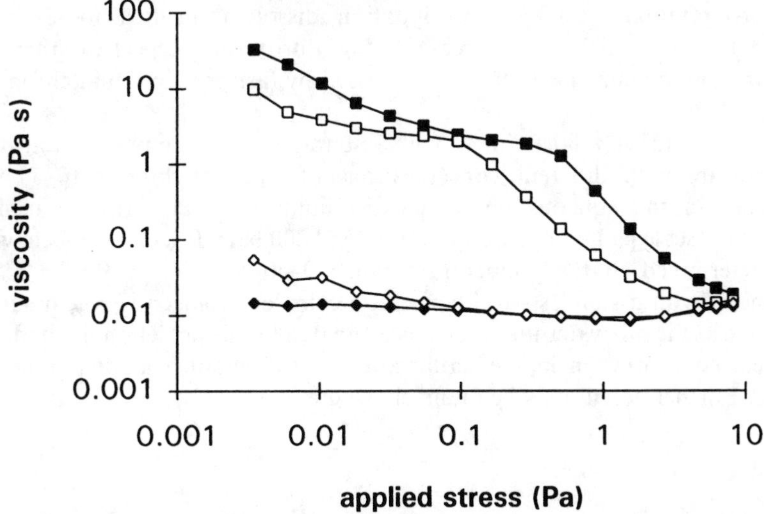

Figure 3 *Influence of ionic calcium present before emulsification on shear rheology of sodium caseinate-stabilized emulsions (4 wt% protein, 35 vol% oil, 0.05 M phosphate, pH 6.8, 30 °C). Apparent viscosity is plotted against shear stress:* ■, *no added* $CaCl_2$; □, *2 mM* $CaCl_2$; ◆, *5 mM* $CaCl_2$; ◇, *8 mM* $CaCl_2$

homogenization addition of calcium ions (at a concentration below that causing casein precipitation) leads to a very substantial improvement in creaming stability due to elimination of the flocculated network structure, as demonstrated by rheological data (Figure 3). This is perhaps an unexpected result, since increased calcium ion content is often associated with destabilization of caseinate-containing dairy emulsions.[29,30]

Below the critical $CaCl_2$ concentration causing caseinate precipitation (10–12 mM), we can envisage the association of sub-micelles into larger protein aggregates. Such stable casein aggregates of size up to 200 nm (*i.e.* approaching the dimensions of whole casein micelles) have been reported[6] in light-scattering studies of caseinate solutions on increasing the Ca^{2+} content from 6 to 10 mM. On theoretical grounds[22,23] such large protein particles (of similar size to emulsion droplets) would not be expected to induce depletion flocculation. Another factor favouring the enhanced stabilization of emulsions at ionic calcium contents of 5–8 mM is the lower concentration of protein available for inducing depletion flocculation because of the higher surface coverage of the caseinate in its more aggregated calcium-bound state.[31]

3 β-Casein Emulsions Containing Added Calcium Ions

Materials and Methods

Bovine β-casein was obtained from the Hannah Research Institute (Ayr, Scotland). It had been carefully prepared from fresh skim milk by a process of acid precipitation, washing, reprecipitation, dissolving in urea, ion-exchange chromatography, dialysis and freeze drying. Purity with respect to other milk protein contaminants (<1–2%) was assessed by fast protein liquid chromatography (FPLC).

Emulsions (5 wt% β-casein, 45 vol% *n*-tetradecane) were prepared at room temperature with different concentrations of calcium chloride (0–30 mM) dissolved in the aqueous phase (0.02 M imidazole/HCl, pH 7.0) using a laboratory-scale jet homogenizer (operating at 300 bar). Rheological behaviour was determined in the temperature range 0–40 °C using a Bohlin CS-50 rheometer with a stainless-steel concentric cylinder geometry. Protein surface coverage at the oil–water interface was estimated by the depletion method. The β-casein concentration in the serum after centrifugation was determined by FPLC. Further details may be found elsewhere.[32]

Results

Figure 4 shows that the non-Newtonian rheology of the calcium-free β-casein-stabilized emulsion (5 wt% protein, 45 vol% oil, d_{32} = 0.63 μm) is substantially dependent on temperature. The behaviour in the range 5–20 °C is qualitatively similar to that for the sodium caseinate-stabilized emulsions (see Figure 3), with a shoulder in the viscosity/stress plot possibly indicating the

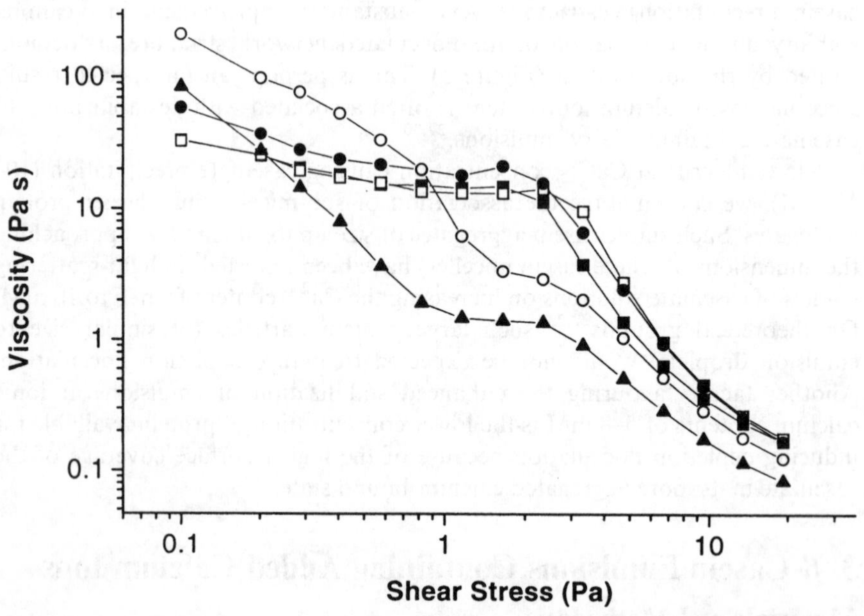

Figure 4 *Effect of temperature on shear rheology of β-casein-stabilized emulsion (5 wt% protein, 45 vol% oil, 0.02 M imidazole, pH 7.0). Apparent viscosity is plotted against shear stress:* ■, *5 °C;* □, *10 °C;* ●, *20 °C;* ○, *30 °C;* ▲, *40 °C*

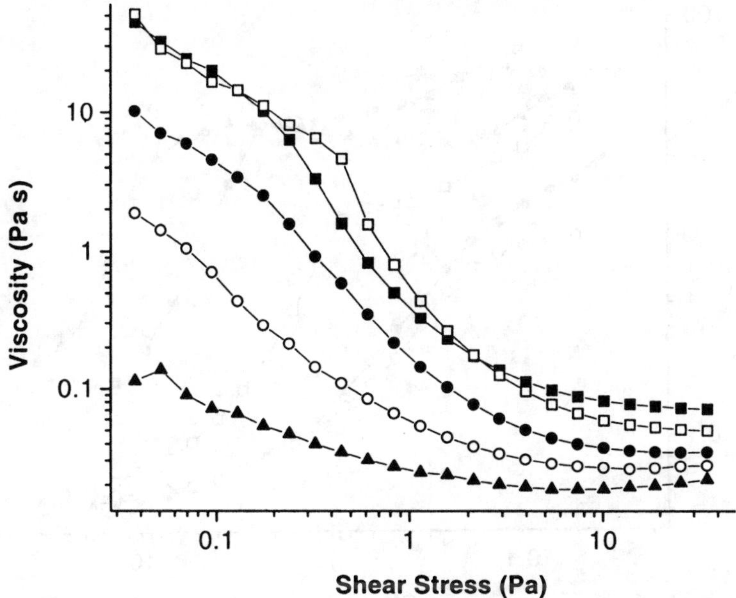

Figure 5 *Effect of temperature on shear rheology of β-casein-stabilized emulsion (5 wt% protein, 45 vol% oil, 0.02 M imidazole, pH 7.0) containing 15 mM $CaCl_2$ added before emulsification. Apparent viscosity is plotted against shear stress:* ■, *0 °C;* □, *10 °C;* ●, *20 °C;* ○, *30 °C;* ▲, *40 °C*

breakdown of a flocculated network structure above a certain critical yield stress.[32] Figure 4 shows that this shoulder is absent from the data obtained at 30–40 °C, suggestive of a rather different type of flocculation at higher temperatures.

The presence of ionic calcium in the aqueous phase prior to emulsification was found to lead to a greater fraction of protein becoming associated with the surface of the droplets. Addition of 15 mM $CaCl_2$ caused a two-fold reduction in the amount of protein remaining in the serum phase, the inferred surface coverage increasing from $\Gamma = 1.1$ mg m^{-2} for the calcium-free system to $\Gamma =$ 6 mg m^{-2} for the emulsion containing 15 mM $CaCl_2$. Figure 5 shows that this latter emulsion has a lower viscosity than the calcium-free system (Figure 4) and is more strongly dependent on temperature. The almost Newtonian behaviour at 40 °C is similar to that exhibited by the caseinate emulsion containing 8 mM $CaCl_2$ (see Figure 3).

Figure 6 shows a complete set of viscosity/stress plots at 20 °C for β-casein-stabilized emulsions containing 0, 10, 15, 20 and 30 mM $CaCl_2$. We observe that the apparent viscosity gradually decreases with increasing Ca^{2+} content up to 15–20 mM, and then increases again as the Ca^{2+} content increases towards 30 mM.

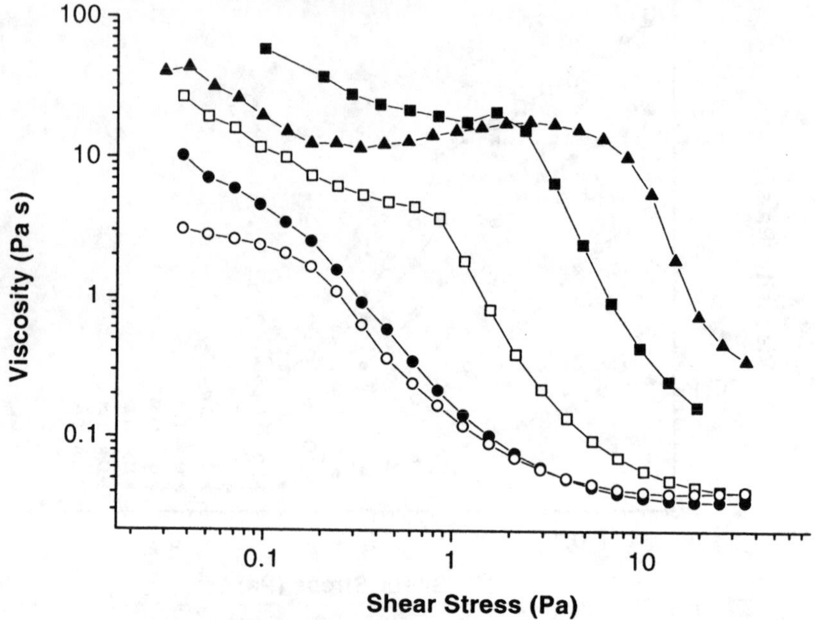

Figure 6 *Influence of ionic calcium present before emulsification on shear rheology of β-casein-stabilized emulsions (5 wt% protein, 45 vol% oil, 0.02 M imidazole, pH 7.0, 20 °C). Apparent viscosity is plotted against shear stress: ■, no added $CaCl_2$; □, 10 mM $CaCl_2$; ●, 15 mM $CaCl_2$; ○, 20 mM $CaCl_2$; ▲, 30 mM $CaCl_2$*

Discussion

These results show that a concentrated β-casein-stabilized emulsion exhibits pseudoplastic flow behaviour with apparent viscosity that decreases with increasing temperature (up to 40 °C) and increasing $CaCl_2$ content (up to 20 mM). The lower sensitivity to Ca^{2+} of the emulsion made with β-casein instead of sodium caseinate is probably due to weaker calcium ion binding of β-casein as compared with α_{s1}-casein. However, aggregation of β-casein at 15 mM Ca^{2+} leads to a large increase in protein surface concentration and consequently a substantial reduction in the non-adsorbed protein fraction. Assuming that flocculation of the β-casein-stabilized emulsion takes place via a similar depletion mechanism to that postulated for sodium caseinate emulsions, the reduced strength of the flocculation with increasing Ca^{2+} content may be attributed to the increased micellar size and/or the diminished amount of protein available in the continuous phase. Another alternative explanation might be that flocculation is caused by a weak (reversible) bridging mechanism involving the phosphoserine moieties in the tails of adsorbed β-casein molecules. Calcium ion binding to these phosphoserines would tend to inhibit network formation by blocking the potential bridging sites.

Precise shapes of the non-Newtonian flow curves at low stresses in Figures 4–6 should be treated with some caution because of wall slip effects. Flocculated suspensions and emulsions are known[33] to be susceptible to wall slip (depletion) arising from displacement of dispersed phase from the vicinity of the solid boundary and the associated formation of a lubricating continuous phase layer near the wall of the rheometer. Evidence for some wall slip in our casein-based emulsions at low shear stresses (<1 Pa) has been indicated by shifts in flow curves on changing the standard stainless-steel rheometer cell (C14) for one of identical dimensions but made with roughened glass surfaces. The two sets of data in Figure 7 are identical at high stresses (down to ~ 0.5 Pa), but it is clear that the low-stress viscosities obtained with the roughened glass surface are considerably higher at the lowest stresses. This is indicative of a wall slip effect,[33] which means that the absolute viscosity values recorded for flocculated casein-stabilized emulsions with the stainless-steel rheometer cell are probably generally underestimated at the lowest stresses. However, the viscosities at higher stresses ($\geqslant 1$ Pa) remain unchanged, and the qualitative trends in the shapes and positions of the non-Newtonian flow curves as a function of temperature, calcium ion content and casein type are expected to be unaffected.

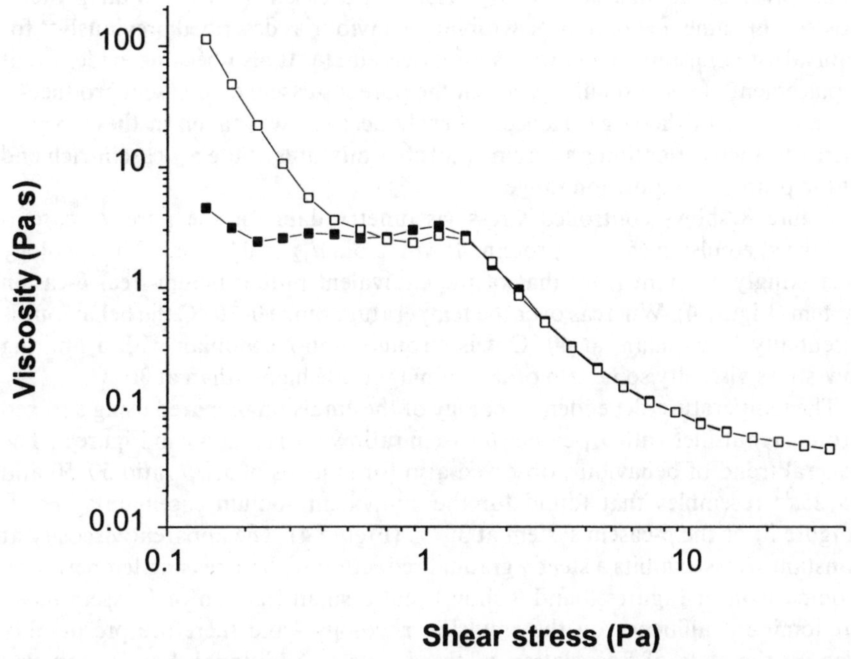

Figure 7 *Test for wall slip effect in experimental viscosity* versus *shear stress data for sodium caseinate-stabilized emulsion (5 wt% protein, 45 vol% oil, 0.02 M imidazole, pH 7.0, 30 °C) investigated with rheometer cells made of different materials:* ■, *stainless steel (smooth);* □, *roughened glass*

4 Emulsions Made with Mixtures of α_{s1}-Casein + β-Casein

Materials and Methods

Pure bovine α_{s1}-casein was obtained as a freeze-dried sample from the Hannah Research Institute (Ayr, Scotland).

Oil-in-water emulsions (5 wt% total protein, 45 vol% *n*-tetradecane, no added calcium) were prepared at room temperature using a laboratory-scale jet homogenizer (operating at 300 bar). The effect of protein composition in the aqueous phase (0.02 M imidazole/HCl, pH 7.0) was investigated by considering various α_{s1}-casein/β-casein ratios (by weight): 0:100, 50:50, 75:25, 90:10, 95:5, 98:2 and 100:0. The rheological behaviour was determined in the temperature range 0–40 °C using the Bohlin CS-50 rheometer with a stainless steel cell. Protein surface coverage and composition at the oil–water interface were determined by the depletion method with the proteins assayed by FPLC. Further details may be found elsewhere.[32]

Results

Preliminary experiments indicated that concentrated oil-in-water emulsions made with 50:50 mixtures of α_{s1}-casein + β-casein (5 wt% total protein) possess the same sort of non-Newtonian behaviour as described previously[21] for equivalent emulsions made with sodium caseinate. It also became evident that replacement of just a small fraction of the pure α_{s1}-casein by β-casein produces a large change in rheology. Hence, an early decision was taken in these experiments to focus attention on casein emulsifier mixtures at the α_{s1}-casein-rich end of the protein composition range.

Figure 8 shows controlled stress viscometry data for the pure α_{s1}-casein-stabilized emulsion (5 wt% protein, 45 vol% oil, d_{32} = 0.93 μm). The rheology is strikingly different from that of the equivalent pure (calcium-free) β-casein system (Figure 4). Whereas over the temperature range 0–30 °C the behaviour is essentially Newtonian, at 40 °C it is strongly non-Newtonian with a limiting low-stress viscosity some two orders of magnitude higher than at 30 °C.

The temperature-dependent rheology of the emulsion prepared using a mixed protein emulsifier with α_{s1}-casein/β-casein ratio 90:10 is shown in Figure 9. The general trend of behaviour, observed also for systems of α_{s1}/β ratio 50:50 and 75:25,[32] resembles that found for the equivalent sodium caseinate system[26] (Figure 3) or the β-casein system at 30 °C (Figure 4). The apparent viscosity at constant stress exhibits a steady gradual reduction with increasing temperature. Comparison of Figures 8 and 9 shows that a small fraction of β-casein has a predominant influence on the emulsion rheology—and therefore presumably also on the state of flocculation of the droplets. Additional data for systems containing even smaller proportions of β-casein (not shown here) indicate[32] that the presence of just 2–5% β-casein can significantly affect the rheology of the pure α_{s1}-casein-stabilized emulsion.

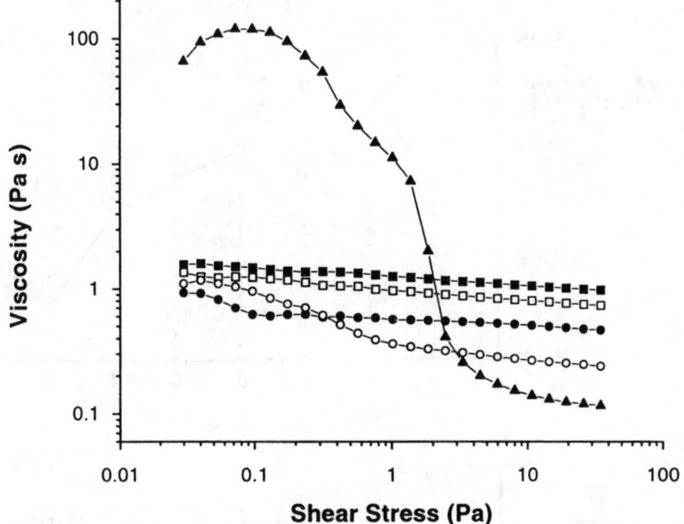

Figure 8 *Effect of temperature on shear rheology of α_{s1}-casein-stabilized emulsion (5 wt% protein, 45 vol% oil, 0.02 M imidazole, pH 7.0). Apparent viscosity is plotted against shear stress:* ■, *0 °C;* □, *10 °C;* ●, *20 °C;* ○, *30 °C;* ▲, *40 °C*

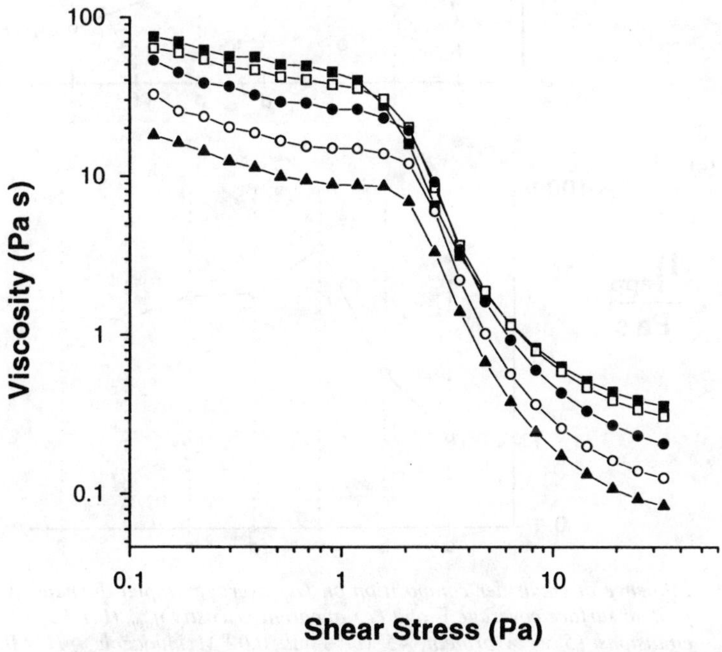

Figure 9 *Effect of temperature on shear rheology of emulsion (5 wt% protein, 45 vol% oil, 0.02 M imidazole, pH 7.0) containing 90 : 10 mixture of α_{s1}-casein + β-casein. Apparent viscosity is plotted against shear stress:* ■, *5 °C;* □, *10 °C;* ●, *20 °C;* ○, *30 °C;* ▲, *40 °C*

(a)

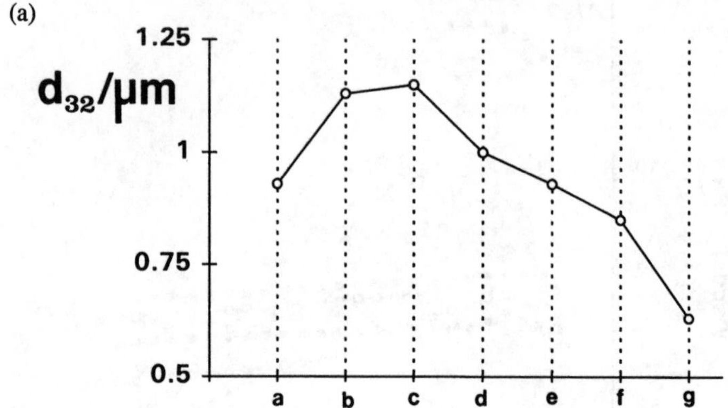

(b)

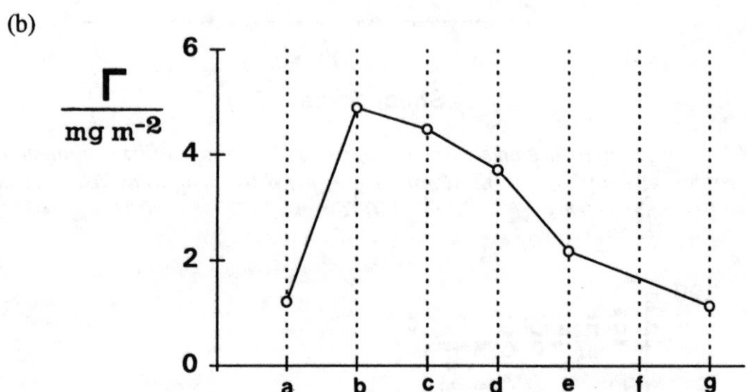

(c)

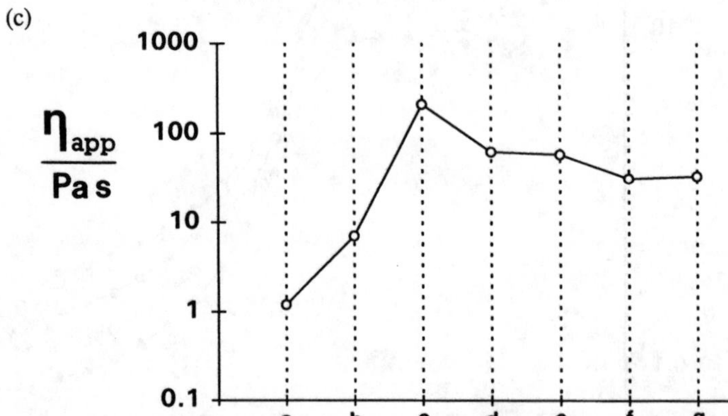

Figure 10 *Influence of emulsifier composition on (a) average droplet diameter* d_{32}*, (b) protein surface coverage* Γ *and (c) apparent viscosity* η_{app} *(0.1 Pa, 10 °C) of emulsions (5 wt% protein, 45 vol% oil, 0.02 M imidazole, pH 7.0) with different* α_{s1}/β *ratios: a, 100:0; b, 98:2; c, 95:5; d, 90:10; e, 75:25; f, 50:50; g, 0:100*

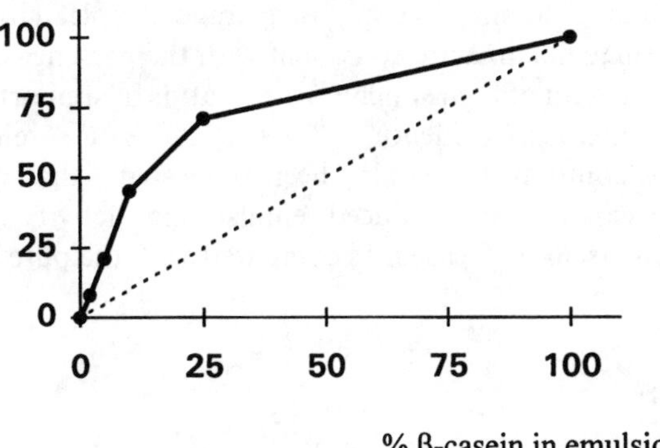

Figure 11 *Proportion of β-casein in adsorbed layer as a function of overall emulsifier composition in emulsions (5 wt% protein, 45 vol% oil, 0.02 M imidazole, pH 7.0) made with different mixtures of α_{s1}-casein + β-casein*

Figure 10 shows the effect of α_{s1}/β ratio (by weight) on values of (a) average droplet size d_{32}, (b) estimated total protein surface coverage Γ and (c) apparent viscosity η_{app} at 0.1 Pa and 10 °C. A larger mean droplet size for the pure α_{s1}-casein emulsion than for the pure β-casein emulsion is consistent with a greater effectiveness of β-casein as an emulsifier. The rather higher d_{32} values (1.0–1.15 μm) in Figure 10(a) for the 98:2, 95:5 and 90:10 compositions implies a lower effectiveness of the binary casein mixture as an emulsifier, possibly due to a greater degree of aggregation.[34] Figure 10(b) shows that the estimated surface coverage in the mixed casein emulsions is much higher than in the pure protein emulsions. This is consistent with some form of complexation between the two different caseins leading to protein multilayer formation at the surface of the droplets. Figure 11 indicates that the β-casein is preferentially adsorbed at the oil–water interface, in agreement with previous work on emulsions of lower protein/oil ratio.[14] The strength of the preferential adsorption tendency of β-casein was indicated by the complete absence of any detectable β-casein in the continuous phases of emulsions prepared at α_{s1}/β ratios of 98:2, 95:5 or 90:10.

Discussion

It is clear from the above that an emulsion made from α_{s1}-casein + β-casein containing considerable unadsorbed protein exhibits rheology that depends in a complex way on temperature and protein composition. This behaviour is presumably related to effects of these variables on the self-assembly properties of the proteins[4–9] and on their distribution between the aqueous continuous phase and the surface of the emulsion droplets.[14] Figure 10(c) shows that replacing just 5% of the α_{s1}-casein by β-casein in a pure α_{s1}-casein-stabilized emulsion leads to an increase in the low-stress apparent viscosity by two orders of magnitude. There is an approximate correlation between the emulsifier composition corresponding to the maximum viscosity [in Figure 10(c)] and that corresponding to maxima in (a) average droplet size and (b) surface coverage. All three maxima are consistent with the presence of some complexes of α_{s1}-casein–β-casein of high stoichiometric ratio. In support of this interpretation is existing literature evidence[35–38] for α_{s1}-casein–β-casein complexation in dilute aqueous solution. It has also been suggested[34] that α_{s1}-casein–β-casein complexation explains the reduced emulsifying activity index for binary mixtures of α_{s1}-casein + β-casein as compared with the pure proteins.

References

1. E. Dickinson, *J. Dairy Sci.*, 1997, **80**, 2607.
2. E. Dickinson, *J. Dairy Res.*, 1989, **56**, 471.
3. H. E. Swaisgood, in 'Developments in Dairy Chemistry—1', ed. P. F. Fox, Applied Science, London, 1982, p. 1.
4. D. G. Schmidt, in 'Developments in Dairy Chemistry—1', ed. P. F. Fox, Applied Science, London, 1982, p. 61.
5. P. H. Stothart, *J. Mol. Biol.*, 1989, **208**, 635.
6. B. Chu, Z. Zhou, G. Wu, and H. M. Farrell, Jr., *J. Colloid Interface Sci.*, 1995, **170**, 162.
7. A. Thurn, W. Burchard, and R. Niki, *Colloid Polym. Sci.*, 1987, **265**, 653.
8. E. Leclerc and P. Calmettes, *Phys. Rev. Lett.*, 1997, **78**, 150.
9. A. Thurn, W. Burchard, and R. Niki, *Colloid Polym. Sci.*, 1987, **265**, 897.
10. E. Dickinson, R. H. Whyman, and D. G. Dalgleish, in 'Food Emulsions and Foams', ed. E. Dickinson, Royal Society of Chemistry, London, 1987, p. 40.
11. E. Pearson, B.Sc. Research Project Report, University of Leeds, 1998.
12. E. Dickinson, M. G. Semenova, and S. A. Antipova, *Food Hydrocolloids*, 1998, **12**, 227.
13. H. Casanova and E. Dickinson, *J. Agric. Food Chem.*, 1998, **46**, 72.
14. E. Dickinson, S. E. Rolfe, and D. G. Dalgleish, *Food Hydrocolloids*, 1988, **2**, 397.
15. D. G. Dalgleish, in 'Food Proteins and their Applications', ed. S. Damodaran and A. Paraf, Marcel Dekker, New York, 1997, p. 199.
16. E. Dickinson, D. S. Horne, V. J. Pinfield, and F. A. M. Leermakers, *J. Chem. Soc. Faraday Trans.*, 1997, **93**, 425.
17. E. Dickinson, V. J. Pinfield, D. S. Horne, and F. A. M. Leermakers, *J. Chem. Soc. Faraday Trans.*, 1997, **93**, 1785.
18. E. Dickinson, in 'Food Colloids: Proteins, Lipids and Polysaccharides', ed. E. Dickinson and B. Bergenståhl, Royal Society of Chemistry, Cambridge, 1997, p. 107.
19. E. Dickinson and M. Golding, *Food Hydrocolloids*, 1997, **11**, 13.
20. E. Dickinson, M. Golding, and M. J. W. Povey, *J. Colloid Interface Sci.*, 1997, **185**, 515.
21. E. Dickinson and M. Golding, *J. Colloid Interface Sci.*, 1997, **191**, 166.
22. S. Asakura and F. Oosawa, *J. Chem. Phys.*, 1954, **22**, 1255.
23. Y. Mao, M. E. Cates, and H. N. W. Lekkerkerker, *Physica A*, 1995, **222**, 10.
24. M. J. W. Povey, in 'New Physico-Chemical Techniques for the Characterization of Complex Food Systems', ed. E. Dickinson, Blackie, Glasgow, 1995, p. 196.
25. V. J. Pinfield, E. Dickinson, and M. J. W. Povey, *Ultrasonics*, 1995, **33**, 246.

26. E. Dickinson and M. Golding, *Colloids Surf. A*, in press.
27. T. van Vliet and P. Walstra, in 'Food Colloids', ed. R. D. Bee, P. Richmond, and J. Mingins, Royal Society of Chemistry, Cambridge, 1989, p. 206.
28. A. Parker, P. A. Gunning, K. Ng, and M. M. Robins, *Food Hydrocolloids*, 1995, **9**, 373.
29. E. Dickinson, S. K. Narhan, and G. Stainsby, *J. Sci. Food Agric.*, 1989, **48**, 225.
30. D. G. Dalgleish and S. O. Agboola, *J. Food Sci.*, 1995, **60**, 399.
31. D. M. Mulvihill and P. C. Murphy, *Int. Dairy J.*, 1991, **1**, 13.
32. H. Casanova and E. Dickinson, *J. Colloid Interface Sci.*, 1998, **207**, 82.
33. H. A. Barnes, *J. Non-Newtonian Fluid Mech.*, 1995, **56**, 221.
34. P. Cayot, J.-L. Courthaudon, and D. Lorient, *J. Agric. Food Chem.*, 1991, **39**, 1369.
35. T. A. J. Payens and H. Nijhuis, *Biochim. Biophys. Acta*, 1974, **336**, 201.
36. M. Yoshikawa, E. Sugimoto, and H. Chiba, *Agric. Biol. Chem.*, 1975, **39**, 1843.
37. L. Pepper and H. M. Farrell, Jr, *J. Dairy Sci.*, 1982, **65**, 2259.
38. T. Ono, S. Odagiri, and T. Takagi, *J. Dairy Res.*, 1983, **50**, 37.

Structure and Rheology of Simulated Particle Gel Systems

By Christopher M. Wijmans, Martin Whittle, and Eric Dickinson

PROCTER DEPARTMENT OF FOOD SCIENCE, UNIVERSITY OF LEEDS, LEEDS LS2 9JT, UK

1 Introduction

Many soft solid-like food colloids (*e.g.*, yoghurt, cheese, margarine) are particle gels.[1] The constituent components of these gels can be fat crystals or protein 'particles' which range in size from individual globular protein molecules (*ca.* 2 nm) to protein-coated emulsion droplets (*ca.* 2 μm). The gelation of these systems can be induced in several ways, such as by heat or enzyme treatment, by lowering the pH, or by addition of divalent counter-ions.[2] The rate of gelation and the texture and rheology of the gel that is formed tend to depend strongly on the experimental conditions. It is clearly of practical advantage to gain a better insight into this dependence. At the same time, it is a great challenge from a purely scientific point of view to try and understand how the solution properties during gelation determine the structure and rheology of a gelled system.

The complexity of the systems we are interested in automatically rules out any simulation that is realistic on the atomic scale. Using such an approach, it would not even be feasible to simulate one single particle from a gel in detail. We have therefore chosen to develop and apply a computer simulation model in which the protein particles are represented as structureless soft spheres, whose motion in solution is determined by 'free-draining' Brownian dynamics. When two such particles come close together, they have the possibility to form a flexible, but permanent, interparticle bond. Such bonds represent the strong attractive interactions (such as covalent linkages, hydrophobic bonds, ion bridges, *etc.*) that occur between particles during the gelation process and that lead to a coherent network structure being formed. The model allows continuous aggregate rearrangement to take place during the gelation process and development of new cross-links during the ageing of the network structure. The experimental conditions during the gelation are most strongly reflected in the sign and magnitude of the interparticle interaction. In addition the reactivity of the particles (that is, the probability that two particles which are in close contact form a bond) has an important influence on the gelation process.

A simulation approach to gel formation has been taken up by Lodge and Heyes.[3] However, their Lennard-Jones-like potentials are completely sphero-symmetric and the 'bonding' between particles is reversible. Although in these simulations gel-like characteristics do appear during the phase separation of a colloidal system,[3] the system does not become 'frozen' at any point during the simulation. This approach simulates transient rather than permanent gels. Silbert *et al.*[4] included hydrodynamic forces based on the resistance pair-drag terms in this sort of simulation and found good agreement with experimental data on depletion flocculated systems.

2 Simulation Method

Basic Model

The simulation algorithm has been described in detail elsewhere.[5] Here we only give a general outline of the simulation model.

We consider a system of size $V = L^3$ with periodic boundary conditions. This system contains N particles of diameter σ, which interact via a steeply repulsive spherical core potential ϕ_C,

$$\phi_C = \varepsilon \left(\frac{\sigma}{r_{ij}} \right)^{36}, \tag{1}$$

where ε is an energy parameter, and $r_{ij} = |\boldsymbol{r}_i - \boldsymbol{r}_j|$ is the interparticle distance between particles i and j. In addition, we introduce a long-range interaction potential which causes a force of magnitude ε_{LR}/σ between two particles at a separation smaller than the long-range cut-off distance $r_c = 1.5\sigma$ (this force is attractive for $\varepsilon_{LR} < 0$ and repulsive for $\varepsilon_{LR} > 0$). Distance, temperature and time are expressed in units of σ, k_B/T and $\sigma(m/\varepsilon)^{1/2}$, respectively, where k_B is Boltzmann's constant. In presenting the results, we use a relaxation time τ_r defined as the time taken for a particle of mass m to diffuse a distance equivalent to one particle radius in a medium of viscosity η_s at infinite dilution: $\tau_r = (3\pi\sigma^3\eta_s)/(4k_BT)$. For $\sigma = 1\ \mu$m the value of τ_r is 0.58 s; for $\sigma = 10$ nm we have $\tau_r = 0.58\ \mu$s.

When two particles approach to within a distance b_1, there is a probability P_B that a bond will be formed between these two particles during a timestep Δt. Bond nodes are defined on the particle surfaces (at $r = \sigma/2$) at points collinear with the particle centres. The subsequent bond potential ϕ_B is defined as

$$\phi_B = \varepsilon_B \left(\frac{b_{ij} - b_1}{b_0} \right)^2 \quad (\text{for } b_{max} > b_{ij} > b_1), \tag{2}$$

where b_{ij} is the inter-node distance. We use the following default parameter values: $\varepsilon_B = 1.0$; $b_1 = 0.1$; $b_0 = 0.1$; $b_{max} = 1.0$. We only allow one bond to be formed between any two particles. The alignment of the bonds collinear with the

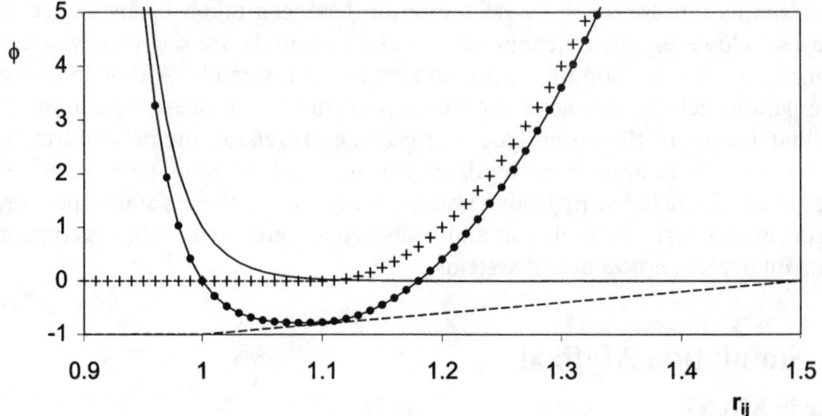

Figure 1 *Interaction potential $\phi(r_{ij})$ between two bonded particles i and j in a collinear configuration with $\varepsilon_{LR} = -1$. The bonding interaction $\phi_B(+)$, the hard-core interaction ϕ_C (full curve) and the long-range interaction ϕ_{LR} (dashed curve) are shown together with the total interaction potential (curve with filled circles)*

particle centres is soon lost due to the diffusive motion of the particles. If the bond length b_{ij} becomes greater than b_{max}, the bond breaks. Figure 1 shows the total interaction potential between two bonded particles with an attractive long-range interaction.

The translational update algorithm is written as a function of the interparticle forces:

$$\boldsymbol{r}_i(t+\Delta t) - \boldsymbol{r}_i(t) = \boldsymbol{F}_i(t)\frac{\Delta t}{\xi} + \boldsymbol{R}_i(t, \Delta t). \tag{3}$$

Here $\boldsymbol{F}_i$ is the force due to all interparticles interactions, ξ is the Stokes friction coefficient, and $\boldsymbol{R}_i(t,\Delta t)$ is the familiar Gaussian random displacement. An analogous rotational update algorithm is also used.[5]

The small-deformation shear rheology of the system can be derived from the interparticle stress tensor. The complex shear modulus is then calculated as the Fourier transform of the stress time correlation function, or it can be derived by separately simulating an oscillatory shearing experiment.[5]

Although we have implied above that all particles have the same size and interactions, the model can be straightforwardly extended to mixtures of different particles.[6]

The Algorithm

The timestep Δt must be chosen small enough to ensure that the force on a particle does not change significantly during one timestep. Because of the steepness of the hard-core potential ϕ_C, Δt cannot be chosen larger than 0.1. This means that the algorithm becomes a lot slower than, for example, the two-dimensional hard-disk gelation model of Dickinson,[7] which Mellema *et al.*[8] extended to hard spheres. In the latter model no forces need to be calculated during the simulation. Instead, one simply corrects for overlap during each timestep by moving overlapping particles apart. One corrects for bonds that become too strongly stretched in a similar manner. Consequently, one can easily use a timestep as large as 5. The disadvantage of this method is that without continuous potentials one cannot easily define the stress tensor. However, by combining both approaches one can benefit from the advantages of either model.

Figure 2 shows that the kinetics of the gelation process are virtually indistinguishable, regardless of which model is applied. No systematic structural differences can be detected between a network formed using continuous potentials and one formed with hard spheres (checked by comparing the pair-distribution functions for both cases). We can therefore use a hard sphere system to form a network, and (after a short relaxation period) use the full continuous potentials to investigate rheological features of the gel. Especially when one wants to generate a large set of gelled systems with different parameters, this can give a considerable gain in computational speed.

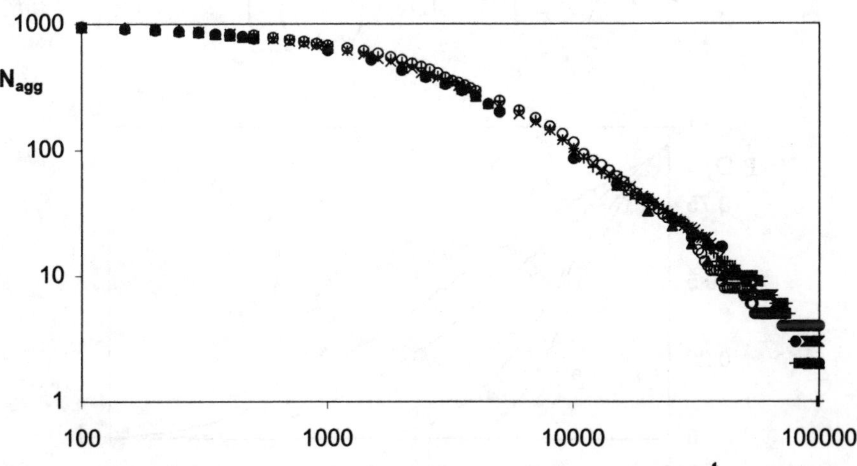

Figure 2 *Comparison between the gelation kinetics using the full continuous interaction potential (symbols ×, + and ○) and the hard sphere model (● and ▲). In the former case we have* $\Delta t = 0.1$*; in the latter case* $\Delta t = 5.0$*. The number of aggregates* N_{agg} *is plotted against time* t*. The different symbols are for independent simulation runs. Parameters:* $N = 1000$*,* $\phi = 0.05$*,* $P_B = 10^{-3}$ *(for* $\Delta t = 0.1$*),* $\varepsilon_{LR} = 0$

3 One-Component Gels

Structural Characterization

A thorough study was previously made[5] of the structure and small-deformation rheology of gels consisting of one type of particle only. The structure of these gels can vary profoundly, depending mainly on the nature of the long-range interaction and the bonding probability. For example, a gel formed with parameters $P_B = 10^{-4}$ and $\varepsilon_{LR} = -2$ has a very coarse structure, the particles being clustered together in dense aggregates. In contrast, the particles are distributed far more evenly throughout the system for $\varepsilon_{LR} = +1$. In the absence of bonding the presence of net attractive interactions would eventually lead to a macroscopic phase separation.[3] The coarse structure is a consequence of this incipient phase separation, which is arrested by the bonding that takes place between the particles. On the other hand, repulsive interactions favour chain formation and lead to a fine structure. This clearly shows how the experimental solution conditions (which determine the interparticle interactions) can very strongly influence the texture of a gelled structure.

In order to give a quantitative description of these coarse and fine structures we introduce $P(D)$ as the probability that a randomly inserted test particle with diameter D does not overlap with any of the particles in the system. For infinitely small spheres ($D \to 0$) this fraction is, of course, $1 - \phi = 0.95$, where ϕ is the particle volume fraction. Figure 3 shows the function $P(D)$ for several gel structures. The curves in this figure can all be fitted very well by the equation

$$P(D) = 0.95 - AD\exp\left(-\left(\frac{D}{\eta}\right)^m\right), \tag{4}$$

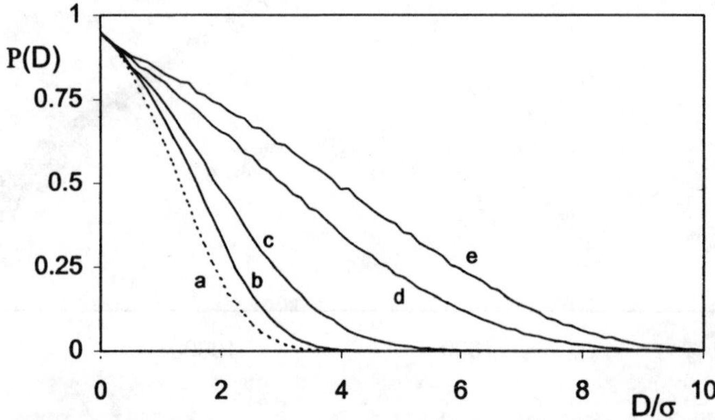

Figure 3 *Probability* P *that a randomly inserted spherical test particle with diameter* D *does not give any overlap with the particles in the gel network. Results are shown for several systems. Positive values of ε_{LR} indicate a repulsive interaction. (a) (------) Initial unbonded system; (b) $\varepsilon_{LR} = +2.0$; (c) $\varepsilon_{LR} = +1.0$; (d) $\varepsilon_{LR} = 0.0$; (e) $\varepsilon_{LR} = -1.0$. In all cases* $P_B = 10^{-4}$

Table 1 *Average pore sizes (defined as the value of* D *for which* P(D) = *0.475) of the systems a–e shown in Figure 3*

system	*a*	*b*	*c*	*d*	*e*
D/σ	1.37	1.65	2.01	3.14	4.15

where A, m and η are fitting parameters. In the initially random (non-gelled) system $P(D)$ falls off most quickly. For the network with a very fine structure ($\varepsilon_{LR} = +2$) $P(D)$ falls off only slightly less slowly with increasing D. However, as the gel structure becomes coarser (as ε_{LR} decreases and becomes negative), larger contiguous volumes of free space occur, leading to far higher values of P at large D. We use the width of the distributions in Figure 3 at $P = \frac{1}{2}(1 - \phi)$ as a measure of the average pore size. Table 1 lists the values of D for which $P(D) = 0.475$ for the systems shown in Figure 3.

Another way to describe the gel structure is by using the fractal concept.[5] The fractal dimension d_f can be determined from the radial distribution function $g(r)$. The logarithm of the integrated distribution function $n(r)$ for these systems is linear in $\log(r)$ over an approximate range of $0.4 < \log(r/\sigma) < 0.7$. The gradient of $\log[n(r)]$ against $\log(r)$ over this limited range is identified as d_f. As the long-range interaction becomes more attractive, the fractal dimension of the network decreases. For example, for $P_B = 10^{-4}$ we find a fractal dimension $d_f = 2.3$ for $\varepsilon_{LR} = +1$, and $d_f = 1.7$ for $\varepsilon_{LR} = -1$. These simulated values of the fractal dimension are comparable to those derived from experimental work on casein particle gels.[9]

Again another way to characterize a gel is in terms of the (average) mobility of its constituent particles. This is an essential parameter that determines the dynamic light scattering behaviour of a system.[10] For a free particle in solution (at infinite dilution) the mean-square displacement $\langle \Delta r^2(t) \rangle$ is $6Dt$, where D is the particle self-diffusion coefficient. The line in Figure 4 represents this behaviour. As particles become bonded during the gelation process, their mobility decreases. In Figure 4 values for $\langle \Delta r^2(t) \rangle$ are also shown at two different points in time during the gelation process (for $t_g = 10^4$ and $t_g = 10^5$) for a gel with $\varepsilon_{LR} = 0$ (and bonding probability $P_B = 10^{-3}$). Furthermore, values have been included for $\varepsilon_{LR} = -1$ ($t_g = 10^5$) and for $\varepsilon_{LR} = +1$ ($t_g = 2 \times 10^5$). For $t_g = 10^5$ and $\varepsilon_{LR} = 0$, $\langle \Delta r^2(10^4) \rangle$ is only 4% of the value for a free particle. For a gel with attractive long-range interactions ($\varepsilon_{LR} = -1$, and $t_g = 10^5$), $\langle \Delta r^2(10^4) \rangle$ is 1/3 smaller than the value found for the gel with $\varepsilon_{LR} = 0$, whereas for the network with repulsive interactions ($\varepsilon_{LR} = +1$, and $t_g = 2 \times 10^5$) it is twice as large. Attractive interparticle forces promote the formation of bonds, thus decreasing the particle mobilities. These results can be used to simulate the dynamic light scattering by particle gels.[11]

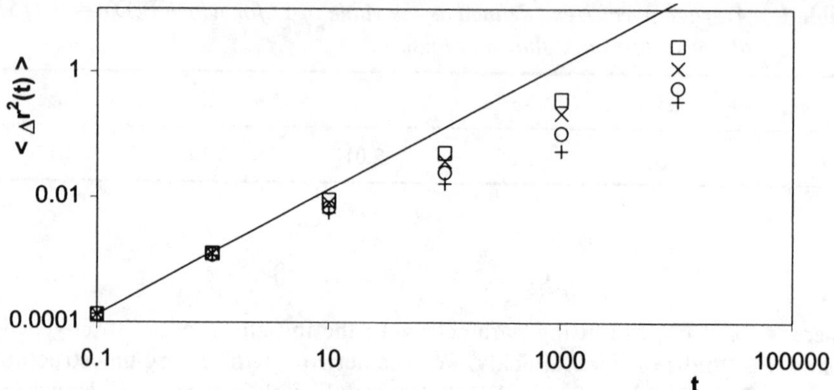

Figure 4 *Average mean-square displacements* $\langle \Delta r^2(t) \rangle$ *of particles in a gel with* $P_B = 10^{-3}$, $\varepsilon_{LR} = 0$ *and gelation time* $t_g = 10^4$ (□), $t_g = 10^5$ (○). *In addition data are shown for gels with attractive long-range interactions* (+, $\varepsilon_{LR} = -1$, $t_g = 10^5$) *and repulsive long-range interactions* (×, $\varepsilon_{LR} = +1$, $t_g = 2 \times 10^5$). *The line shows the mean-square displacement of a freely diffusing particle*

Rheology

From a rheological point of view the gel character is most clearly reflected in the stress correlation function $C_s(t)$.[5] As the interaction between the particles becomes less repulsive and/or more attractive, the stress correlation starts to display a significant long-time tail due to the slower stress relaxation processes in the dense close-packed regions. However, the frequency-dependent storage and loss moduli, $G'(f)$ and $G''(f)$, are relatively insensitive to the network structure. When normalized by their infinite frequency modulus G_∞, the functions $G'(f)$ and $G''(f)$ show only slight variations among networks with very different structural characteristics. The high frequency modulus itself does depend strongly on the network structure. It appears that G_∞ scales linearly with the number of bonds n_B in the system (Figure 5). The average number of bonds per particle is larger in gels with close packed regions (ε_{LR} negative) than in gels with a much finer structure that are formed under the influence of repulsive interactions.

The description of the rheology given above is valid for small deformations. For small strains (*e.g.* $\gamma = 0.05$) the stress is a linear function of the strain, and the system can be subjected to an oscillatory shear strain without any particle bonds being broken. When the gels are subjected to larger strains,[12] the aggregate first starts to tear (bonds break) and then it breaks up into smaller aggregates. If bond formation is included in the simulation, the combination of bond breakage and formation reaches an equilibrium state and the maximum aggregate size is limited depending on the strain-rate. Larger strain-rates give smaller maximum aggregate sizes.

In Figure 6 the stress S has been plotted as a function of time for several (constant) strain rates $\dot{\gamma}$. We have furthermore introduced the dimensionless

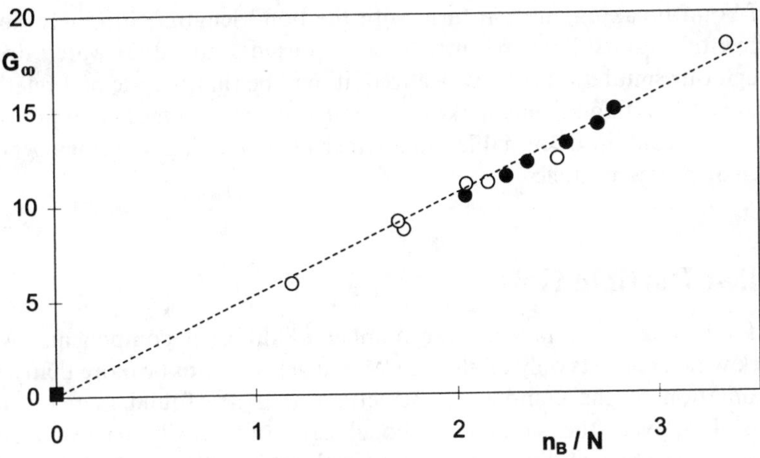

Figure 5 *The high frequency modulus* G_∞ *plotted against the number of bonds per particle* n_B/N. *Parameters:* $P_B = 10^{-4}$ (●); $P_B = 10^{-3}$ (○); *unbonded* (■)
(Reproduced with permission from ref. 5)

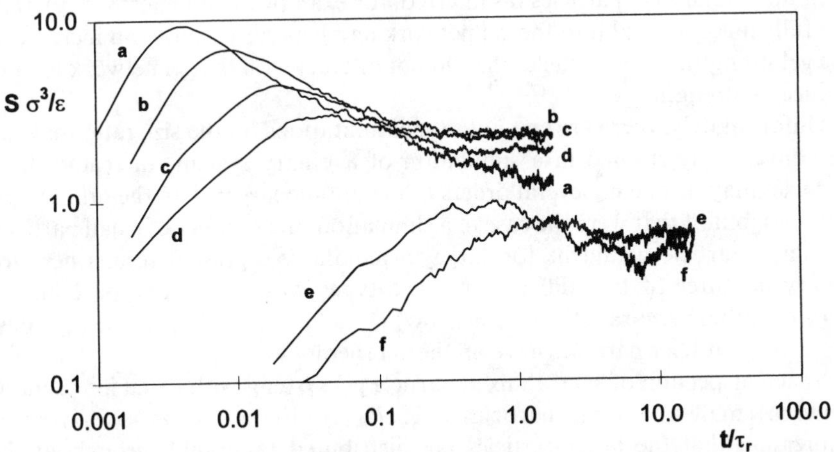

Figure 6 *Log–log plot of the simulated stress response* S *to strain plotted against time* t *at a series of strain rates corresponding to a Peclet number* Pe*: (a) 56.6; (b) 28.3; (c) 11.3; (d) 5.66; (e) 0.566; (f) 0.283*
(Reproduced with permission from ref. 12)

Peclet number $Pe = \dot{\gamma}\tau_r/2$, which represents the relative contributions of shear flow and Brownian motion. The stress overshoot seen in Figure 6 is characteristic of many real viscoelastic systems exhibiting shear-induced network fragmentation. These include margarine,[13] flour-water dough[14] and gelatinized corn starch systems.[15] The overshoot maximum occurs at roughly the same strain irrespective of strain-rate, and is strongly correlated with the maximum rate of bond breaking, and, in turn, with the bond length.[5] Strikingly similar experimental results have recently been reported[14] for flour-water dough. Although our simulation is very idealized, it may be that the use of a relatively long-ranged bonding scheme makes the model relevant to more complex food systems that contain deformable biopolymer chains acting as bridging bonds between dispersed particles.

4 Filled Particle Gels

Most food systems contain a large number of different components. A gel network which consists only of all identical particles cannot be more than a first approximation to the complicated structures that are found in many food colloids. However, the simulation model can quite easily be extended to incorporate particles of different sizes and with different interactions.[6] Here we just consider binary systems where a second component is added to a system of particles that on its own forms a gel. Experimentally a lot of work has been done to investigate the effect of such 'filler' particles on the rheological properties of the gel (see the references cited in ref. 6). When the filler particles are compatible with the primary gel particles (as inferred, for example, from microscopy), they are fully incorporated into the gel network and generally lead to an increase of the gel strength. Filler particles that do not interact with the gel network tend to reduce its strength.

Unfortunately, there are computational limitations on the size ratio that can be conveniently studied in a simulation of a binary system. In reality, filler particles may be one or several orders of magnitude larger than the primary gel particles; but, if that were the case in a simulation, the number of small particles per large particle would be too large to handle. We present results here for binary mixtures of two different particle types with size ratios of 2 and 3. Although these size ratios are rather low, we believe that they do clearly show the effect that filler particles have on the gel rheology.

Visual inspection of such a mixed particle gel system, with equal interactions between small and large particles (*e.g.* $P_B = 10^{-3}$, $\varepsilon_{LR} = 0$), gives the impression that the large particles are distributed randomly throughout the simulated network. This is confirmed by the radial distribution functions.[6]

Figure 7 shows the effect of the filler particles on the shear moduli of the network. A comparison is made between the shear moduli of the mixed system and that of a gel without the filler particles. The filled squares show the ratio of $G'(f)$ in the binary system of size ratio 3 and $G'(f)$ in the equivalent single-particle gel. The filled circles show the ratio of $G''(f)$ in the binary system and $G''(f)$ in the single-particle gel. Over the whole frequency range the storage modulus is larger in the binary system. At low frequencies the loss modulus also increases due to the filler particles. However, at high frequencies (where the value of G'' is small) G'' shows a small reduction. The value of the phase angle $\delta = \tan^{-1}[G''(f)/G'(f)]$ increases due to the presence of filler particles. The open symbols in Figure 7 are for a binary system with a size ratio of 2 (N_1 =

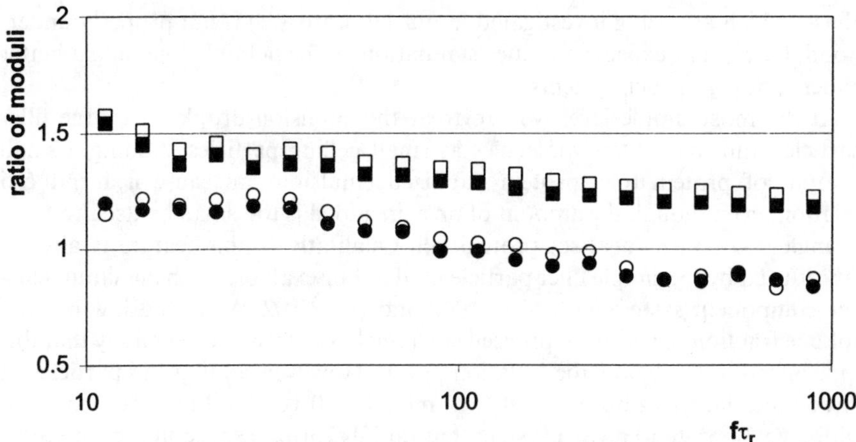

Figure 7 *Ratios of the storage modulus (squares) and loss modulus (circles) in a binary system (size ratio 2 or 3) and a one-component gel. The filled squares show the ratio of* G′(f) *in the binary system of size ratio 3 and* G′(f) *in the equivalent single-particle gel. The filled circles show the ratio of* G″(f) *in the binary system of size ratio 3 and* G″(f) *in the single-particle gel. The open squares show the ratio of* G′(f) *in the binary system of size ratio 2 and* G′(f) *in the equivalent single-particle gel. The open circles show the ratio of* G″(f) *in the binary system of size ratio 2 and* G″(f) *in the single-particle gel. All data points are averages over the sampling intervals*
(Reproduced with permission from ref. 6)

1000 and N_2 = 125, so that the volume fraction of both particle types is still the same). The smaller filler particles have a slightly stronger influence than the larger ones, although qualitatively the effects are the same. Furthermore, a one-component gel with N_1 = 2000 (which has the same overall particle volume fraction as the binary gels) has substantially larger values for the shear modulus than the binary systems.

The data presented in Figure 7 are in good agreement with the wide range of experimental data which show that filler particles that are compatible with the gel network strengthen the gel.[6] In contrast, we have not been able to find a disruptive effect in the simulations for the case when the filler particles do not interact with the gel-forming particles. In this case we probably would need far larger filler particles (that is, large in comparison with the naturally occurring void size in the one-component gel) to disturb the intrinsic network structure and lower the gel strength.

5 Emulsion Gels

Mixed particle gels containing emulsion droplets and protein are good examples of gels that play an important role in food systems. Even the behaviour of a relatively simple experimental model system (such as, for example, heat-set β-lactoglobulin emulsion gels containing *n*-tetradecane or soya oil as the dispersed phase, which are being investigated in our laboratory[16]) is not properly understood. One might expect computer simulations to be helpful in gaining a better understanding of such systems.

At the most simple level we can treat the emulsion droplets as large filler particles and the protein molecules as small gelling particles. Adding a small amount of protein to a protein-stabilized emulsion can cause heat-induced gelation, even though the amount of protein added is too small by itself to form a single-component heat-set protein gel. Qualitatively this feature is already reproduced by the simple filler particle model. For example, we have simulated a one-component system with $N = 1000$ and $\phi = 0.02$. At such a low particle volume fraction the gelation proceeds extremely slowly and one can say that this concentration lies below the 'gelation point'. However, adding 185 particles of three times larger radius ($\phi = 0.1$, to model a 10 vol% oil-in-water emulsion with 2 vol% protein) gives a system that rapidly forms a space-filling network.

In order to make the above model more realistic we can increase the emulsion droplet radius to 15 times that of the protein particles. Although a size ratio of 15 is still rather small, it does at least reflect the fact that the volume of an emulsion droplet is substantially larger than that of a protein particle. However, because of the large number of small particles involved, we can only take one emulsion droplet into account in the simulation. As we are using periodic boundary conditions, we are still able to mimic an infinitely large ensemble of emulsion droplets, albeit in a slightly artificial (ordered) way. A second improvement we can now make to the simple filler particle model concerns the nature of the interaction between the emulsion droplet and the protein. The emulsion itself is normally stabilized by protein. The protein that adsorbs at the oil–water interface is typically the same protein that causes gelation. We therefore consider a system with one large emulsion droplet and 1000 randomly distributed protein particles. We introduce a strong attractive long-range interaction between the emulsion droplet and the particle ($\varepsilon_{LR} = -20$). Now we let the particles adsorb onto the droplet surface while no bonding takes place ($P_B = 0$). After time $t = 3 \times 10^5$, about two thirds of the particles have become adsorbed to the droplet surface, forming a monolayer of protein. This configuration represents a system of protein-stabilized emulsion droplet + protein solution, and we use it as the starting configuration for the idealized emulsion gelation simulation (with a bonding probability between protein particles of $P_B = 10^{-3}$).

Figure 8 shows the configuration that is reached after a gelling time $t_g = 10^5$. The adsorbed particles, which now are bonded together, form a spherical cavity around the emulsion droplet. In addition, the non-adsorbed particles form branches sticking out into the solution, although not all these particles are yet incorporated into one aggregate. If branches from different emulsion droplets bond to each other, bridges will be formed between the droplets leading to a percolating gel network. The emulsion droplets pictured in Figure 8 form a regular array, which is of course a spurious artefact of the periodic boundary conditions and the fact there is only one emulsion droplet in the central simulation box. The total number of inter-particle bonds is roughly the same

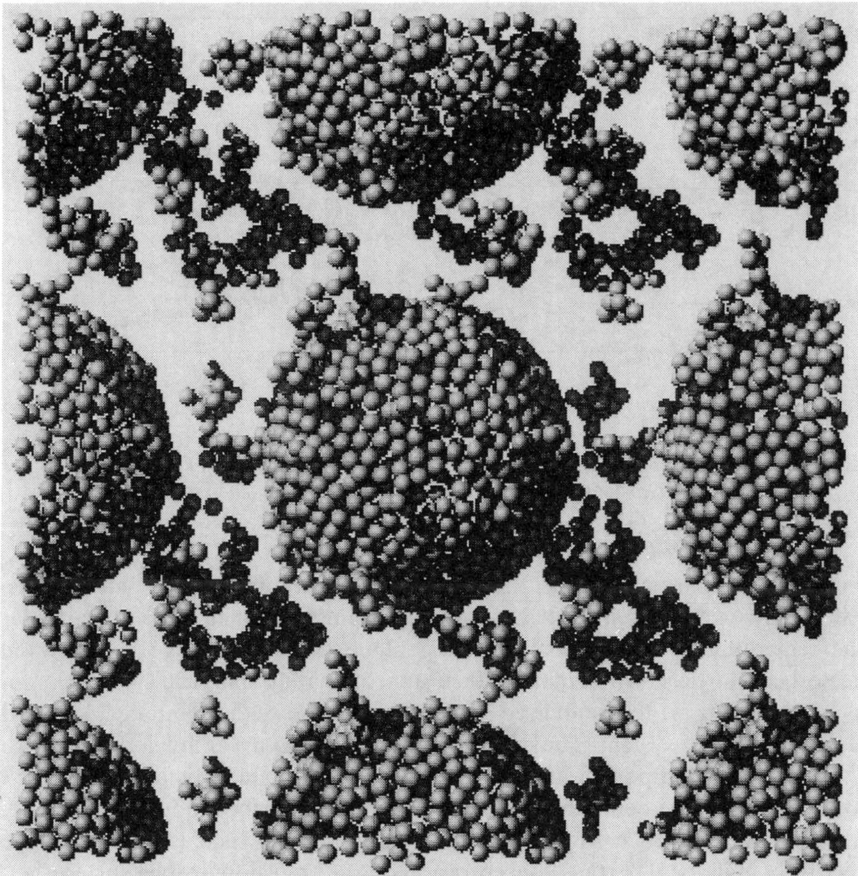

Figure 8 *Snapshot of a simulated emulsion gel as discussed in the text. Only the protein particles (small spheres) are shown*

(n_B = 2322) as that in a single-component gel with 1000 particles at ϕ = 0.05. As the adsorbed particles effectively form a two-dimensional network, one might have expected a smaller number of bonds per particle than in a three dimensional gel since a two-dimensional lattice has a lower coordination number than a similar three-dimensional one. However, due to the adsorption, the local particle concentration is greater, which should lead to a larger number of bonds. Anyhow, these two effects seem roughly to cancel each other.

The shear modulus of the system shown in Figure 8 is very much dominated by the interaction between the adsorbed particles and the emulsion droplet. However, this is largely an artefact of the model, which leads to a very large stress due to the long-range character of the interaction between the protein particles and the large droplet. We can therefore consider the opposite situation that the emulsion droplet does not contribute directly to the stress tensor. As an emulsion droplet is not a rigid body, it seems reasonable to assume that the

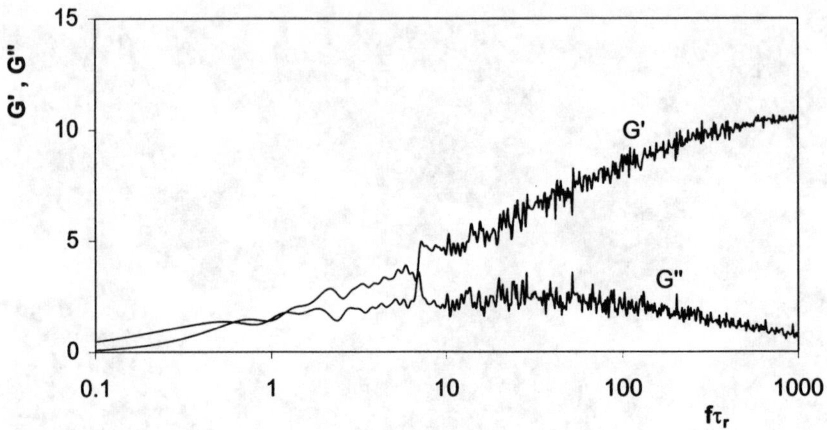

Figure 9 *Shear storage and loss moduli,* G' *and* G''*, as a function of frequency* f *for the protein network (small spheres) in the simulated emulsion gel of Figure 8*

emulsion droplet can relax the stress by undergoing shape fluctuations. Effectively, we can then remove the droplet from the system and only consider the remaining protein particle network. Due to the extensive bonding the 'adsorbed' particles will remain in their approximately spherical configuration, as in a globular protein-stabilized foam. Figure 9 shows the computed $G'(f)$ and $G''(f)$ for such a system. The overall shapes of these curves are very similar to those of a one-component gel. For G_∞/n_B we find a value which is about 10% lower than in a one-component gel. The lower value for the modulus is probably due, on average, to a slightly smaller bond length, which is caused by the relatively high local particle concentration on the emulsion droplet surface.

6 Concluding Remarks

We have seen how the simulation model predicts different gel structures depending on the interactions between the primary particles, and how this can affect the rheological characteristics of the gel. We have also demonstrated how simulated small-deformation rheological properties of bidisperse aggregated particle networks successfully reproduce the mechanical properties of real filled gels. Furthermore, we have made a first attempt to model the much more complicated system of a gel formed from a protein-stabilized emulsion. In this case the representation of an emulsion droplet as a rigid, non-deformable sphere limits the applicability of the model. Although the introduction of deformable particles would severely increase the complexity of the simulation algorithm, one could introduce some internal structure to a droplet by treating it as a three-dimensional array of small particles all bonded to one other. The mechanical properties of an individual emulsion droplet would then be determined by its internal bond density and the stiffness of these internal bonds. Another possible extension of the model is its application towards simulating adsorbed protein films at oil–water interfaces. The rheology of such interfaces is a crucial factor affecting the stability behaviour of emulsions. By modelling a well-defined flat oil–water interface, it is possible to investigate both its shear and its dilatational rheology. This approach will enable comparison with experimental work carried out in a Langmuir trough on macroscopic model monolayers of adsorbed proteins and lipids.

Acknowledgements

This work is financially supported by the Biotechnology and Biological Sciences Research Council of Great Britain and by Contract FAIR-CT96-1216 of the EU Framework IV Programme.

References

1. E. Dickinson, *Chem. Ind.*, 1990, 595.
2. E. Dickinson, in '1st International Symposium on Food Rheology and Structure', ed. E. J. Windhab and B. Wolf, Vincentz Verlag, Hannover, 1997, p. 50.
3. J. F. M. Lodge and D. M. Heyes, *J. Chem. Soc. Faraday Trans.*, 1997, **93**, 437.
4. L. E. Silbert, J. R. Melrose, and R. C. Ball, *Phys. Rev. E*, 1997, **56**, 7067.
5. M. Whittle and E. Dickinson, *Mol. Phys.*, 1997, **90**, 739.
6. C. M. Wijmans and E. Dickinson, *J. Chem. Soc. Faraday Trans.*, 1998, **94**, 129.
7. E. Dickinson, *J. Chem. Soc. Faraday Trans.*, 1994, **90**, 173.
8. M. Mellema, J. van Opheusden, and T. van Vliet, this volume, p. 176.
9. D. S. Horne, *Faraday Discuss. Chem. Soc.*, 1987, **83**, 259.
10. D. A. Weitz and D. J. Pine, in 'Light Scattering: Principles and Development', ed. W. Brown, Clarendon Press, Oxford, 1996, p. 652.
11. D. S. Horne, in 'New Physico-Chemical Techniques for the Characterization of Complex Food Systems', ed. E. Dickinson, Blackie, Glasgow, 1995, p. 240.
12. M. Whittle and E. Dickinson, *J. Chem. Phys.*, 1997, **107**, 10191.
13. W. Kloek, T. van Vliet, and P. Walstra, in 'Food Colloids: Proteins, Lipids and Polysacharides', ed. E. Dickinson and B. Bergenståhl, Royal Society of Chemistry, Cambridge, 1997, p. 168.
14. N. Phan-Thien, M. Safari-Ardi, and A. Morales-Patiño, *Rheol. Acta*, 1997, **36**, 38.
15. A. S. Navarro, M. N. Martino, and N. E. Zaritsky, *J. Texture Stud.*, 1997, **28**, 365.
16. J. Chen and E. Dickinson, *J. Agric. Food Chem.*, 1998, **46**, 91.

Rheology and Physical Stability of Low-Calorie Salad Dressings

By Carlota Pascual, M. Carmen Alfaro, and José Muñoz*

DEPARTAMENTO DE INGENIERÍA QUÍMICA, UNIVERSIDAD DE SEVILLA, C/P. GARCÍA S/N, 41007 SEVILLA, SPAIN

1 Introduction

The increasing trend towards nutritional and health awareness is leading to a demand from consumers for reduced-oil-content food emulsions and other modified spreads. Take, for instance, the market for mayonnaise in mainland Spain and the Balearic islands: this involves sales of more than 37 million kg per year with an annual growth rate of about 5%. The low-calorie mayonnaise market accounts for roughly 17% of total sales, with a growth rate per year around 4%.[1] Similar or higher figures may be observed for other European and North American countries. For these reasons, it is an important challenge for the food industry to optimize the performance of low-calorie formulations, which must mimic the macroscopic properties of full oil emulsions in terms of physical appearance, mouthfeel, texture, long-term stability, flavour and other organoleptic properties.[2]

The objective of this work was to carry out a rheological characterization and a study of the long-term physical stability of model oil-in-water emulsions, similar to commercial low-calorie salad dressings, and formulated to take into account the standard expectations of consumers. Two emulsifiers are compared: salted liquid egg yolk and spray-dried egg yolk. Emulsions containing spray-dried egg yolk were prepared as a reference system for investigating further egg yolk treatments, like cholesterol extraction.

2 Materials

The following materials were used to manufacture the low-calorie mayonnaise samples: pasteurized liquid salted egg yolk (LEY) provided by Hijos de Ybarra, SA. (Seville, Spain); spray-dried egg yolk (SDEY) provided by Ovosec (Valladolid, Spain); modified starch (MS) from Cerestar (Barcelona, Spain);

*To whom correspondence should be addressed

xanthan gum (XG) for laboratory use from Sigma (St. Louis, USA); sunflower oil; vinegar (equivalent acetic acid concentration: 10% wt); sugar; salt; preservatives (sodium benzoate and potassium sorbate, technical quality); and deionized water.

3 Methods

Manufacture of Low-Calorie Salad Dressing

Different preliminary procedures were compared and the best one found by trial and error. The best procedure was selected to avoid blockage of the oil-feeding valve, slurries, and lump occurrence.

The procedure of emulsion manufacture consisted of two main steps. In the first step an aqueous dispersion of polysaccharide or polysaccharides + sugar was prepared in a manner depending on whether XG was present or not. The required amounts of MS and sugar were blended and readily dispersed in deionized water at room temperature. Then the dispersion was kept in a bath at about 80 °C for 7 minutes so as to achieve a suitable gelation of the MS. Finally, the gel formed was allowed to cool to room temperature before being stored at *ca.* 4 °C. When formulating with XG, a weak gel dispersion of xanthan was first prepared.

The second step involved the actual manufacture of the low-oil salad dressing. An Ultraturrax T50 homogenizer (Ika) was used at rotational speeds ranging from 4000 to 9000 rpm with an emulsifying time of *ca.* 10 minutes. Emulsion manufacture was carried out in a bath with a set temperature of 5°C.

The reduced-calorie oil-in-water emulsions were prepared on the basis of equal egg yolk solid content. The qualitative composition was chosen following well-established industrial formulations for low-calorie mayonnaise. Important commercial qualities were taken into account. The emulsion compositions are displayed in Table 1.

Table 1 *Composition of emulsions made with liquid egg yolk (LEY) and spray-dried egg yolk (SDEY)*

	LEY (wt%)	*SDEY (wt%)*
Sunflower oil	34	34
Egg yolk	3.4	1.28
Modified starch	4	(3.3–4)
Xanthan gum	–	(0.7–0)
Sugar	4.2	4.2
Vinegar	4.7	4.7
Salt	1.4	1.74
Preservatives	0.1	0.1
Additional water	48.2	49.98

A Malvern Mastersizer was used to determine the droplet-size distribution (polydisperse model) at room temperature of the emulsions prepared with MS as the only polysaccharide. However, emulsions manufactured with MS + XG showed a high concentration of lumps when diluted for characterization by laser light scattering, despite using nonionic surfactants as dispersant. This prevented these samples from being characterized. The appearance of lumps when using XG is believed to be associated with its viscoelasticity and the type of homogenizer used. We consider that the higher energy input provided by a colloid mill would overcome this problem.

Rheological Characterization

All the rheological tests were carried out at 20 °C on emulsions that had been stored at 4 °C for 3 days.

A Rotovisco RV20 rheometer (Haake) with measuring head CV100 was first used to find out whether the emulsions prepared exhibited slip effects when sheared. In order to achieve this goal, several flow curves determined using different sensor systems (different geometries) were compared. The following thermo-mechanical history was applied: equilibration time to room temperature = 2 h; equilibration time after sample loading = 20 min; shear-rate range = 0.01 to 200 s^{-1} with a logarithmic variation; steady-state criteria = 90 s per point/slope for steady state approximation (0.01). Several sensor systems were used: Mooney-Ewart, ME-15 (radii ratio = 1.078) and ME-16 (radii ratio = 1.037); smooth cone & plate, CP45 (diameter = 45 mm, angle = 4°, gap = 0.175 mm); serrated plate & plate, Q20 (diameter = 19.25 mm, gap = 1 mm).

A Bohlin CS-50 controlled stress rheometer with a serrated plate & plate sensor system (diameter = 20 mm, gap = 1 mm) was used to conduct small-amplitude oscillatory shear experiments (SAOS).

4 Results and Discussion

Characterization of Slip Effects

As a preliminary task, the sensor systems used in this study were calibrated with a Newtonian oil and with a non-Newtonian standard fluid, A1 (a solution of polyisobutylene in decalin).[3] The results obtained with the CR rheometer were satisfactory provided the raw data of the serrated plate & plate sensor system were suitably corrected for non-Newtonian behaviour[4] (see Figure 1).

Slip effects (wall depletion phenomena) occur in the flow of two-phase systems because of the displacement of the disperse phase away from solid boundaries. One of the ways in which wall depletion manifests itself is the lack of reproducibility of viscosity measured in different sensor systems with different size geometries. Slip effects in rotational rheometers may be overcome by physically modifying the surface of sensor systems. For instance, the use of roughened plate & plate sensor systems is strongly recommended.[5]

Results obtained showed that viscosity values were not influenced by the

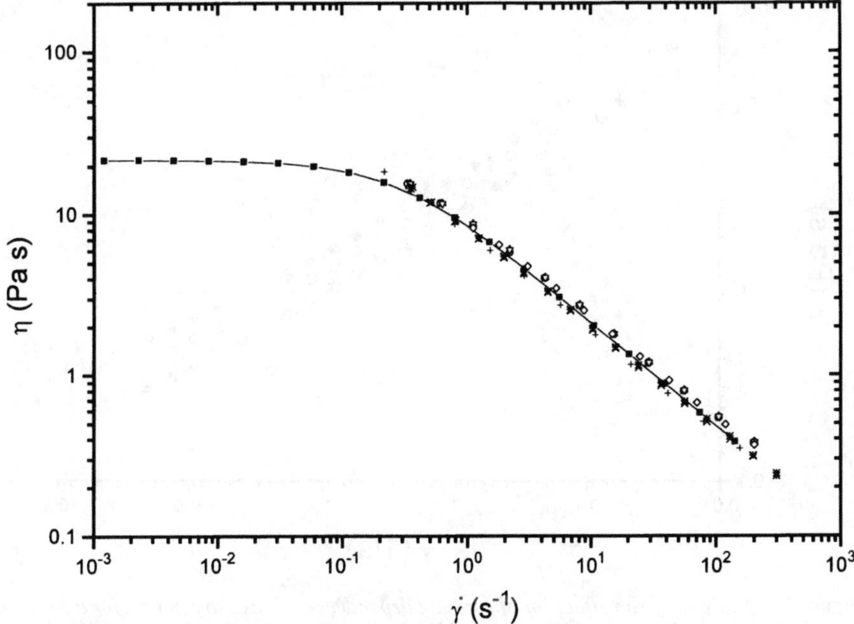

Figure 1 *Shear-rate dependence of viscosity, η ($\dot{\gamma}$), for A1 non-Newtonian standard fluid using different sensor systems. Mooney-Ewart ME-15 (×), Mooney-Ewart ME-16 (∗), smooth cone & plate (+), serrated plate & plate: run 1 (□), run 2 (○), run 3 (△), run 4 (▽) and run 5 (◇), reference 3 (■)*

distance between the tip of the ME-15 Mooney-Ewart sensor system and the bottom of its beaker. Neither were viscosities significantly affected by a 50% gap reduction in Mooney-Ewart sensor systems, as demonstrated by the fact that results shown by ME-16 fell on those of ME-15 (Figure 2). The roughened plate & plate sensor system yielded higher viscosities than the smooth surface cone & plate sensor system, and the latter in turn yielded higher viscosities than the Mooney-Ewart sensor systems, as shown in Figure 2.

These results indicate that slip effects are likely to influence the rheological measurements unless special sensor systems are used. Therefore, only roughened plate & plate sensor systems were used to carry out the more detailed rheological characterization.

Influence of Homogenization Conditions on Emulsion Properties

The effect of the rotor/stator homogenizer rotational speed was studied in order to search for the optimum manufacturing procedure. Preliminary results showed that the dispersion homogenized at 4000 rpm did not have a good appearance on mere visual inspection. Emulsions prepared at between 5000 and 7000 rpm exhibited a similar viscous consistency, but at higher speeds (8000 and

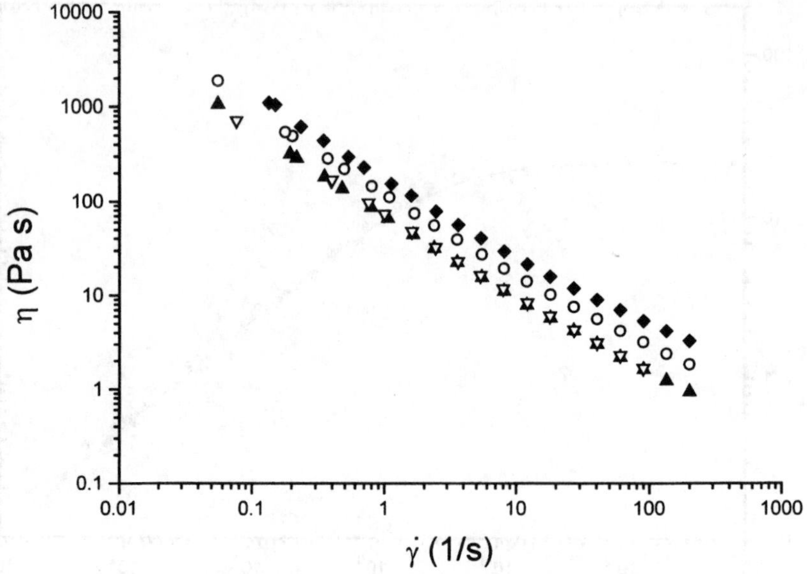

Figure 2 *Analysis of slip effects on shear-rate dependence of viscosity, $\eta\ (\dot{\gamma})$, for a 34 wt% oil salad dressing containing SDEY, determined using four different sensor systems. Mooney-Ewart ME-15 (▲), Mooney-Ewart ME-16 (▽), smooth cone & plate (○), and serrated plate & plate (◆)*

9000 rpm) the viscous consistency of emulsions was reduced. For this reason, this study is focused on emulsions manufactured at between 5000 and 7000 rpm.

The critical shear stress for the onset of nonlinearity did not significantly vary between 5000 and 7000 rpm in either SDEY or LEY emulsions. While the latter exhibited an average critical shear stress of 3 Pa, the former showed values more than twice as high as those shown by the salad dressings containing LEY. It must be pointed out that G'' initially increased with shear stress at the onset of nonlinearity. This behaviour has been also reported[6] for concentrated oil-in-water emulsions. However, if these results are replotted as a function of strain, no substantial effects in the extension of the linear viscoelastic domain could be observed upon changing the emulsifier. These results are consistent with some differences in the absolute magnitude of the complex modulus.

The mechanical spectra showed typical properties of weak gels, with G' one order of magnitude higher than G''. The frequency dependence of G' is small, and it exhibits a definite tendency to decrease with frequency in the low frequency range. This demonstrates a lack of the so-called equilibrium modulus G_e, typical of strong gels. In other words, fluid-like properties can be predicted to be dominant at long relaxation times. No clear effect of the rotational speed on the mechanical spectra of either LEY or SDEY emulsions was observed.

A comparison of the viscoelastic spectra of emulsions containing LEY and SDEY reveals that the latter are slightly more viscoelastic than the former as demonstrated by the lower values of the loss tangent and the slower G' decay as

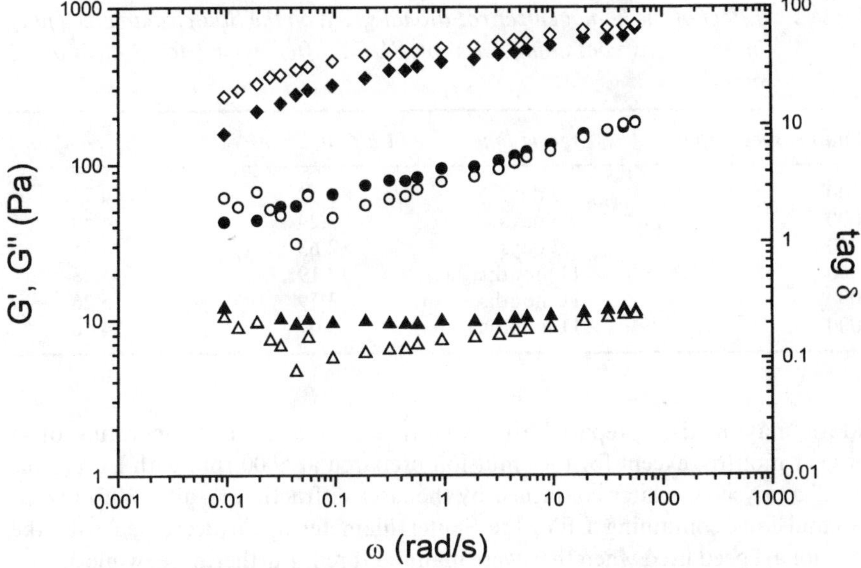

Figure 3 *Loss tangent (tag δ) (▲,△), storage modulus (G') (◆,◇), and loss modulus (G") (●,○) plotted against frequency on a log–log scale for salad dressings prepared at 6000 rpm containing LEY (filled symbols) or SDEY (open symbols)*

frequency decreases (Figure 3). In other words, the SDEY emulsions are characterized by a longer relaxation mechanism. The small differences found in the linear viscoelastic properties of low-oil-content salad dressings manufactured with either LEY or SDEY may be attributed to a certain degree of protein denaturation during the atomization process.[7] Indeed, a more extensive unfolding of proteins in SDEY emulsions as a consequence of the thermal treatment, as well as a more extensive disruption of LDL micelles and egg yolk granules, as demonstrated by optical microscopy, are likely to occur.

Emulsions containing LEY, as well as emulsions with SDEY, showed a clear drop in the Sauter diameter with increasing rotational speed (Table 2). The droplet-size distributions were not dramatically affected by the rotational speed. The fact that the Sauter diameter for emulsions containing SDEY were found to be higher than those of emulsions containing LEY may be due to the fact that the liquid egg yolk can be transported to the oil–water interface more rapidly than the solid egg yolk, and also to a change in the adsorption properties of some egg yolk components on the oil–water interface. These factors may also be responsible for the different droplet-size distribution. This was narrower for emulsions containing SDEY than for emulsions containing LEY.

Physical Stability

On visual observation, no phase separation occurred for any of the reduced-

Table 2 *Effect of the homogenizer rotational speed on the Sauter diameter (μm) in emulsions containing LEY or SDEY after ageing for 3 days or 11 months*

Rotation speed (rpm)	*Ageing time*	*LEY emulsions*	*SDEY emulsions*
5000	3 days	3.51	4.73
6000	3 days	3.24	4.49
7000	3 days	2.63	4.35
5000	11 months	13.19	5.28
6000	11 months	3.79	5.26
7000	11 months	3.07	4.46

calorie mayonnaises prepared after a storage time at room temperature of at least 11 months, except for the emulsion prepared at 5000 rpm with LEY. This visual analysis was later confirmed by the laser diffraction results. With regard to emulsions containing LEY, the Sauter diameter again decreased with the rotational speed used when they were manufactured. Furthermore, while ageing only provoked a slight increase in the Sauter diameter for the emulsions prepared at 6000 and 7000 rpm, a marked increase was evident for that prepared at 5000 rpm. On the other hand, the Sauter diameter of emulsions prepared with SDEY hardly increased at all after ageing for 11 months, demonstrating their excellent shelf-life (Table 2).

Effect of XG Concentration on Emulsions Containing SDEY

The mechanism causing non-linear properties was the same as that for the emulsions manufactured without XG. However, the G' drop was less sharp and the critical shear stresses were much higher (Table 3). In fact, the critical shear stress showed a tendency to increase with the XG/MS ratio. It is noteworthy that such a high value is exhibited by the system containing 0.15% XG. Once more, a plot of G' *versus* strain reveals that most of the systems essentially exhibit the same critical strain for the onset of non-linear response, emphasizing that the differences can be attributed to their corresponding complex moduli.

The shape of the mechanical spectra does not dramatically change upon

Table 3 *Critical shear stress for the onset of non-linear viscoelastic behaviour (τ_c°), the plateau modulus (G_N°) and the slope of the plateau zone in the relaxation spectrum (n) as a function of XG concentration*

Concentration (wt%)	*0*	*0.15*	*0.3*	*0.4*	*0.5*	*0.6*	*0.7*
τ_c° Pa	7	20	12	18	22.5	25	30
G_N° Pa	500	1000	675	700	850	775	1300
n	−0.17	0.26	0.03	−0.03	0.39	−0.03	0.09

increasing the XG concentration, since the data basically lie along the so-called plateau zone. The values of G'' usually show a clear minimum at an intermediate frequency which locates the plateau shear modulus in the G' curve: $G_N{}^\circ$. However, $G_N{}^\circ$ can be better located on the basis of a loss tangent *versus* frequency plot.[8] The occurrence of such a minimum or a plateau zone in the G'' curves has been related[9] in polymer rheology to the existence of entanglements between the macromolecular chains. Likewise, this behaviour in concentrated emulsions can be associated with strong interactions between oil droplets resulting in extensive flocculation. Furthermore, the actual value of $G_N{}^\circ$ may be related to the density of entanglements in the network structure formed.[10]

As the XG concentration increases a local maximum in $G_N{}^\circ$ is found at 0.15% XG followed by a dramatic decrease at 0.3% XG. However, above this concentration $G_N{}^\circ$ steadily increases. Similar results have been reported by Pawlosky and Dickinson[11] for BSA/dextran sulphate emulsions. These authors attributed the occurrence of a maximum in G^* (complex modulus) to a bridging flocculation mechanism. Invoking such an interpretation is not straightforward for these low-oil-content salad dressings due to the large number of ingredients involved and to the fact that low concentrations of a non-adsorbing polysaccharide like xanthan gum have been reported to induce a depletion mechanism.[12] Furthermore, no evidence of enhanced creaming was observed in these emulsions. Apart from that, bulk viscosity measurements of SDEY–XG suspensions did not show significant complexation, although the relevant concentrations in the interface must be rather different. Finally, steric stabilization can be assumed to be playing an increasingly important role as XG concentration increases.

The conversion of oscillatory data into a relaxation spectrum provides the key for interpreting viscoelastic properties in the time domain. Discrete relaxation spectra were determined by fitting the SAOS results to the generalized Maxwell model,

$$G' = \sum_{i=1}^{N} G_i\omega^2\lambda_i^2/(1+\omega^2\lambda_i^2) \tag{1}$$

$$G'' = \sum_{i=1}^{N} G_i\omega\lambda_i/(1+\omega^2\lambda_i^2) \tag{2}$$

using an iterative regression method.[13]

The corresponding discrete relaxation spectra (that is to say plots of G_i *versus* λ_i) were analysed using the empirical Madiedo equation,[14] *i.e.*,

$$H(\lambda) = \frac{\alpha\lambda^m + \beta\lambda^n}{1+(\lambda/\lambda_t)^p}, \tag{3}$$

where m, n and $n - p$ are related to characteristic slopes for the transition, plateau and terminal zones, respectively. The quantity λ_t is the pseudoterminal relaxation time. When the LEY emulsion spectrum was compared to that derived for the SDEY emulsion, the latter exhibited a slightly wider plateau zone and did not show a clear pseudo-terminal region.

The characteristic slopes of the plateau regions of the SDEY emulsions containing XG are higher than for the XG-free system. Discarding the unexpectedly high value shown by the system containing 0.5% XG, the higher slopes are also found for the 0.15% and 0.7% XG systems, that is to say, those which show the higher plateau modulus values (Table 3).

5 Concluding Remarks

Low-calorie salad dressings have been formulated with either pasteurized salted liquid egg yolk or spray-dried egg yolk using a rotor/stator homogenizer. The best results were obtained between 6000 and 7000 rpm according to the physical stability and rheological studies carried out, whose results resembled those shown by commercial high-oil content salad dressings.[15] In fact, these emulsions were found to be stable for at least 11 months, on the basis of the Sauter average diameter and droplet-size distribution hardly changing within this period of time.

The fact that slip effects are likely to occur unless serrated sensor systems are used has been proved. Additionally, the need for a non-Newtonian correction when using plate & plate sensor systems was evident.

Changes in homogenizer rotational speed between 5000 and 7000 rpm did not significantly influence the linear viscoelastic properties of these low-oil salad dressings after 3 days storage. The fact that SDEY emulsions turn out to be more viscoelastic than those made with LEY has been attributed to a more extensive unfolding of proteins due to the atomization process. The maximum in the viscoelastic properties found at low XG concentration could be associated with a bridging flocculation mechanism, though the possibility of a depletion mechanism should not be completely discarded. Further increase in XG concentration leads to stronger steric stabilization.

Acknowledgements

This work is part of a project sponsored by the CICYT, Spain (research project ALI-96 0892). The authors also acknowledge financial support from a Spain–United Kingdom Acciones Integradas project (HB 96-0232). Technical information provided by Hijos de Ybarra, S.A. (Sevilla), Ovosec, S.A. (Valladolid) and Cerestar, S.A. (Barcelona) is acknowledged.

References

1. A. C. Nielsen Company, private communication, 1997.
2. K. Wendin, K. Aaby, A. Edris, M. R. Ellekjaer, R. Albin, B. Bergenståhl, L. Johansson, E. P. Willers, and R. Solheim, *Food Hydrocolloids*, 1997, **11**, 87.

3. N. E. Hudson, private communication, 1997.
4. J. Ferguson and Z. Kemblowski, 'Applied Fluid Rheology', Elsevier Science, Cambridge, 1991, p. 112.
5. H. A. Barnes, *J. Non-Newtonian Fluid Mech.*, 1995, **56**, 221.
6. Th. F. Tadros, *Langmuir*, 1990, **6**, 28.
7. A. F. Guerrero and H. R. Ball, Jr., *J. Texture Stud.*, 1994, **25**, 363.
8. S. Wu, *J. Polym. Sci.*, 1989, **27**, 723.
9. M. E. De Rosa and H. H. Winter, *Rheol. Acta*, 1994, **33**, 221.
10. J. M. Franco, A. Guerrero, and C. Gallegos, *Rheol. Acta*, 1995, **34**, 513.
11. K. Pawlowsky and E. Dickinson, in 'Food Colloids: Proteins, Lipids and Polysaccharides', ed. E. Dickinson and B. Bergenståhl, Royal Society of Chemistry, Cambridge, 1997, p. 258.
12. E. Dickinson, in 'Proceedings of Second World Congress on Emulsion', Bordeaux, 23–26 September 1997, vol. 4, p. 83.
13. F. Martínez, PhD. Thesis, University of Seville, 1996.
14. J. M. Madiedo, J. Muñoz, and C. Gallegos, in 'Rheology and Fluid Mechanics of Nonlinear Materials', ed. D. A. Siginer and S. G. Advani, American Society of Mechanical Engineers, Atlanta, 1996, p. 151.
15. J. Muñoz and P. Sherman, *J. Texture Stud.*, 1990, **21**, 411.

Microstructure in Relation to the Textural Properties of Mayonnaise

By Maud Langton, Annika Åström, and Anne-Marie Hermansson

SIK—THE SWEDISH INSTITUTE FOR FOOD AND BIOTECHNOLOGY, BOX 5401, SE 402 29 GÖTEBORG, SWEDEN

1 Introduction

Mayonnaise is an oil-in-water emulsion, with a relatively high content of oil. The size and shape of the oil droplets, and the connection between them, in combination with the viscosity of the aqueous phase, have an impact on the stability and textural properties of the product and on optical aspects such as colour and gloss.

The distribution of the phases can vary according to the ingredients and processing conditions used. There are few published results on the microstructure of mayonnaise in particular or on the sizes of droplets in undiluted emulsions in general. Undiluted samples can be visualized by cryotechniques and confocal laser scanning microscopy (CLSM). Previously, cryo-scanning electron microscopy (cryo-SEM) has been used to characterize mayonnaise[1] as well as CLSM,[2] light microscopy (LM)[1,3] and embedded sections for transmission electron microscopy (TEM).[1,4,5] Difficulties with estimating the sizes of emulsion droplets in general have been discussed earlier.[6–8] One relatively new method is to apply image analysis to confocal images of undiluted samples.

The effect of process conditions on the storage modulus G' has previously been investigated with reference to light and full-fat mayonnaises.[9] Windhab *et al.* found a general increase in the storage modulus for full-fat emulsions when the energy input was increased. In order to optimize the quality properties of food, we have to quantify the process–rheology–structure–texture relationships. The perception of texture and mouth-feel characteristics is a highly dynamic process in which physicochemical properties of the food are continuously altered by chewing, salivation, and, potentially, adjustment to body temperature.[10]

Previously we have characterized the microstructure of mayonnaises using CLSM and TEM and quantified microstructural parameters with the help of image analysis.[11] In addition, we have analysed the sensory properties of

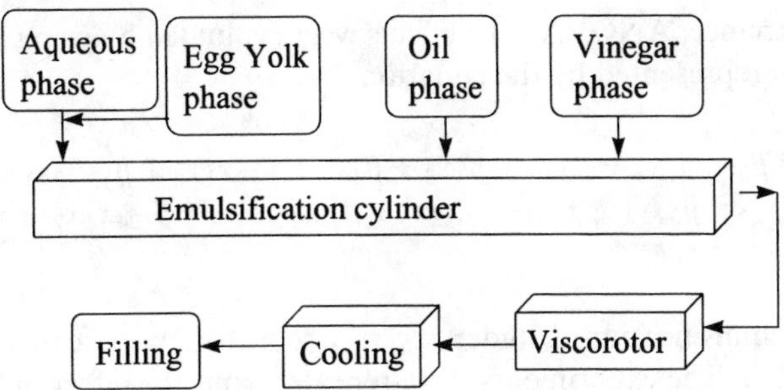

Figure 1 *Schematic drawing of the process*

mayonnaises.[12] Changes in the properties were achieved by varying the process and formulation according to factorial designs. The aim of this paper is to relate microstructural parameters, oil droplet size and distribution of egg yolk to rheological or textural properties.

2 Materials and Methods

The mayonnaise was produced by a cold process line, using pilot plant equipment from Schröder & Co, Lübeck, at Palsgaard Industri A/S, Juelsminde Denmark. The line is shown schematically in Figure 1, and the aqueous phase was first formed by dissolving the dry components in the water and then mixing with the egg yolk. The oil was added to this mixture in the first emulsifying cylinder, where the vinegar was also added. The emulsion was then passed through the colloid mill (viscorotor) to produce the final emulsion, which was cooled before filling.

The oils were supplied by Aarhus Oliefabrik A/S, under the trade names Shogun for the soybean oil and Colzao for the rape seed oil (a refined and deodorized rape seed oil of low erucic acid content). The water phase was stabilized by PALSGAARD KP 94-159 (KP) and egg yolk, which also functions as an emulsifier. The KP product from Palsgaard Industri is a mixture of polysaccharides, guar gum (E412) and xanthan gum (E415). The pasteurized egg yolk was purchased from Honum A/S, Denmark.

The emulsion fat content was 70 wt%, and the egg yolk content varied between 2.5 and 6 wt%. The added sweetening was saccarose. To all mayonnaises were also added NaCl, sodium benzoate, vinegar, colour and aroma, at a constant level.

A fully two-level factorial design was used, including two centre points, and with four design variables: speed of emulsification cylinder, speed of viscorotor, exit temperature, and content of egg yolk, thus forming 18 samples. From this design eight sample points, comprised of a full two-level factorial design with three variables, were repeated later with different batches of the raw materials for a more detailed investigation. The responses were analysed by means of analysis of variance, ANOVA. The effects were evaluated by using the response surface model represented by the equation

$$\mathbf{Y} = \mathbf{X}\beta + \varepsilon = \beta_0 + \beta_1 x_1 + \beta_2 x_2 + \beta_3 x_3 + \beta_4 x_4 + \beta_{12} x_1 x_2 + \beta_{13} x_1 x_3 + \beta_{14} x_{14} + \beta_{23} x_2 x_3 + \beta_{24} x_{24} + \beta_{34} x_3 x_4 + \beta_{123} x_1 x_2 x_3 + \beta_{234} x_2 x_3 x_4 + \beta_{124} x_1 x_2 x_4 + \beta_{134} x_1 x_3 x_4 + \beta_{1234} x_1 x_2 x_3 x_4 + e, \quad (1)$$

where x_1 = emulsification cylinder, x_2 = viscorotor, x_3 = exit temperature, and x_4 = batch. The experiments were repeated, and the effect of variation in batch x_4 was analyzed. The emulsification cylinder speed was set at 550, (759) and 1000 rpm, the viscorotor was set at 0, (350) and 700 rpm, the exit temperature of the cooling device was set at 10 °C, (17.5 °C) and 25 °C, and the content of the egg yolk was set at 2.5%.

Two different microscopy techniques were used to study the mayonnaise: confocal laser scanning microscopy (CLSM) and transmission electron microscopy (TEM) which both visualize undiluted samples.[11] The CLSM samples were stained with Nile blue and observed under the microscope at 15 °C. All preparations were made under chilled conditions. Small pieces were quickly frozen for TEM preparation. Frozen samples were fractured, etched and rotary shadowed. The replica was washed, dried and observed under the TEM at an accelerating voltage of 80 kV.

The average oil droplet size and the average particle size of the egg yolk were estimated using a stereological approach. All measurements were performed on CLSM images. The object size was estimated by using the so-called star volume defined by

$$v^* = \frac{\pi}{3}\overline{l_0^3}, \quad (2)$$

where l_0 is the line intercept length.[11,13] The star volume is an estimate of the volume-weighted mean volume, and is used in this study as the estimated mean size.

Viscoelastic measurements were performed with a Bohlin VOR Rheometer (Bohlin Instruments Ltd., Chichester, UK). The measuring system was a plate–plate (PP30) with 1 or 2 mm gap, and a strain of 2×10^{-3}. The strain used was within the linear region, which was confirmed by measurements of modulus *versus* strain. The frequency was 1 Hz. The samples were equilibrated for 5 min at 15 °C. The viscoelastic properties were measured while increasing the temperature by 1.5 °C min^{-1} from 15 to 80 °C.

The sensory characterization was performed by quantitative descriptive analysis, using a trained sensory panel with documented ability to generate valid and reproducible data.[12] During several training sessions the sensory panel developed and defined the most appropriate attributes. During the training sessions the panel also agreed on the procedure for quantitative evaluation. Each attribute was evaluated by marking on an unstructured intensity scale. A 100 mm unstructured line scale was used, anchored 10 mm from each end with the terms *slight* and *much*. The data were collected with a computerized system (PSA Systems/3, Oliemans Punter & Partners, The Netherlands) with each panellist marking the intensity of each attribute on the computer screen by a click with the mouse. The samples were evaluated at a temperature of 15 °C.

3 Results

The microstructure of undiluted samples was characterized at different structural levels by both CLSM and TEM. The nature of the fat phase, the size of the oil droplets, the nature of the aqueous phase, and the distribution of the egg yolk will be described. The effect of the droplet size and the size of egg yolk particles on the texture, as measured instrumentally and perceived sensorially, will be discussed.

Oil Droplets

At low magnification using CLSM a large field of view is obtained; an example of a full-fat mayonnaise is shown in Figure 2(a). Many droplets of ~5 μm in diameter can be seen in the micrograph. At this level of magnification the largest oil droplets can be detected. A higher magnification is achieved by TEM, and a freeze-etched sample is shown in Figure 2(b). At this level, a network of oil droplets can be confirmed. In Figure 2(b) the connections between droplets can be observed as well as individual deformed droplets. The most spherical droplets were found when the emulsification cylinder was operated at the highest speed.

Droplets forming a network can have an influence on the texture. An even higher magnification was used in Figure 2(c), which shows that one of the connections is formed of a crystalline structure and another of a very small droplet. The smallest droplet in this image cannot be resolved by CLSM. Many very small droplets, finely dispersed between large droplets, can also have an impact on the texture. In full fat mayonnaise there is a large span in droplet size, from very small (~0.2 μm) up to 30 μm.

The theoretical limit for producing droplets of equal size is a volume fraction of around 74%.[14] At a higher oil content the droplets have to be deformed, or a fraction of very small droplets must be present between the large droplets. We have recently shown[11] that when reducing the oil content from 80% to 70% the amount of very small droplets is reduced; thus the effect of very small droplets in 70% mayonnaise was assumed to be negligible. The density of connections between droplets forming a network was also reduced in 70% mayonnaise. Figure 3 shows the smallest and largest droplets achieved by varying the processing condition, keeping the formulation the same. The volume-weighted mean volume of droplets was estimated from CLSM images by using a stereological approach, and this was used as an effective measure of the droplet size. Eight different conditions were evaluated, with replication. The storage

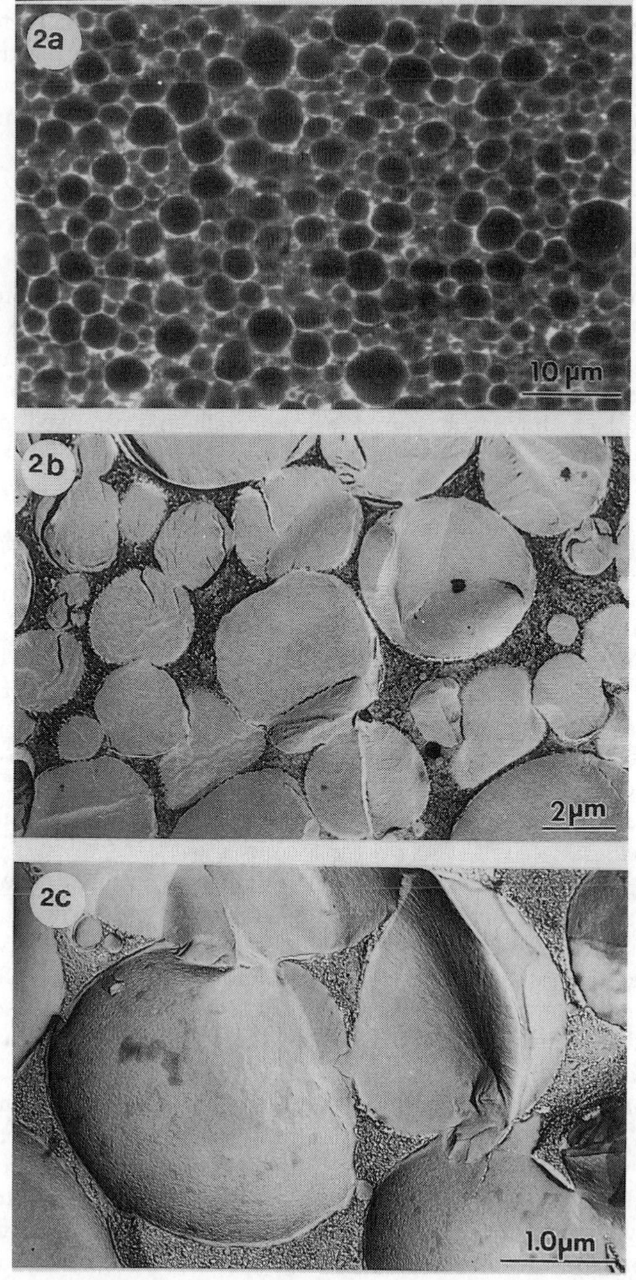

Figure 2 *The microstructure of full-fat mayonnaise at different structural levels as visualized by (a) CLSM at low magnification, (b) TEM of a freeze-etched sample of mayonnaise, at medium magnification showing network of droplets and deformed droplets, and (c) TEM at high magnification showing connections between droplets*

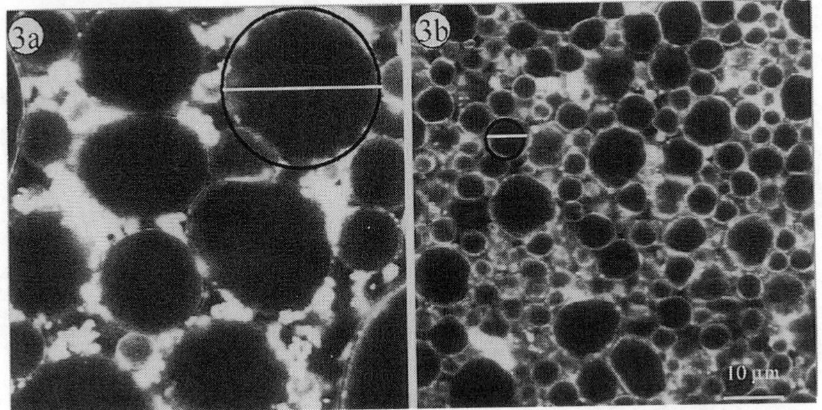

Figure 3 *CLSM micrograph of 70% mayonnaise. (a) A sample with large oil droplets formed with emulsification cylinder at 55 rpm, viscorotor at 700 rpm and exit temperature at 10 °C. (b) A sample with small oil droplets formed with emulsification cylinder at 1000 rpm, viscorotor at 700 rpm and exit temperature at 25 °C. Circles indicate the diameter corresponding to the sphere calculated from the estimated mean volume, by image analysis, for the two samples*

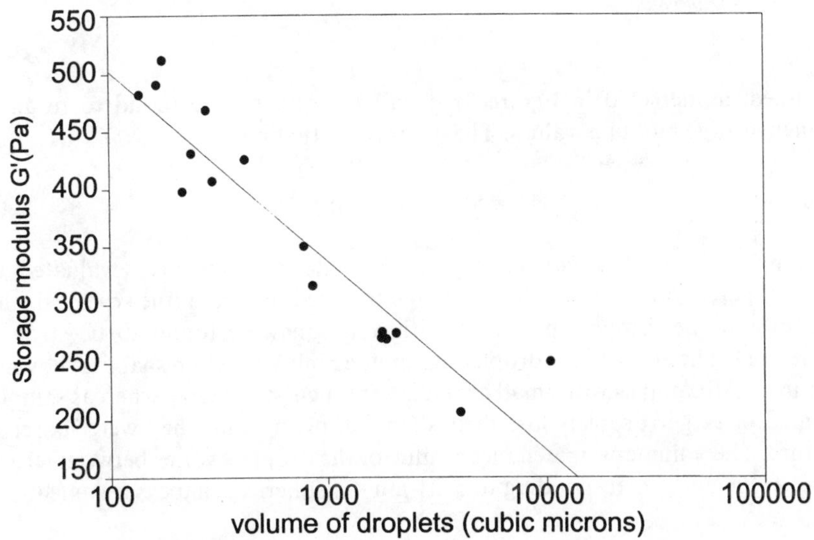

Figure 4 *The storage modulus* G′ *of mayonnaise as a function of the size of the oil droplets, estimated as the volume-weighted mean value of the volume (in cubic micrometres)*

modulus G' was measured for the same samples. Figure 4 shows the storage modulus as a function of the logarithm of the mean volume of the droplets. The smallest mean droplets volume was 140 μm^3 and the largest 1.05×10^4 μm^3, corresponding to diameters of 6 and 27 μm respectively (if a spherical shape was

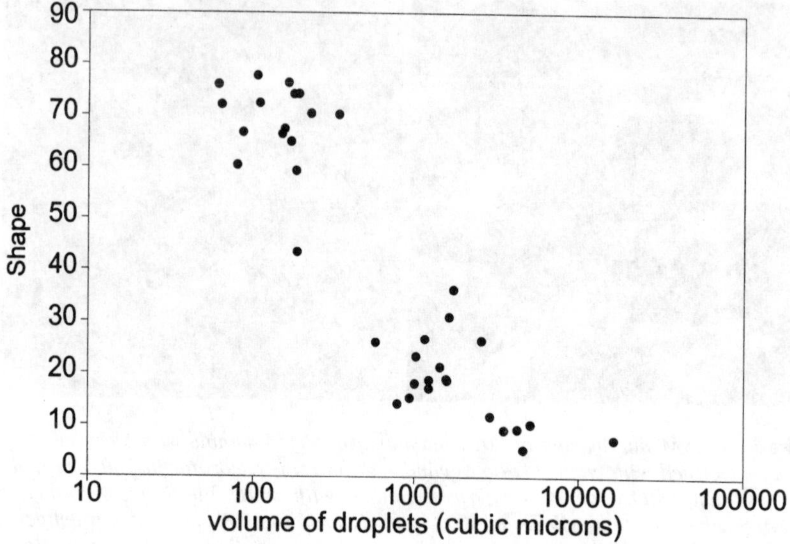

Figure 5 *The ability of mayonnaise samples to retain shape as a function of oil droplet size. The shape was evaluated by the trained sensory panel, and the size was estimated as the star volume, in cubic micrometres, which is a volume-weighted mean value*

assumed, as depicted in Figure 3). Smaller droplets were found to result in higher storage modulus values. The data were fitted by

$$G'/\text{Pa} = 832 - 165 \log_{10} (v^*/\mu\text{m}^3). \tag{3}$$

Samples from an expanded experimental design were also evaluated by sensory perception. How well the sample retained its shape (the so-called star-shape) and the firmness perceived with a spoon were found to be strongly correlated. The effect of the droplet size on the ability to retain shape is shown in Figure 5. All samples with small droplets kept their star-shape, whereas samples formed of larger droplets lost their shape, implying that they were looser in texture. The volume-weighted mean value of the droplets varies between 60 and 1.63×10^4 μm^3, corresponding to 5–31 μm if a spherical shape is assumed.

Effect of Egg Yolk Particles in 70% Mayonnaise

The aqueous phase was analyzed by using TEM, and density variations were noted. Figure 6(a) shows inhomogeneities in the aqueous phase which originate from unevenly dispersed egg yolk. The same recipe used with different processing gives the mayonnaise shown in Figure 6(b), which has the egg yolk evenly distributed in the water phase. Aggregates of egg yolk are also detected in the CLSM micrograph, where they appear as bright particles (cf. Figures 3 and 6).

The size of the egg yolk particles was estimated and evaluated by means of

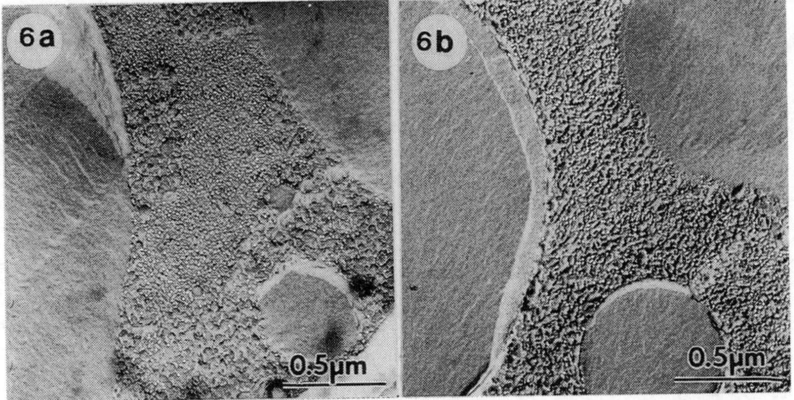

Figure 6 *TEM micrograph showing 70% mayonnaise formed at (a) 10 °C exit temperature when aggregates of small globular particles of egg yolk in the water phase were detected, and (b) 25 °C exit temperature, when the egg yolk was more evenly distributed in the water phase*

analysis of variance. The main effect, according to the response surface model, on the size of egg yolk particles is shown in Figure 7. Two of the processing parameters, namely the emulsification cylinder speed and the exit temperature, are shown to have a strong effect on size of egg yolk particles, whereas the speed of the viscorotor has no major effect on the size. An effect of batch number is found, dependent on the quality of the raw material, the pasteurized egg yolk. No batch effect was found on the oil droplet size; thus the size of the droplets could be repeated using the same process conditions. The volume-weighted mean value of the egg yolk particles was found to vary between 1.8 and 180 μm^3, corresponding to diameters at between 1.5 and 7 μm if a spherical shape is assumed. Figure 8 shows the storage modulus as a function of temperature.

The samples with a high exit temperature show an immediate increase in the storage modulus, whereas the samples with a low exit temperature have the same G' values up to higher temperatures.

4 Discussion

This work presents the feasibility of characterizing different features in an important food emulsion, mayonnaise. The microstructure has been characterized at different structural levels. At the ultra-structural level, small droplets of around 0.2 μm were found which were impossible to detect by CLSM. In order to find a relationship between microstructure, texture and rheology, the fat content was kept constant at 70%, which is the condition where equal-sized oil droplets can be formed.

The effect of droplet size on the storage modulus has been quantified, as well as the influence of droplet size on the perceived texture. The storage modulus is negatively correlated to the logarithm of the mean volume of the droplets, *i.e.*,

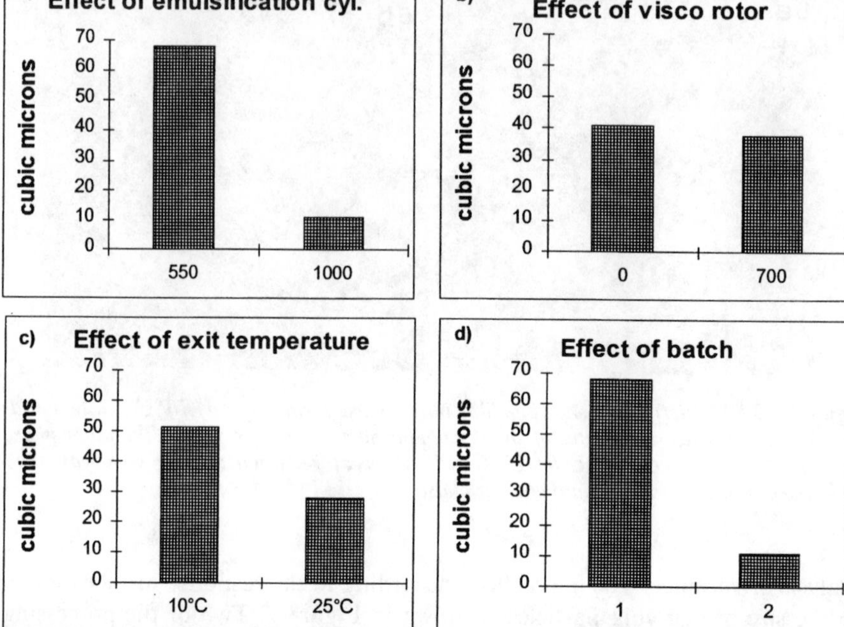

Figure 7 *Main effect of the dependent variables on the size of the egg yolk particle, in cubic micrometres, according to equation (3). Plot (a) illustrates β_1, the main effect of the emulsification cylinder; plot (b) illustrates β_2, the main effect of the viscorotor; plot (c) illustrates β_3, the main effect of the exit temperature; and plot (d) illustrates β_4, the main effect of the batch of the raw material*

increasing droplets size resulted in lower values of the storage modulus. Mayonnaises formed of smaller droplets were also perceived to keep their shape better. We also found previously[15] that the perceived texture is logarithmically related to microstructural parameters for whey protein gels.

The distribution of egg yolk is affected by both the processing conditions and the quality of the raw material used. In this study, the low speed of the emulsifier cylinder was found not to be sufficient to disperse the egg yolk evenly in the aqueous phase. The system could therefore be interpreted as an under-processed emulsion. The setting temperature and exit temperature of the emulsion influences the storage modulus as a function of temperature for the final product. Thus, the processing conditions affect both the droplet size and the distribution of the egg yolk, which in turn affect the storage modulus and the perceived texture of the final product. Microstructure determination in combination with image analysis is confirmed to be a powerful tool for establishing correlations of the type structure–rheology–texture–process. The knowledge of the relationship between microstructure and texture should lead to optimized food production processes as well as to the development of new products with desired properties.

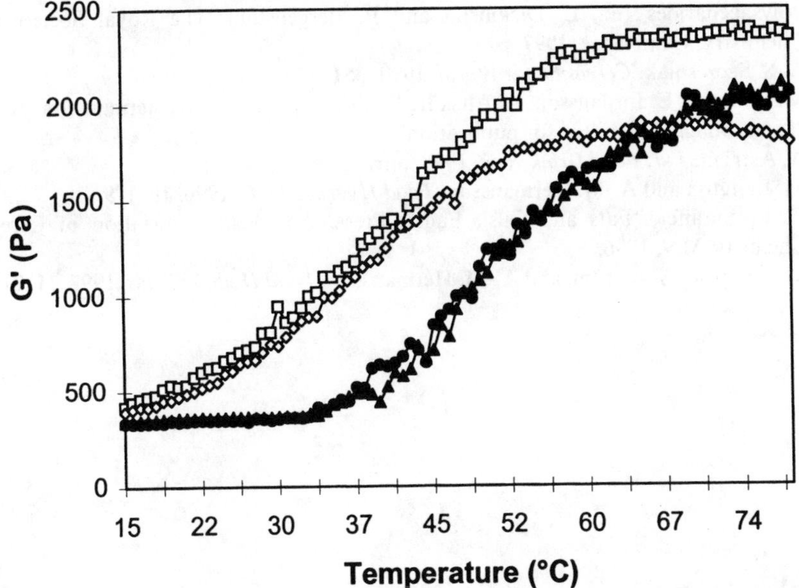

Figure 8 *The storage modulus G′ of samples as a function of temperature:* ■*, exit temperature of 10 °C and viscorotor at 0 rpm;* ▲*, exit temperature of 10 °C and viscorotor at 700 rpm;* □*, exit temperature of 25 °C and viscorotor at 0 rpm;* ◇*, exit temperature of 25 °C and viscorotor at 700 rpm*

Acknowledgements

This work is part of the NordFood project No P93131, 'Quality of Emulsion Products', and support from the Nordic Industrial Fund, Mills AD, and Palsgaard is gratefully acknowledged.

References

1. M. Tanaka and H. Fukuda, *Can. Inst. Food Sci. Technol. J.*, 1976, **9**(3), 130.
2. I. Heerjte, P. van der Vlist, J. C. G. Blonk, H. A. C. Hendricks, and G. J. Brakenhoff, *Food Microstruct.*, 1987, **6**, 115.
3. L. D. Ford, R. Borwanker, R. W. Martin, and D. N. Holcomb, in 'Food Emulsions', 3rd Edn., ed. S. E. Friberg and K. Larsson, Marcel Dekker, New York, 1997, p. 361.
4. C. M. Chang, W. D. Powrie, and O. Fennema, *Can. Inst. Food Sci. Technol.*, 1972, **5**(3), 134.
5. M. A. Tung and L. J. Jones, *Scanning Electron Microscopy*, 1981, **III**, 523.
6. D. G. Dalgleish, in 'Emulsions and Emulsion Stability', ed. J. Sjöblom, Marcel Dekker, New York, 1996, p. 287.
7. P. Walstra, H. Oortwijn, and J. J. de Graaf, *Neth. Milk Dairy J.*, 1969, **23**, 12.
8. W. Buchheim and P. Dejmek, in 'Food Emulsions', 2nd Edn., ed. K. Larsson and S. Friberg, Marcel Dekker, New York, 1990, p. 203.
9. E. J. Windhab, B. Wolf, and E. Byekwaso, in 'Food Colloids—Proteins, Lipids and

Polysaccharides', ed. E. Dickinson and B. Bergenståhl, The Royal Society of Chemistry, Cambridge, 1997, p. 3.
10. A. S. Szczesniak, *Cereal Foods World*, 1990, **351**, 1201.
11. M. Langton, E. Jordansson, A. Altskär, C. Sørensen, and A.-M. Hermansson, *Food Hydrocolloids*, submitted for publication.
12. A. Åström, *Eur. Food Drink Rev.*, 1998, Spring, 43.
13. M. Langton and A.-M. Hermansson, *Food Hydrocolloids*, 1996, **10**, 179.
14. C. E. Stauffer, 'Fats and Oils', Eagan Press, American Association of Cereal Chemists, MN, 1996.
15. M. Langton, A. Åström, and A.-M. Hermansson, *Food Hydrocolloids*, 1997, **11**, 217.

Subject Index